W9-AHL-459

Medical Informatics

Practical Guide for Healthcare and Information Technology Professionals

Fourth Edition

Robert E. Hoyt MD FACP

Editor

Ann Yoshihashi MD FACE

Associate Editor

Medical Informatics
Practical Guide for Healthcare and Information Technology Professionals

---

Disclaimer

Every effort has been made to make this book as accurate as possible but no warranty is implied. The information provided is on an "as is" basis. The authors and the publisher shall have neither liability nor responsibility to any person or entity with respect to any loss or damages arising from the information contained in this book. The views expressed in this book are those of the author and do not necessarily reflect the official policy or position of the University of West Florida, Department of the Navy, Department of Defense, nor the U.S. Government.

Copyright © 2010

First Edition: June 2007

Second Edition: August 2008

Third Edition: November 2009

All rights reserved. No part of this book may be reproduced or transmitted in any form, by any means, electronic or mechanical, including photocopying, recording, or by any information storage and retrieval system, without written permission from the publisher, except for inclusion of brief excerpts in connection with reviews or scholarly analysis.

ISBN-13: 978-0-557-60808-9

Published by: Lulu.com

Editors

Robert E. Hoyt MD, FACP
Co-Director Medical Informatics
Instructor, School of Allied Health and Life Sciences
University of West Florida
Pensacola, FL
Assistant Professor of Medicine
Adjunct Assistant Professor of Family Medicine
The Uniformed Services University of the Health Sciences
Bethesda, MD

Ann K. Yoshihashi MD, FACE
Guest Lecturer, School of Allied Health and Life Sciences
Medical Informatics
University of West Florida
Medical Analyst
Naval Operational Medicine Institute
Pensacola, FL

Contributors

Elmer V. Bernstam MD, MSE
Professor of Biomedical Informatics and
Internal Medicine
School of Biomedical Informatics and
Medical School
The University of Texas Health Science
Center at Houston
Houston, TX

Robert W. Cruz
Manager, Technical Development
Computer Programs and Systems, Inc
Mobile, AL

Brent Hutfless MS, CISSP, GSLC
Manager, Information Security
Naval Operational Medicine Institute
Pensacola, FL

Todd R. Johnson PhD
Professor of Biomedical Informatics
School of Biomedical Informatics
The University of Texas Health Science
Center at Houston
Houston, TX

Justice Mbizo Dr PH
Masters of Public Health Program
School of Allied Health and Life Sciences
University of West Florida
Pensacola, FL

M. Hassan Murad MD, MPH
Assistant Professor of Medicine
Mayo Clinic College of Medicine
Rochester, MN

Jane A. Pellegrino, MSLS, AHIP
Head, Library Services Department
Naval Medical Center Portsmouth
Portsmouth, VA

Steve Steffensen, MD, LCDR
Chief Medical Information Officer
Telemedicine and Advanced Technology
Research Center
Fort Detrick, MD

Fred Trotter
Director, Liberty Health Software Foundation
Houston, TX

Brandy G. Ziesemer, RHIA, CCS
Health Information Manager and Associate
Professor
Lake-Sumter Community College,
Leesburg, FL

Preface to the Fourth Edition

With this fourth edition of the book we continue our goal of helping healthcare and information technology professionals to keep up to date on the key topics in the rapidly evolving field of medical informatics. This edition includes important program updates by the Office of the National Coordinator for Health Information Technology (ONC) and other offices and agencies of the Department of Health and Human Services (HHS). The American Recovery and Reinvestment Act (ARRA) of 2009 and specifically the HITECH Act have had a major impact on promoting health information technology (HIT). It is important for the average reader to understand the implications of such sweeping legislation on the implementation of HIT into the practice of medicine. We have re-written numerous chapters to reflect the changes brought about by ARRA. The newly updated chapters will include several final rules posted in the July 2010 time frame dealing with electronic health records, particularly Meaningful Use objectives and measures.

We have made every attempt to provide the most up-to-date information about medical informatics by constantly reviewing the medical and lay literature. Our goal is to present the most recent changes, the most interesting concepts and both sides of each controversy. Our book is intended to be an introduction to the field of Medical informatics that will entice individuals to go further in their education.

In our experience, individuals from very diverse backgrounds seem to be interested in this relatively new field. We hope our book will better educate technology workers about clinical issues and better educate clinical workers about technology issues.

Given the newness of this field, research may be lacking or inadequate in many areas of medical informatics. We have to rely on surveys and expert opinions where the medical literature is weak or non-existent. Frequently, this lack of knowledge can result in more noise than signal and more hype than fact. We are dedicated to presenting the issues fairly and objectively. Approximately 1300 medical literature references and web links are included in this book that help direct readers to additional information. In our fourth edition we have added new content to most chapters and created several new chapters.

While we are vendor-agnostic we are not opposed to presenting interesting hardware and software, including open source, we think will be of interest to our readers. One of the goals of this book is to promote and disseminate innovations that might help healthcare workers as well as technology developers. The fact we mention specific hardware or software or web-based applications does not mean we endorse the vendor; instead, it is our attempt to highlight an interesting concept that might lead others in a new direction.

Our book's emphasis will always be on the medical issues that the average clinician or hospital faces, with potential solutions that are easy to understand. We won't be reluctant to present the obstacles new innovations face or negative publications written on the subject. If we can introduce the reader to a single new concept or tool that improves patient care or efficiency, we will consider our work successful.

We appreciate feedback regarding how to make this book as user friendly and educational as possible. We welcome your input so we can continue to publish the most accurate and up-to-date information.

Please note that all proceeds will be donated to support the advancement of medical informatics education.

Robert E. Hoyt MD FACP
Ann Yoshihashi MD FACE

Table of Contents

1

Overview of Medical Informatics

ROBERT E. HOYT
ELMER V. BERNSTAM
TODD R. JOHNSON

Learning Objectives

After reading this chapter the reader should be able to:

- State the definition and origin of medical informatics
- Identify the forces behind medical informatics
- Describe the key players involved in medical informatics
- State the potential impact of the ARRA and HITECH Act on medical informatics
- List the barriers to health information technology (HIT) adoption
- Describe the educational and career opportunities in medical informatics

"During the past few decades the volume of medical knowledge has increased so rapidly that we are witnessing an unprecedented growth in the number of medical specialties and subspecialties. Bringing this new knowledge to the aid of our patients in an economical and equitable fashion has stressed our system of medical care to the point where it is now declared to be in a crisis. All these difficulties arise from the present, nearly unmanageable volume of medical knowledge and the limitations under which humans can process information"

Marsden S. Blois, *Information and Medicine: The Nature of Medical Descriptions, 1984*

Medical informatics began as a new field of study in the 1950s-1960s time frame but only recently gained recognition as an important component of many aspects of healthcare. Its emergence is partly due to the multiple challenges facing the practice of medicine today. As the 1984 quote above indicates, the growth in the volume of medical knowledge and patient information that has occurred due to better understanding of human health has resulted in more treatments and interventions that produce more information. Likewise, the increase in specialization has also created the need to share and coordinate patient information. Furthermore, clinicians need to be able to access medical information expeditiously, regardless of location or time of day. Technology has the potential to help with each of those areas. With the advent of the Internet, high speed computers, voice recognition, wireless and mobile technology, healthcare professionals today have many more tools available at their disposal. However, in general, technology is advancing faster than healthcare professionals can assimilate it into their practice of medicine. One could also argue that there is a critical limitation of current information technology that manages data and not information. Thus, there is a mismatch between what we need (i.e. something to help us manage meaningful data = information) and what we have (ineffective ways to manage information).

Additionally, given the volume of data and rapidly changing technologies, there is a great need for ongoing Informatics education of all healthcare workers.

In this chapter we will present an overview of Medical informatics with emphasis on the factors that helped create and sustain this new field and the key players involved.

Definitions and Concepts

Informatics is the science of information and the blending of people, biomedicine and technology. Individuals who practice Informatics are known as informaticians or informaticists, such as, a nurse informaticist.

There is an information hierarchy that is important in the information sciences, as depicted in the pyramid in figure 1.1. Notice that there is much more data than information, knowledge or wisdom. As data are consumed and analyzed the amount of knowledge and wisdom produced is much smaller. The following are definitions to better understand the hierarchy:

- **Data** are symbols or observations reflecting differences in the world. Data are the plural of datum (singular). Thus, a datum is the lowest level of abstraction, such as a number in a database (e.g. 5), or packets sent across a network (e.g. 10010100). Note that there is no meaning associated with data; the "5" could represent 5 fingers, 5 minutes or have no real meaning at all. Modern computers process data accurately and rapidly

- **Information** is meaningful data or facts from which conclusions can be drawn by humans or computers. For example, "5 fingers" has meaning in that it is the number of fingers on a normal human hand. Modern computers do not process information, they process data. This is a fundamental problem and challenge in informatics

- **Knowledge** is information that is justifiably considered to be true. For example, a rising prostate specific antigen (PSA) level suggests an increased likelihood of prostate cancer

- **Wisdom** is the critical use of knowledge to make intelligent decisions and to work through situations of signal versus noise. For example, a rising PSA could mean prostate infection and not cancer.

Health information technology provides the tools to generate information from data that humans (clinicians and researchers) can turn into knowledge and wisdom. [1-2] Thus, enabling and improving human decision making is a central concern of informaticians

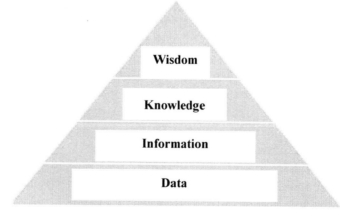

Figure 1.1 Information hierarchy

Another important concept to understand about data is that there are different levels of data (figure 1.2). Paper forms would be considered level 1 with serious limitations, in regards to sharing, storing and analyzing. Level 2 data could be scanned-in documents. Level 3 data are entered into a computer and are data that are structured and retrievable, but not computable between different computers. Level 4 data are computable data. That means the data is electronic, capable of being stored in data fields and computable because it is in a format that disparate computers can share (interoperable) and interpret (analyzable).

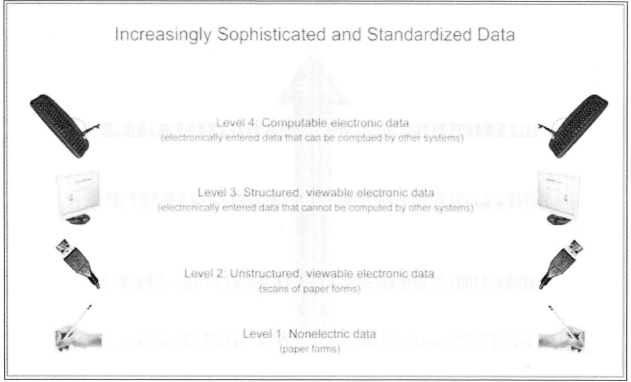

Figure 1.2 Levels of data (courtesy Government Accounting Office)

Therefore, the information sciences tend to promote data in formats that can be rapidly transmitted, shared and analyzed. Paper records and reports do not allow this, without a great deal of manual labor.

With the advent of electronic health records, health information exchanges and multiple hospital electronic information systems we have the ability and the need to collate and analyze large amounts of data to improve clinical and financial decisions. We have developed enterprise systems that: integrate disparate information; archive data; provide the ability to data mine using business intelligence and analytic tools. Figure 1.3 demonstrates a typical enterprise data system.

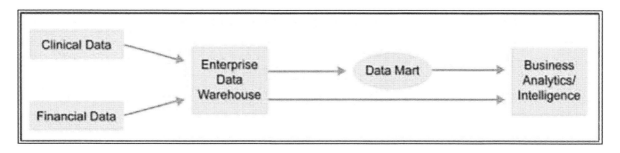

Figure 1.3 Enterprise data warehouse and data mining

Medical informatics is the field of information science concerned with management of healthcare data and information through the application of computers and computer technologies. In reality, it is more about applying information in the healthcare field than it is about technology per se. Technology merely facilitates the collection, storage, transmission and analysis of data. This field also includes data standards (such as HL7) and controlled medical vocabularies (such as SNOMED) we will cover in the chapter on data standards.

The definition of medical informatics is dynamic because the field is relatively new and rapidly changing. The following are four definitions frequently cited:

> *"science of information, where information is defined as data with meaning. Biomedical Informatics is the science of information applied to, or studied in the context of biomedicine. Some, but not all of this information is also knowledge"*[3]

> *"scientific field that deals with resources, devices and formalized methods for optimizing the storage, retrieval and management of biomedical information for problem solving and decision making"* [4]

> *"application of computers, communications and information technology and systems to all fields of medicine - medical care, medical education and medical research"* [5]

> *"understanding, skills and tools that enable the sharing and use of information to deliver healthcare and promote health"* [6]

Medical informatics is also known as *health informatics, clinical informatics* and *bioinformatics* in some circles. However, the consensus is that *bioinformatics* involves the integration of biology and technology and can be defined as the:

> *"analysis of biological information using computers and statistical techniques; the science of developing and utilizing computer databases and algorithms to accelerate and enhance biological research"* [7]

Some prefer *biomedical informatics* because it encompasses bioinformatics and medical, dental, nursing, public health, pharmacy, medical imaging and veterinary informatics.[8] As we move closer to integrating human genetics into the day-to-day practice of medicine this more global definition may gain traction. We have chosen to use medical informatics throughout the book for consistency.

Health information technology (HIT or healthIT) is defined as the "application of computers and technology in healthcare settings".

Health information management (HIM) traditionally focused on the paper medical record and coding. With the advent of the electronic health record HIM specialists now have to deal with a new set of issues, such as privacy and multiple new concepts such as voice recognition. For a discussion of the definition, concepts and implications (e.g. distinguishing from other related fields) we refer you to a 2010 article by Bernstam, Smith and Johnson [3] and a 2009 article by Hersh. [9]

Background

Given the fact that most businesses incorporate technology into their enterprise fabric, one could argue that it was just a matter of time before the tectonic forces of medicine and technology collided. As more medical information was published and more medical data became available as a result of computerization, the need to automate, collect and analyze data escalated. Also, as new technologies such as electronic health records appeared, ancillary technologies such as disease registries, voice recognition and picture archiving and communication systems arose to augment functionality. In turn, these new technologies prompted the need for expertise in health information technology that spawned new specialties and careers.

Medical informatics emphasizes *information brokerage;* the sharing of a variety of information back and forth between people and healthcare entities. Examples of medical information that needs to be shared include: lab results, x-ray results, vaccination status, medication allergy status, consultant's notes and hospital discharge summaries. Medical informaticians harness the power of information technology to expedite the transfer and analysis of data, leading to improved efficiencies and knowledge. The field also interfaces with other fields such as the clinical sciences, computer sciences, biomedical engineering, biology, library sciences and public health, to mention a few.

Information drives health information technology (HIT) and interacts with many important functions in healthcare organizations and serves as a common thread that facilitates these functions (figure 1.4). This is one of the reasons the Joint Commission created the management of information standard for hospital certification.[10]

Many aspects of medical informatics noted in figure 1.4 are interconnected. To accomplish data collection and analysis there are hospital information systems (HISs) that collect financial, administrative and clinical information and subsystems such as the laboratory (LISs) and radiology information systems (RISs). As an example, a healthcare organization is concerned that too many of its diabetics are not well controlled and believes it would benefit by offering a diabetic web portal. With a portal, diabetics can upload blood sugars and blood pressures to a central web site so that diabetic educators and/or clinicians can analyze the results and make recommendations. The following technologies and issues are involved with just this one initiative and we will cover each in other chapters:

- The web based portal involves consumer (patient) informatics and telemedicine
- Management of diabetes requires online medical resources, evidence based medicine, clinical practice guidelines and an electronic health record with a disease registry
- If the use of the diabetic web portal improves diabetic control, clinicians may be eligible for improved reimbursement, known as pay-for-performance

There are multiple forces driving the adoption of health information technology, but the major ones are the need to: increase the efficiency of healthcare (i.e. decrease medical costs and improve outcomes), improve the quality of healthcare, resulting in improved patient safety. Over the past 40 years, there has been increasing recognition that some variation in practice cannot be justified. For example, patients in some areas of the United States are undergoing more invasive procedures than similar patients in other areas. Thus, there has been a movement to standardize the care of common (and expensive) conditions, such as coronary artery disease, congestive heart failure and diabetes. Computerized clinical practice guidelines are one way to provide advice at the point of care.

In this book we will discuss each driving force and their inter-relationships. In addition to these three forces, the natural diffusion of technology also exerts an influence. In other words, as technologies such as wireless and voice recognition become more common place, easier to use and less expensive, new applications will arise that have applicability to the field of medicine. Technological innovations appear at a startling pace as demonstrated by Moore's Law.

"The number of transistors on a computer chip doubles every 1.5 years "[11]
Gordon Moore, co-founder Intel Corporation 1965

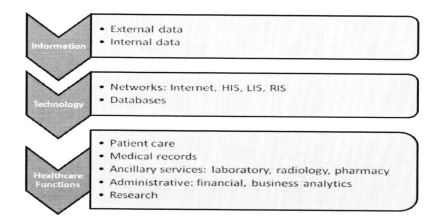

Figure 1.4 Information, information technology and healthcare functions

Moore's Law describes the exponential growth of transistors in computers. Technology will continue to evolve at a rapid rate but it is important to realize that it often advances in an asynchronous manner. For example, laptop computers have advanced greatly with excellent processor speed and memory but their utility is limited by a battery life of roughly 4-6 hours. This is a significant limitation given the fact that most nurses now work 8-12 hour shifts, so short battery life is one factor that currently limits the utility of laptop computers in healthcare.

The electronic health record (EHR), covered in chapter two, could be considered the centerpiece of medical informatics with its potential to improve patient safety, medical quality, productivity and data retrieval. Although the current adoption rate is low, EHRs will likely become the focal point of all patient encounters in the future. Multiple resources that are currently standalone programs are being incorporated into the EHR, e.g. electronic prescribing, physician/patient education, genetic profiles and artificial intelligence. It is anticipated that EHR use will eventually be shown to improve patient outcomes like morbidity and mortality as a result of decision support tools that decrease medication errors and standardize care with embedded clinical guidelines. However, at present, because EHRs do not adequately support clinicians' information needs and workflow, they do little to improve patient care and in some cases have been shown to reduce the quality of care.[12] Informaticians will play a major role in reversing this trend.

It is also important to realize that one of the outcomes of EHRs will be voluminous healthcare data. As pointed out by Steve Balmer, the CEO of Microsoft, there will be an "explosion of data" as a result of automating and digitizing multiple medical processes.[13] Adding new technologies such as electronic prescribing and health information exchanges will produce data that heretofore has not been available. This explains, in part, why technology giants such as Microsoft, Intel, IBM and Google are now entering the healthcare arena. As we begin mining medical data from entire regions or organizations we will be able to make much better evidence based decisions. As we will point out in other chapters, large organizations such as Kaiser Permanente have the necessary information technology tools and large patient population to be able to make evidence based decisions. Pooling data is essential because most practices in the United States are small and do not provide enough information on their own to show the kind of statistical significance we need to alter the practice of medicine.[14]

The federal government understands the importance of data in every sector of our economy. For that reason, they developed www.data.gov to make datasets from the federal agencies available to a multitude of interested parties, such as healthcare organizations, developers, researchers, etc. Datasets are available through several categories: "raw data", special tools and a geodata catalog.

As a result of this initiative, a variety of applications, mashups and visualizations have been developed. Similarly, the Department of Health and Human Services is working on a Health Indicators Warehouse that is projected to be functional in late 2010. Current available datasets include hundreds of health indicators that will help measure progress towards the Health People 2020 program. One of the HHS public-private projects is known as Community Health Data Initiative that will release healthcare data for anyone to use, to include software developers. The downloadable data will cover healthcare performance at the local, state and national level. As an example, prevalence of disease, quality cost and utilization data from Medicare will be available. Importantly, this initiative will be working with technology companies, researchers and others to develop applications and initiatives to improve healthcare. [15-18] The chapter on Public Health Informatics will discuss how Microsoft and Google have taken advantage of these datasets.

The introduction of information technology into the practice of medicine has been tumultuous for many reasons. Not only are new technologies expensive, they affect workflow and require advanced training. Unfortunately, this type of training rarely occurs during medical or nursing school or after graduation. More healthcare professionals who are "bilingual" in technology and medicine will be needed to realize the potential of new technologies. Vendors, insurance companies and governmental organizations will also be looking for the same expertise.

Historical Highlights

Information technology has been pervasive in the field of Medicine for only about three decades but its roots began in the 1950s. [19] Since the earlier days we have experienced astronomical advances in technology to include personal computers, high resolution imaging, the Internet, mobile technology and wireless, to mention only a few. In the beginning there was no strategy or vision as to how to advance healthcare using information technology. Now, we have the involvement of multiple federal and private agencies that are plotting future healthcare reform, partly based on health information technology. The following are some of the more noteworthy developments in health information technology:

- **Computers**. The first general purpose computer (ENIAC) was released in 1946 and required 1000 sq ft of floor space. Primitive computers such as the Commodore and Atari appeared in the early 1980s along with IBM's first personal computer, with a total of 16 K of memory.[20] Computers were first theorized to be useful for medical diagnosis and treatment by Ledley and Lusted in the 1950's.[21] They reasoned that computers could archive and process information more rapidly than humans. The programming language known as MUMPS was developed in Octo Barnett's lab at Massachusetts General Hospital in the 1970s. MUMPS exists today in the popular electronic health record know as VistA, used by the Veterans Affairs medical system [22]
- **Origin of medical informatics.** It is thought that the origin of the term medical informatics dates back to the 1960's in France ("Informatique Medicale") [23]
- **MEDLINE.** In the mid-1960s MEDLINE and MEDLARS were created to organize the world's medical literature. For older clinicians who can recall trying to research a topic using the multi-volume text *Index Medicus*, this represented a quantum leap forward
- **Artificial Intelligence.** Artificial intelligence (AI) medical projects such as MYCIN (Stanford University) and INTERNIST-1 (University of Pittsburg) appeared in the 1970s and 1980s. [24] Since 1966 AI has had many periods where research flourished and where it floundered, known as "AI winters" [12]
- **Internet.** The development of the Internet began in 1969 with the creation of the government project ARPANET.[25] The World Wide Web (WWW or Web) was conceived by

Tim Berners-Lee in 1990 and the first web browser Mosaic appeared in 1993.[26-27] The Internet is the backbone for digital medical libraries, health information exchanges and web based medical applications, to include electronic health records. Although the terms Web and Internet are often used interchangeably, the Internet is the "network of networks" consisting of hardware and software that connects computers to each other. The Web is a set of protocols (particularly related to HyperText Transfer Protocol or HTTP) that are supported by the Internet. Thus, there are many Internet applications (e.g., email) that are not part of the Web

- **Electronic Health Record.** The electronic health record has been discussed since the 1970's and recommended by the Institute of Medicine in 1991 [28]

- **Mobile technology.** The PalmPilot PDA appeared in 1996 as the first truly popular handheld computing device.[29] Personal Digital Assistants (PDAs) loaded with medical software became standard equipment for residents in training. They have been quickly supplanted by smartphones like the iPhone. PDAs and smartphones will be discussed in more detail in the chapter on mobile technology

- **Human Genome Project.** In 2003 the Human Genome Project (HGP) was completed after thirteen years of international collaborative research. Mapping all human genes was one of the greatest accomplishments in scientific history. Finalizing a draft of the genome is the first step. What remains is making sense of the data. In other words, we need to understand the difference between data (the code), information (what the code means) and knowledge (what we do with the information). [30] Data from mega-databases will likely change the way we practice medicine in the future. The HGP will be discussed in the chapter on Bioinformatics

- **Nationwide Health Information Network.** The concept was developed in 2004 as the National Health Information Infrastructure and renamed the Nationwide Health Information Network (NHIN). The goal of the NHIN is to connect all electronic health records, health information organizations and government agencies in one decade.[31] Achieving interoperability among all healthcare systems and workers in the United States will be a monumental challenge. This will be discussed in more detail in several other chapters

Key Players in Health Information Technology

Health information technology (HIT) is important to multiple players in the field of medicine. The common goals of these different groups are outlined in table 1.1

Table 1.1 Goals of HIT

Goal	Process
Improve	Communication and continuity of care Quality of care Patient outcomes Clinician productivity Return on investment
Reduce	Medical errors and resultant litigation Duplication of tests
Standardize	Medical care by individuals and organizations
Accelerate	Access to care and administrative transactions
Protect	Privacy and ensure security

In the next section we list the key players in HIT and how they utilize health information technology (adapted from *Crossing the Quality Chasm*). [32]

Patients

- Online searches for health information
- Web portals for storing personal medical information, making appointments, checking lab results, e-visits, etc
- Research choice of physician, hospital or insurance plan
- Online patient surveys
- Online chat, blogs, podcasts, vodcasts and support groups and Web 2.0 social networking
- Personal health records
- Limited access to electronic health records
- Telemedicine and home telemonitoring

Clinicians

- Online searches with MEDLINE, Google and other search engines
- Online resources and digital libraries
- Patient web portals, secure e-mail and e-visits
- Physician web portals
- Clinical decision support, e.g. reminders and alerts
- Electronic health records (EHRs)
- Personal digital assistants (PDAs) and smartphones loaded with medical software
- Telemedicine and telehomecare
- Voice recognition software
- Online continuing medical education (CME)
- Electronic prescribing
- Disease management registries
- Picture archiving and communication systems (PACS)
- Pay for performance (P4P)
- Health information organizations (HIOs)
- E-research
- Electronic billing and coding

Support staff

- Patient enrollment
- Electronic appointments
- Electronic coding and billing
- EHRs
- Web based credentialing
- Telehomecare monitoring
- Practice management software
- Secure patient-office e-mail communication
- Electronic medication administration record (e-Mar)
- Online educational resources and CME
- Disease registries

Public Health

- Incident reports
- Syndromic surveillance as part of bio-terrorism program
- Establish link to all public health departments
- Geographic information systems to link disease outbreaks with geography
- Telemedicine
- Disease registries as part of EHRs or health information exchanges
- Remote reporting using mobile technology

Government

- Nationwide Health Information Network
- Financial support for EHR adoption and health information exchange
- Development of standards, services and policies for HIT
- Information technology pilot projects and grants
- Disease management
- Pay for performance
- Electronic health records and personal health records
- Electronic prescribing
- Telemedicine
- Broadband adoption
- Health information organizations
- Regional extension centers

Medical Educators

- Online medical resources for clinicians, patients and staff
- Online CME
- MEDLINE searches
- Telehealth via video teleconferencing, podcasts, etc

Insurance Companies

- Electronic claims transmission
- Trend analysis
- Physician profiling
- Information systems for P4P
- Monitor adherence to clinical guidelines
- Monitor adherence to preferred formularies
- Promote claims based personal health records and information exchanges
- Reduce litigation by improved patient safety through fewer medication errors
- Alerts to reduce test duplication

Hospitals

- Electronic health records
- Electronic coding and billing
- Information systems to monitor outcomes, length of stay, disease management, etc

- Bar coding and radio frequency identification (RFID) to track patients, medications, assets, etc
- Wireless technology
- E-intensive care units
- Patient and physician portals
- E-prescribing
- Member of health information organizations (HIOs)
- Telemedicine
- Picture archiving and communication systems (PACS)

Medical Researchers

- Database creation to study populations, genetics and disease states
- Online collaborative web sites e.g. CaBIG
- Electronic case report forms (eCRFs)
- Software for statistical analysis of data e.g. SPSS
- Literature searches with multiple search engines
- Randomization using software programs
- Improved subject recruitment using EHRs and e-mail
- Online submission of grants

Technology Vendors

- Applying new technology innovations in the field of medicine: hardware, software, genomics, etc
- Data mining
- Interoperability

Organizations involved with HIT

Academic Organizations

Institute of Medicine (IOM). One of the leading organizations in the United States to promote health information technology is the Institute of Medicine. It was established in 1970 by the National Academy of Sciences with the task of evaluating policy relevant to healthcare and providing feedback to the Federal Government and the public. In their two pioneering books *To Error is Human* (1999) and *Crossing the Quality Chasm* (2001), they reported that approximately 98,000 deaths occur yearly due to medical errors. It is their contention that an information technology infrastructure will help the six aims set forth by the IOM: safe, effective, patient centered, timely, efficient and equitable medical care. The infrastructure would support *"efforts to re-engineer care processes, manage the burgeoning clinical knowledge base, coordinate patient care across clinicians and settings over time, support multidisciplinary team functioning and facilitate performance and outcome measurements for improvement and accountability"*. They also stress *"the importance of building such an infrastructure to support evidence based practice, including the provision of more organized and reliable information sources on the Internet for both consumers and clinicians and the development and application of decision support tools"*.

Two of the IOM's twelve executive recommendations directly relate to information technology:

> *Recommendation 7: "improve access to clinical information and support clinical decision making"*

> _Recommendation 9:_ "_Congress, the executive branch, leaders of health care organizations, public and private purchasers and health informatics associations and vendors should make a renewed national commitment to building an information infrastructure to support health care delivery, consumer health, quality measurement and improvement, public accountability, clinical and health services research, and clinical education. This commitment should lead to the elimination of most handwritten clinical data by the end of the decade_"

The IOM cites twelve information technology applications that might narrow the quality chasm. Many of these will be discussed in other chapters:

1. Web based personal health records
2. Patient's access to hospital information systems to access their lab and x-ray reports
3. Access to general health information via the Internet
4. Electronic medical records with clinical decision support
5. Pre-visit online histories
6. Inter-hospital data sharing (health information exchange), e.g. lab results
7. Information to manage populations using patient registries and reminders
8. Patient-physician electronic messaging
9. Online data entry by patients for monitoring, e.g. glucose results
10. Online scheduling
11. Computer assisted telephone triage and assistance (nurse call centers)
12. Online access to clinician or hospital performance data. [33-34]

The Association of American Medical Colleges (AAMC). For more than twenty years the AAMC has been an advocate of incorporating informatics into medical school curricula and promoting medical informatics in general. In their _Better Health 2010 Report_ they made the following recommendations:

- Optimize the health and healthcare of individuals and populations through best practice information management
- Enable continuous and life-long performance based learning
- Create tools and resources to support discovery, innovation and dissemination of research results
- Build and operate a robust information environment that simultaneously enables healthcare, fosters learning and advances science [35]

Public-Private Organizations

Bridges to Excellence. This organization consists of employers, physicians, health plans and patients. They currently have four programs incentivized by bonuses:

- Adoption of information technology systems to improve patient care
- Adoption of national guidelines for diabetes care
- Adoption of guidelines for cardiac care
- Adoption of guidelines for spine care [36]

eHealth Initiative. This is a non-profit organization promoting the use of information technology to improve quality and patient safety. Its membership includes virtually all stakeholders involved in the delivery of healthcare. This organization created the "_Connecting Communities for Better Health Program_" that provides seed money to support and connect disparate healthcare communities. They also offer the *"Connecting Communities Toolkit"* that provides guidance how to create and sustain health information organizations. They also provide an annual survey of HIOs. In 2009 they added an extensive section on navigating the ARRA (stimulus package), discussed later in this chapter.[37]

Center for Information Technology Leadership. CITL was chartered in 2002 as a research arm of the Partners Healthcare System in Boston. They make recommendations to other healthcare systems and vendors based on their research. The research areas are: telehealth, diabetes, health information exchanges and ambulatory computerized physician order entry. [38]

Leapfrog. Leapfrog is a consortium of over one hundred and seventy major employers seeking to purchase the highest quality and safest healthcare. Voluntary reporting by hospitals has made hospital comparisons possible and the results are reported on their website. They also have a hospital rewards program to provide incentives to hospitals that show they deliver quality care. One of their patient safety measures is the use of inpatient computerized physician order entry (CPOE) that will be covered in several other chapters. [39]

National Alliance for Health Information Technology. Created in 2002, this non-profit organization is comprised of senior healthcare leaders who work towards a consensus on multiple HIT issues. Members are from prominent healthcare organizations, technology vendors, the Joint Commission and PricewaterhouseCoopers. In 2008 they were funded by the Office of the National Coordinator of Health Information Technology (ONC) to develop definitions for commonly used but controversial terms, such as the electronic medical record. These new definitions will be included in several chapters. This group disbanded in August 2009.[40]

Connecting For Health. This organization is a public-private collaboration operated by the Markle Foundation and funded partially by the Robert Wood Johnson Foundation. With over 100 stakeholders, its primary mission is to promote interoperable HIT. They published *Common Framework: Resources for Implementing Private and Secure Health Information Exchange* that helps organizations exchange information in a secure and private manner, with shared policies and technical standards. The Common Framework with 9 policies guides and 7 technical guides is available free for download on their web site.[41]

National eHealth Collaborative (NeHC). This government-civilian-consumer collaborative has a board of directors of 18 members and 3 federal liaisons. It took over in early 2009 when the American Health Information Community (AHIC) was dissolved. They are charged with prioritization of HIT standards to promote interoperability. They create "value cases" and refer those for harmonization of standards and once accepted they will be adopted by the certification organizations such as the Certification Commission for Health Information Technology (CCHIT).[42]

Healthcare Information Technology Standards Panel (HITSP). This panel is a public-private partnership established in 2005 by the Department of Health and Human Services (DHHS). It is also sponsored by the American National Standards Institute (ANSI), in cooperation with HIMSS, ATI and Booz Allen Hamilton. HITSP was charged by the ONC to harmonize standards based on "use cases" derived from AHIC requirements. They have worked with 15 standards development organizations (SDOs) to identify the best standards. The HITSP recommends standards to the Secretary of HHS who accepts them and officially recognizes them one year later after a period of review. Ironically, the private sector is not required to comply with these federal standards, but nevertheless they will likely become the industry-wide standards. Specific standards are available as downloadable pdf documents on the HITSP web site. It should be pointed out that the specifications are system architecture neutral. Each interoperability specification is a suite of documents that provides a roadmap of how standards and specifications will answer the requirements of the use case. For instance, specifics of the standard for using the Continuity of Care Document (CCD) were released as C32 in March 2008 with a detailed explanation of the technical aspects. The CCD is discussed further in the chapter on data standards. The interoperability specifications (ISs) are as follows: (IS 01) EHR lab results reporting; (IS 02) Biosurveillance; (IS 03) Consumer empowerment; (IS 04) Emergency responders EHR; (IS 05) Consumer empowerment and access to clinical information via media; (IS 06) Quality and (IS 07) Medication management. It is likely this Panel will be dissolved, now that ONC has created a HIT Standards Panel. Their contract with the government was terminated in April 2010. [43]

The Certification Commission for Healthcare Information Technology (CCHIT) was created by HIMSS, AHIMA and Alliance and now includes the American College of Physicians, American Academy of Family Physicians, American Academy of Pediatrics, California HealthCare Foundation, Hospital Corporation of America, McKesson, Sutter Health, United Health Foundation and Wellpoint. Its goals are to: reduce the risk of health information technology (HIT) investment by physicians; ensure interoperability of HIT; enhance the availability of HIT incentives and accelerate the adoption of interoperable HIT. They work with HITSP to adopt standards for EHRs. Their initial step was to certify ambulatory electronic health records. By early 2010 they certify the following categories: ambulatory EHRs, inpatient EHRs, Health Information Exchanges, Emergency EHRs, Cardiovascular Medicine EHRs, Child Health EHRs and E-prescribing. Later in 2010 they intend to add Oncology, Woman's Health, Behavioral Health, Clinical Research, Dermatology and Long Term and Post-Acute Care EHRs. EHRs that have received certification are listed on the web site. The Commission consists of 20 commissioners from a variety of backgrounds and numerous volunteers in their work groups. Substantial changes have been made to the Commission, in part due to the effect of the ARRA on projected EHR use. In June 2009, after two town hall meetings, CCHIT decided they would offer three different levels of EHR certification so more EHRs would qualify for Medicare or Medicaid reimbursement under ARRA: 1) EHR-C (CCHIT certified® 2011), a comprehensive certification that would actually exceed federal standards. Certification for ambulatory EHRs could cost about $37,000 and $9,000 yearly. 2) EHR-M or modular or preliminary certification for e-prescribing, PHRs, registries, etc that meet federal standards. Certification standards would likely have to be adjusted to match "Meaningful Use" criteria, discussed in the chapter on EHRs. This category may appeal to the open source groups. Projected cost would be in the $6,000-$33,000 range, depending on the number of modules certified and annual renewal fees of between $1-5,000. 3) EHR-S or site certification would cover self-assembly of non-certified sources and cost $150-$300. It has been estimated that certification of an inpatient EHR could cost between $75,000 and $200,000. [44-45]

National Committee on Vital and Health Statistics (NCVHS) is a public advisory body to the Secretary of Health and Human Services. It is composed of 18 members from the private sector who are subject matter experts in the fields of health statistics, electronic health information exchange, privacy/security, data standards and epidemiology. They have been very involved in advising the Secretary in matters related to the Nationwide Health Information Network (NHIN). [46]

Health Information Security and Privacy Collaborative (HISPC) was established by a research organization that was funded by the Agency for Healthcare Research and Quality. Their goals are to identify best practices and develop solutions for interoperable electronic health information. [47]

US Federal Government

The federal government has maintained that information technology is essential to improving the quality of medical care and containing costs. It is a major financer of health care with the following programs: Medicare/Medicaid, Veterans Health Administration, Military Health System, Indian Health Service and the Federal Employees Health Benefits Program. It is therefore no surprise that they are heavily involved in health information technology and stand to benefit greatly from an interoperable Nationwide Health Information Network. Agencies such as Medicare/Medicaid and the Agency for Healthcare Research and Quality conduct HIT pilot projects that potentially could improve the quality of medical care and/or decrease medical costs. The federal government has recognized the importance of technology in multiple areas and as a result has a new federal chief technology officer and chief technology officer for HHS.

Before specific government agencies are discussed we will outline the new programs included in the American Recovery and Reinvestment Act of 2009 that impact the information sciences.

American Recovery and Reinvestment Act (ARRA). Without a doubt, the most significant recent governmental initiative that affected the field of Informatics was the ARRA. This legislation will impact HIT adoption, particularly EHRs, as well as training and research. ARRA has five broad goals: (a) improve medical quality, patient safety, healthcare efficiency and reduce health disparities; (b) engage patients and families; (c) improve care coordination; (d) ensure adequate privacy and security of personal health information (e) improve population and public health. Title IV and XIII of ARRA, known as the Health Information Technology for Economic and Clinical Health (HITECH) Act was devoted to funding of HIT programs. Table 1.2 summarizes the major pertinent programs that have monies dedicated for these initiatives. The HealthIT web site under the DHHS outlines the details of many of the programs listed in the table. (www.healthit.hhs.gov) In addition to the programs listed in table 1.2 the following are also important initiatives that were part of the ARRA

- Privacy and HIPAA changes; to be discussed in chapter on privacy and security
- $4.7 billion for the National Telecommunications and Information Administration's Broadband Technology Opportunities Program. This will fund the National Broadband plan discussed in the chapter on telemedicine
- $2.5 billion for USDA's Distance Learning, Telemedicine and Broadband Program
- $85 million for Indian Health Services HIT programs
- $500 million for Social Security Administration HIT programs
- $50 million for Veterans Affairs (VA) HIT programs
- Department of Labor released $225 million in February 2010 to support 55 programs in 30 states in healthcare, IT and other high growth fields
- Mandates the Secretary of HHS to conduct a study on the impact of open source electronic health records for "community clinics" and report the findings by October 1 2010 [48]

Table 1.2 Significant ARRA and HITECH programs that impact the information sciences and HIT

Program or Recipient	Amount	Programmatic details
ONC	$2 billion	Discretionary money to develop the support for multiple programs. Establish Privacy Officer, HIT Standards Committee and HIT Policy Committee within ONC
States	$548 million	Support for statewide health information exchanges. As of mid-2010 56 states, territories and other entities have been funded
NIST	$20 million	Develop HIT standards
HRSA	$1.5 billion	Upgrade community health centers to include HIT initiatives, such as EHRs
AHRQ, NIH	$1.1 billion	Develop comparative effective research (CER) programs
Physicians	$19-$30 billion	Medicare and state administered Medicaid will reimburse physicians for meaningful use of certified electronic health records (EHRs). Details outlined in the chapter on EHRs
US States	$647 million	Create 60 Regional Extension Centers to promote HIT, particularly EHRs for primary care physicians in rural areas. Goal is to support 100,000 clinicians in 2 years
HIT Research Center	$50 million	Collect feedback from the regional extension centers, in order to generate "lessons learned"
Beacon Communities	$220 million	Beacon Program will support 15-20 communities that serve as role models because of their early adoption of HIT
Community Colleges	$36 million	Support 5 consortia to rapidly create programs to be used at 70 community colleges. Students (non-degree) are expected to graduate in less than six months and be knowledgeable in IT, healthcare, workflow, redesign of practices, change strategies and quality improvement techniques.
Community Colleges	$10 million	Support 5 curriculum development centers that create material to be used by community colleges. The curriculum must cover 20 topics included in **Appendix A** (end of chapter). Four short term positions have been identified: implementation support specialist, practice workflow and information management redesign specialist, clinician consultant and implementation manager.
Competency Exam Program	$6 million	Support one center to create a competency exam. There will be no charge for the first 10,000 students to take the exam
University Based Training	$32 million	Support 8 institutions to develop programs for HIT professionals requiring university level training
Strategic HIT Advanced Research Projects (SHARP)	$60 million	Awarded to four centers in 2010. Four focus areas are: HIT security to reduce risk and cultivate technologies of trust, support clinicians to align patient centered care with their practice, improve architectures and applications to exchange information accurately and securely and secondary use of EHR data to improve quality, population health and clinical research

US Department of Health & Human Services (HHS) is the department that serves as an umbrella for most of the important government agencies that impact HIT. The Office of the National Coordinator for Health Information Technology reports directly to the Secretary of HHS and is not an agency. The following are some of the agencies under HHS:
- Agency for Healthcare Research & Quality
- Centers for Medicare & Medicaid Services
- Centers for Disease Control & Prevention
- Health Resources & Services Administration
- National Institute of Health [49]

Office of the National Coordinator for Health Information Technology (ONC). The most significant goal of (ONC) is the creation of a universal interoperable electronic health record by the year 2014. To accomplish this goal they are working to harmonize data standards to ensure interoperability and to facilitate health information exchange. ONC reorganized in December 2009, resulting in the following offices: Office of Economic Modeling and Analysis, Office of the Chief Scientist, Office of the Deputy Coordinator for Programs and Policy, Office of the Deputy National Coordinator for Operations and Office of the Chief Privacy Officer. The following is their 2010 mission statement: [50]

- Promoting development of a nationwide HIT infrastructure that allows for electronic use and exchange of information that:
    - Ensures secure and protected patient health information
    - Improves health care quality
    - Reduces health care costs
    - Informs medical decisions at the time/place of care
    - Includes meaningful public input in infrastructure development
    - Improves coordination of care and information among hospitals, labs, physicians, etc
    - Improves public health activities and facilitates early identification/rapid response to public health emergencies
    - Facilitates health and clinical research
    - Promotes early detection prevention and management of chronic diseases
    - Promotes a more effective marketplace
    - Improves efforts to reduce health disparities
- Providing leadership in the development, recognition and implementation of standards and the certification of HIT products
- Health IT policy coordination
- Strategic planning for HIT adoption and health information exchange
- Establishing governance for the Nationwide Health Information Network

In summary, ONC is responsible for coordinating all aspects of health information technology in the United States. They are involved with the adoption, standards harmonization, interoperability, privacy/security and certification of electronic health records. In addition they are coordinating the efforts to create the Nationwide Health Information Exchange (NHIN). They participate with and support multiple private and public health information technology initiatives.

In 2009 Dr. David Blumenthal was selected as the National Coordinator by President Obama. The next two federal advisory committees discussed are part of ONC and were created as part of the ARRA.

Health IT Policy Committee (HITPC). The main goal of this committee is to set priorities regarding what standards are needed for information exchange and establish the policy framework for the development and adoption of national health information exchange. The committee has 20 multi-disciplinary members with the chair being Dr. Blumenthal. In 2010 three working groups focused on: Meaningful Use; the adoption and the certification process of EHRs and information exchange. They will also evaluate eight HIT areas that require policy and prioritization and these areas are noted in the transcripts from their meetings posted on their web site. [50]

Health IT Standards Committee (HITSC). This committee has 23 multi-disciplinary members and chaired by Jonathan Perlin. They are tasked to look at standards, implementation specifications and certification criteria for the exchange of health information. They will likely focus on issues that are prioritized by HITPC. They will use the National Institute of Standards and Technology (NIST) to test standards. Both committees will make recommendations to the National Coordinator. They have established 3 working groups: clinical quality, clinical operations and privacy/security. [50]

Agency for Healthcare Research and Quality (AHRQ). The AHRQ is *"the lead Federal agency charged with improving the quality, safety, efficiency, and effectiveness of health care for all Americans. As one of 12 agencies within the Department of Health and Human Services, AHRQ supports health services research that will improve the quality of health care and promote evidence based decision making"*. This agency sets aside significant grant money to support healthcare information technology (HIT) each year. Since 2004 AHRQ has invested about $280 million in grants to research HIT. The AHRQ also maintains the National Resource Center for HIT and an extensive patient safety and quality section. They also maintain an extensive HIT Knowledge Library with over 6,000 resources. [51]

Centers for Medicare and Medicaid Services (CMS). CMS is responsible for providing care to 44 million Medicare and 48 million Medicaid patients (2007 data). In an effort to improve quality and decrease costs, CMS has information technology pilot projects in multiple areas, to include pay for performance demonstration projects that link payments to improved patient outcomes. They will reimburse for Meaningful Use of certified EHRs. Several projects will be discussed in later chapters. [52-53]

Centers for Disease Control and Prevention (CDC). Although not a primary HIT site, the CDC has used HIT to promote population health-related issues. Among their programs of interest:
- Public Health Information Network (PHIN), covered in the chapter on public health informatics
- Human Genome Epidemiology Network (HuGENET™) correlates genetic information with public health
- Family History Public Health Initiative is a web site that records family history information and encourages saving it in a digital format so it can be shared. Discussed more in chapter on bioinformatics
- Public Health Image Library contains photos, images and videos on medical topics
- Geographic information systems (GIS) are also covered in chapter on public health informatics
- Podcasts, RSS feeds and web widgets on medical topics
- Online Health Library
- Mobile Pilot Project September-December 2009 to text message patients about public health issues [54]

Health Resources and Services Administration (HRSA) is part of HHS with the primary mission of assisting medical care for the underserved and uninsured in the United States. As noted in the section on the ARRA, HRSA will support grants for community health centers to include the installation and upgrades of health information technology. They have been a long term grant supporter of telemedicine. On their site they post a 2008 monograph entitled *"The Underserved and Health Information Technology: Issues and Opportunities"*. [55]

State Governments and HIT

There are a variety of state-based HIT initiatives, evaluating the adoption of technologies such as electronic health records and e-prescribing. State Medicaid offices are anxious to conduct pilot projects aimed at reducing costs and/or improving quality of care. The State Alliance for e-Health was created in 2006 in an attempt to navigate the issues of best practices, policies and adoption obstacles. Support for the Alliance is from ONC as well as a private-public advisory committee. They have three task forces: health information protection, health care practice-health information communication and data exchange taskforces. Their highest priorities are e-prescribing and the privacy and security of health information.[56]

Barriers to Health Information Technology Adoption

According to Anderson, the United States is at least 12 years behind many industrialized nations, in terms of HIT adoption. Total investment in 2005 per capita was 43 cents, compared to $21 for Canada, $4.93 for Australia, $21 for Germany and $192 for the United Kingdom.[57] Healthcare organizations tend to spend only 3-4% of their budget on information technology, which is far less than other information dependent industries.[58] Healthcare information technology adoption has multiple barriers listed below and discussed in later chapters:

- **Inadequate time**. This complaint is a common thread that runs throughout most discussions of technology barriers. Busy clinicians complain that they don't have enough time to read, learn new technologies or research vendors. They are also not reimbursed to become technology experts. They usually have to turn to physician champions, local IT support or others for technology advice

- **Inadequate expertise and workforce**. In order for the United States to experience widespread HIT adoption and implementation, it will require education of all healthcare workers. According to Dr. Blumenthal (National Coordinator for Health Information Technology) the United States will need approximately 51,000 skilled health informaticians over the next 5 years to create, install and maintain HIT.[50] Dr. William Hersh of the Oregon Health and Science University, echoes the need for a work force capable of leading implementation of the electronic health record and other technologies.[59] Educational offerings will need to be expanded at universities, community colleges and medical schools. There is a substantial difference between different medical schools and healthcare organizations, in terms of HIT sophistication. The first Work Force for Health Information Transformation Strategy Summit, hosted by the American Medical Informatics Association (AMIA) and the American Health Information Management Association (AHIMA) made several strategic recommendations regarding how to improve the work force.[60] The American Medical Informatics Association has been the leader in attempting to increase the health information technology workforce with its *AMIA 10x10 Program*.[61] Their goal is to train 10,000 skilled workers in the next 10 years. Clearly, with the new influx of federal government support from the stimulus package there is a great need for health informaticians. In addition to skilled informaticians, we will need to educate residents in training and faculty at medical schools, given the rapidly changing nature of HIT. The APA Summit on Medical Student Education Task Force on Informatics and Technology recommended that instead of CME, we need "longitudinal, skills-based tutoring by informaticians".[62] Family Medicine residency programs are generally ahead of other specialty training programs in regards to IT training. They also recommend a longitudinal approach to IT competencies [63]

- **Cost**. It is estimated that a Nationwide Health Information Network (NHIN) will cost $156 billion dollars over five years and $48 billion annually in operating expenses.[64] Technologies such as picture archiving and communications systems (PACS) and electronic health records are also very expensive. The ARRA will help underwrite the initial purchase of some technologies but long term support will be a different challenge

- **Lack of interoperability**. Electronic health records and the NHIN cannot function until data standards are adopted and implemented nationwide. Interoperability and data standards are covered in more detail in other chapters

- **Change in workflow**. Significant changes in workflow will be required to integrate technology into the inpatient and outpatient setting. As an example, clinicians may be accustomed to ordering lab or x-rays by giving a handwritten request to a nurse who actually places the order. Now they have to learn to use computerized physician order entry (CPOE). As with most new technologies, older users have more difficulty changing their habits, even if it will eventually save time or money. According to Dr Carolyn Clancy, the director of AHRQ:

    *"The main challenges are not technical; it's more about integrating
    HIT with workflow, making it work for patients and clinicians who
    don't necessarily think like the computer guys do"* [65]

- **Privacy**. The Health Insurance Portability and Accountability Act (HIPAA) of 1996 was created initially for the portability, privacy and security of personal health information (PHI) that was largely paper-based. HIPAA regulations were updated in 2009 to better cover the electronic transmission of PHI or (ePHI). This Act has caused healthcare organizations to re-think healthcare information privacy and security. This will be covered in more detail in the chapter on privacy and security. In the past few years there have been a series of privacy breeches and stolen identities in healthcare organizations, thus adding to the angst

- **Legal.** The Stark and Anti-kickback laws prevent hospital systems from providing or sharing technology such as computers and software with referring physicians. Exceptions were made to these laws in 2006, as will be pointed out in other chapters. This is particularly important for hospitals in order to share electronic health records and e-prescribing programs with clinician's offices. Many new legal issues are likely to appear. As an example, there has been discussion of empowering the US Food and Drug Administration to regulate electronic health records and medical software

- **Behavioral change.** Perhaps the most challenging barrier is behavior. In *The Prince* by Machiavelli, it was stated "there is nothing more difficult to be taken in hand, more perilous to conduct, or more uncertain in its success, than to take the lead in the introduction of a new order of things".[66] Dr. Frederick Knoll of Stanford University described the five stages of medical technology acceptance: (1) abject horror, (2) swift denunciation, (3) profound skepticism, (4) clinical evaluation, then, finally (5) acceptance as the standard of care.[67] It is unrealistic to expect all medical personnel to embrace technology. In 1962 Everett Rogers wrote *Diffusion of Innovations* in which he delineated different categories of acceptance of innovation:
    o the innovators (2.5%) are so motivated they may need to be slowed down
    o early adopters (13.5%) accept the new change and teach others
    o early majority adopters (34%) require some motivation and information from others in order to adopt
    o the late majority (34%) require encouragement to get them to eventually accept the innovation
    o laggards (16%) require removal of all barriers and often require a direct order [68]

It is important to realize, therefore, that at least 50% of medical personnel will be slow to accept any information technology innovations and they will be perceived as dragging their feet or being "Luddites". [69] With declining reimbursement and emphasis on increased productivity, clinicians have a natural and sometimes healthy dose of skepticism. They dread widespread implementation of anything new unless they feel certain it will make their lives or the lives of their patients better. In this situation, selecting clinical champions and conducting intensive training are critical to implementation success

- **Health Information Technology: Hype versus Fact.** The Gartner IT Research Group describes 5 phases of the hype-cycle that detail the progression of technology from the "technology trigger" to the "peak of inflated expectations" to the "trough of disillusionment" to the "slope of enlightenment" to the "plateau of productivity".[70] Figure 1.5 shows the hype curve for a variety of IT technologies for 2009.

As already noted, clinicians tend to be leery about new technologies that promise a lot, but deliver little. As a rule, if technology doesn't save time or money physicians are not interested. Importantly, current studies that evaluate HIT are often lacking for multiple reasons, discussed in these articles. [71-72]

Both the RAND Corporation and the Center for Information Technology Leadership reported in 2005 that HIT would save the US about $80 billion annually.[38, 73] The Congressional Budget Office (CBO), on the other hand, refuted this optimistic viewpoint in May 2008. They published a monograph entitled *Evidence on the Costs and Benefits of Health Information Technology* that reviews the evidence on the adoption and benefits of HIT, the costs of implementing, possible factors to explain the low adoption rate and the role of the federal government in implementing HIT. The bottom line for the CBO is that "*By itself, the adoption of more health IT is generally not sufficient to produce significant cost savings.*"[74]

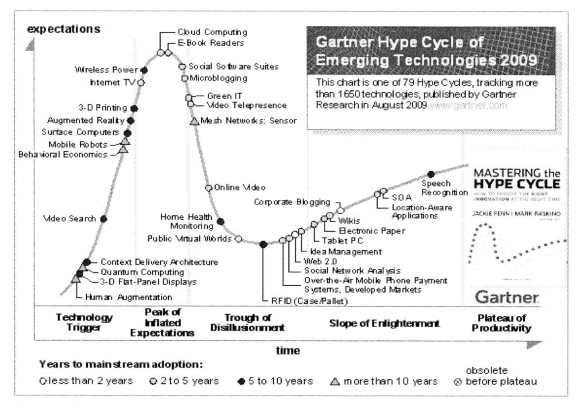

Figure 1.5 Gartner Hype Cycle 2009 (Courtesy www.Gartner.com)

In 2009 Himmelstein et al. reported on an annual survey of computerization of 4000 hospitals from 2003-2007. He concluded that hospital computing modestly improves process quality measures but does not reduce administrative or overall costs. Moreover, hospitals included on the "Most Wired" list performed no better than less wired hospitals, in terms of quality, costs and administrative costs. [75]

There are a significant number of healthcare consultants that don't believe that greater HIT adoption will result in significant cost savings, in spite of supporting these newer technologies.[76] Importantly, Carol Diamond of the Markle Foundation points out that HIT success can't be measured by the number of hospitals that have adopted EHRs but instead whether patient outcomes improve.[77]

Medical Informatics Programs

One of the best sites to review the various medical informatics programs in the United States and overseas can be found on the American Medical Informatics Association's web site. Another excellent site for listing available medical informatics programs in the United States and the United Kingdom is the Biohealthmatics web site. Medical informatics programs can be degree, certificate, fellowship and short courses. Most programs are part of a medical or nursing school and others may be part of a health related organization such as the National Library of Medicine. Courses can be online, taught in a classroom setting or both. Medical informatics degree programs are available as follows: associate degree, undergraduate degree, Master's degree, PhD degree or part of another degree program. Master's degrees may be focused on applied training or readying students for a research career. The following table was extracted (April 2010) from information posted on the AMIA site. This will give the reader an idea of how many programs are available in North America and in which category. In addition, it will provide an idea as to the rapid growth of medical informatics programs in a relatively short period of time. [78-79]

The majority of medical informatics students in the past have come from healthcare fields. With the current economy and the new monies from the ARRA, IT professionals from other industries are enrolling in medical informatics training programs. Often these professionals bring expertise in technology implementation, evaluation and/or user training and programming skills but they often lack clinical experience in healthcare.

Table 1.3 Medical informatics programs listed on the AMIA web site

Program type	2008	2010
Associate degree	1	1
Undergraduate degree	5	9
Masters degree	65	70
PhD degree	26	30
Certificate	36	43
Short courses	13	14
Online courses	28	33
10x10 programs	3	10

Medical Informatics Organizations

The following organizations are considered among the most important and influential in health information technology:

American Medical Informatics Association (AMIA)

- Founded in 1990 by the merger of the American Association for Medical Systems and Informatics, the American College of Medical Informatics and the Symposium on Computer Applications in Medical Care
- In 2006 it became a member of the Council of Medical Specialty Societies
- As of 2009 AMIA has greater than 4000 members from clinical, technical and research sectors
- Members are from 65 countries
- In 2009 Dr. Ed Shortliffe became the President and CEO of AMIA
- Web site includes job exchange, academic programs, fellowships, grants, and an e-newsletter
- Membership includes subscription to the Journal of the American Medical Informatics Association (JAMIA)
- Opportunity to join a working group (20) to discuss issues and formulate white papers
- Annual national symposium in the fall as well as a spring Congress [78]

International Medical Informatics Association

- Began in 1967 but became officially an independent endorsed organization in 1989
- Membership consists of national, institutional, affiliate members and honorary fellows
- AMIA is the US representative to the IMIA
- IMIA supports the triennial World Congress on Medical and Health Informatics, known as Medinfo. Next meeting is September 13-16 2010
- IMIA supports multiple working groups and special interest groups
- *Applied Clinical Informatics* is the new 2010 journal for the IMIA [80]

Healthcare Information and Management Systems Society (HIMSS)

- Founded in 1961
- As of 2007 has over 20,000 members
- 300 corporations are members
- Annual symposium with more than 20,000 attendees
- Professional certification
- Educational publications, books and CD-ROMs
- Web conferences on medical informatics topics
- HIMSS Analytics is a subsidiary that provides data and analytic expertise
- Surveys on multiple topics [81]

American Health Information Management Association (AHIMA)

- Founded in 1928 for medical records librarians
- As of 2010 has more than 59,000 members
- It began as a medical records association but now includes any healthcare worker involved in information management. It offers seven credentials related to four areas: Coding, HIM, privacy and analysis
- *"AHIMA supports the common goal of applying modern technology to and advancing best practices in health information management"*
- AHIMA web site has an excellent HIT resource section

- AHIMA journal is available on their web site at no cost [60]

Alliance for Nursing Informatics (ANI)

- Combines 25 separate nursing informatics organizations
- As of 2007 has more than 3,000 members
- Sponsored by both the AMIA and HIMSS
- Provides a collaborative group for consensus about nursing informatics [82]

American Telemedicine Association

- Established in 1993 to promote telecommunications technology
- Has transitioned to telemedicine, telehealth or eHealth
- Mission is to promote remote access to medical care through telemedicine technology
- Official journal is *Telemedicine and e-Health* [83]

Medical Informatics Careers

The timing is excellent for a career in medical informatics. With the emphasis on increasing adoption of electronic health records and health information exchange coupled with support from the stimulus package there has been tremendous interest in medical informatics. Healthcare organizations and HIT vendors will be looking for workers who are knowledgeable in both technology and medicine. The Department of Labor estimates that there will be 4% growth in the demand for trained health informatics specialist in multiple areas in the private, federal and military sectors. Informaticians will be needed to design, implement and govern many new technologies arriving on the medical scene, as well as train users. It is anticipated that government reimbursement for EHRs and support for health information exchange will only increase the need for skilled HIT workers. The Biohealthmatics, HIMSS, American Nurse Informatics, Health IT News, AHIMA and the AMIA web sites list multiple interesting health IT jobs. Examples include nurse and physician informaticists, systems analysts, information directors, chief information officers (CIOs) and chief medical information officers (CMIOs).[60,78,79,81-82,84] Recruiting organizations also maintain multiple listings for health IT jobs.

The American Medical Informatics Association is in the process of establishing the medical subspecialty of *Clinical Informatics*. It is likely that it will take several years to have this new specialty approved by the American Board of Medical Specialties (ABMS) so individuals can be board certified in this new specialty. In the 2009 March/April issue of the JAMIA, the core content for this new specialty is spelled out.[85-86]

Although physicians can become chief medical information officers in very large organizations, the reality is that nurses and others have the greatest potential to be involved with IT training and implementation at the average hospital or large clinic. Larger, more urban clinics may have the luxury of in-house IT staff, unlike smaller and more rural practices. Notably, nursing already has an informatics specialty certification. Nurse informaticists today are likely to help implement IT initiatives such as EHRs, bar coding and e-prescribing.

Medical Informatics Resources

Because of the rapidly changing nature of technology it is difficult to find resources that are current. It is also difficult to find resources that are not overly technical that would be appropriate for the medical informatics neophyte. There are numerous excellent journals, e-journals and e-newsletters that contain articles that discuss important aspects of health information technology. Because medical informatics is gaining popularity in the field of medicine many excellent articles can also be found in major medical journals that do not normally focus on technology. As an example, *Health Affairs*, a bimonthly journal features web exclusives, blogs and e-newsletters of interest to informaticians.[87] Furthermore, several informatics-related web sites link to the major national and international medical informatics print and online journals.[88-91]

Books:

- Handbook of Biomedical Informatics. Wikipedia Books. 2009 [92]
- Guide to Health Informatics. Enrico Coiera. 2003 [93]
- Biomedical Informatics: Computer Applications in Health Care and Biomedicine. EH Shortliffe and J Cimino 2006 [94]
- Sabbatini RME Medical Informatics: Concepts, Methodologies, Tools and Applications. J Tan. Four Volumes. 2009 [95]

Journals:

- *Journal of the American Medical Informatics Association* is the bimonthly journal of the AMIA. It features peer reviewed articles that run the gamut from theoretical models to practical solutions. The journal is included in the AMIA membership and is most appropriate for medical and IT professionals [96]
- *International Journal of Medical Informatics* is an international monthly journal that covers information systems, decision support, computerized educational programs and articles aimed at healthcare organizations. In addition to standard articles, they publish short technical articles and reviews [97]
- *Journal of Biomedical Informatics* was formally known as *Computers and Biomedical Research*. Its editor is Dr. Ed Shortliffe and the emphasis of this bimonthly journal is bioinformatics [98]
- *Journal of AHIMA* is published 11 months of the year for its members to stay current in health information management-related issues [99]
- *CIN: Computers, Informatics, Nursing* is a bimonthly print journal targeting the nursing professional. Also offers PDA downloads, RSS feeds and a newsletter [100]

E-journals:

- *BMC Medical Informatics and Decision Making* is an open-access free online journal publishing peer-reviewed research articles. This journal is part of BioMed Central, an online publisher of 188 online free full text journals. Because it is an open-access model it allows for much more rapid review and publication, a plus for informatics journals. www.biomedcentral.com/bmcmedinformdecismak/ [101]
- *The Open Medical Informatics Journal* is another open-access free online journal that publishes Medical Informatics research articles and reviews. Bentham Science publishes 89 online and print journals as well as 200 online open-access journals. An abstract is available online and the full text pdf copy is downloadable. www.bentham.org/open/tominfoj/ [102]
- *Journal of Medical Internet Research (JMIR)* is an independent open-access online journal that publishes articles related to medicine and the Internet. The articles are free to read in an html format but there is a cost to download articles in a pdf format or to become a member. http://www.jmir.org/ [103]

- *Electronic Journal of Health Informatics (eJHI)* is an Australian-based international open access electronic journal that offers open access (no fee) to both authors and readers. http://ejhi.net [104]
- *Applied Clinical Informatics* is the fee-based e-journal for the International Medical Informatics Association (IMIA) and the Association of Medical Directors of Information Systems (AMDIS). Its first issue appeared in early 2010. http://www.schattauer.de/index.php?id=558&L=1 [105]
- *Perspectives in Health Information Management* is the open-access research peer-reviewed e-journal for AHIMA, published four times a year. http://perspectives.ahima.org [106]

Informatics-Related E-newsletters:

- *iHealthBeat* is a free daily e-mail newsletter on health information technology published as a courtesy by the California Healthcare Foundation. It is also available through RSS feeds, Twitter and they offer frequent podcasts [107]
- *HealthCareITNews* is available as a daily online, RSS feed or print journal. It is published in partnership with HIMSS and reviews broad topics in HIT. They also publish the online e-journals *NHINWatch, MobileHealthWatch* and *Health IT Blog* [84]
- *eHealth SmartBrief* is a free newsletter e-mailed three times weekly. In addition to broad coverage of HIT, they offer RSS feeds, blogs, reader polls and job postings [108]
- *Health Data Management* offers a free daily e-newsletter, in addition to their comprehensive web site. The web site offers 20 channels or categories of IT information, webinars, whitepapers, podcasts and RSS feeds [109]

Online Resource Sites:

- *University of West Florida Medical Informatics Program Resource Site* augments this book with valuable web links organized in a similar manner as the book chapters. It also includes links to excellent informatics newsletters and journals [110]
- *Agency for Healthcare Research and Quality Knowledge Library* is another excellent resource with over 6,000 articles and other resources that discuss health information technology related issues [111]
- *Family Medicine Digital Resources Library* was created by Dr. Tom Agresta and supported by the Society of Teachers of Family Medicine to promote Informatics education of Family Medicine physicians. In early 2010 they posted 14 presentations that are available to the public [112]

Informatics Blogs
- *HealthIT Buzz* is a new Blog offered on the HHS HealthIT web site [113]
- *Life as a Health CIO* by Dr. John Halamka offers insights from his perspective as CIO of Harvard Medical School and Beth Israel Deaconess Medical Center [114]
- *The Health Care Blog* is hosted by Matthew Holt and considered to be "a free-wheeling discussion of the latest healthcare developments" to include health information technology [115]
- *E-CareManagement* focuses on chronic disease management, technology, strategy, issues and trends. Content is posted by Vince Kuraitis, a HIT consultant for Better Health Technologies [116]
- *Health Informatics Forum* is an international forum dedicated to health informatics professionals and students. Has extensive web links [117]

- *Biological Informatics* was created by Marcus Zillman to compile multiple biomedical informatics sites (100+) into one, as well as a blog [118]
- *HealthTechtopia* compiles the top 50 medical informatics blogs. It is subdivided into General Health Informatics, Anatomy & Physiology, Information Science and Information Technology, Computer Science, Statistics and Radiology and Medical Imaging [119]
- *Biomedexperts* is a free social network for biomedical researchers. They have created groups based on what articles have been published by the scientists involved. The claim to have profiles on 1.8 million biomedical researchers from 190 countries. Profiles were generated from the last 10 years of PubMed. In this manner research networks can be created [120]
- *EMR & HIPAA Blog* hosted by John Lynn that covers EHRs, HIPAA and HIT issues [121]

Key Points

- Medical informatics focuses on the science of information, as applied to healthcare and biomedicine

- Health information technology (HIT) holds promise for improving healthcare quality, reducing costs and expediting the exchange of information

- The 2009 HITECH Act will be a driver of HIT in the immediate future

- Barriers to widespread adoption of HIT include: time, cost, privacy, change in workflow, legal, behavioral barriers and lack of high quality studies

- Many new degree and certificate programs are available in Medical Informatics

Conclusion

Medical Informatics is a new, exciting and evolving field. New specialties and careers are now possible. In spite of its importance and popularity, significant obstacles remain. Health information technology has the potential to improve medical quality, patient safety, educational resources and patient-physician communication, while decreasing cost. Although technology holds great promise, it is not the solution for every problem facing medicine today. As noted by Dr. Safron of the American Medical Informatics Association *"technology is not the destination, it is the transportation"*.[78]

Research in Medical Informatics is being published at an increasing rate so hopefully new approaches and tools will be evaluated more often and more objectively. Better studies are needed to demonstrate the effects of health information technology on actual patient outcomes and return on investment, rather than studies based solely on surveys and expert opinion.

The effects of the multiple programs supported by the HITECH Act will likely be both transformational and challenging for the average practitioner.

Acknowledgements

We would like to thank Dr. Irmgard Willcockson (UT-Houston) for her contributions to the Medical Informatics Programs section.

Appendix A: Workforce Curriculum

Topic	Description
History of HIT in USA	Traces the development of IT systems in health care and public health, beginning with the experiments of the 1950s and 1960s and culminating in the HITECH act. Introduces the concept of meaningful use.
HIM systems	A "theory" component, specific to health care and public health applications. Introduction to health IT standards, health-related data structures, software applications; enterprise architecture in health care and public health organizations
Working with HIT systems	A laboratory component. Students will work with simulated systems or real systems with simulated data. As they play the role of practitioners using these systems, they will learn what is happening "under the hood." They will experience threats to security and appreciate the need for standards, high levels of usability, and how errors can occur. Materials must support hands-on experience in computer labs and on-site in health organizations.
Networking and HIM	More in-depth analysis of data mobility including the hardware infrastructure (wires, wireless, and devices supporting them), the ISO stack, standards, Internet protocol, federations and grids, the NHIN and other nationwide approaches.
Quality improvement	Introduces the concepts of health IT and practice workflow redesign as instruments of quality improvement. Addresses establishing a culture that supports increased quality and safety. Discusses approaches to assessing patient safety issues and implementing quality management and reporting through electronic systems.
Usability and human factors	Discussion of rapid prototyping, user-centered design and evaluation, usability; understanding effects of new technology and workflow on downstream processes; facilitation of a unit-wide focus group or simulation.
Intro to healthcare and public health	A survey of how health care and public health are organized and services delivered in the U.S. Covers public policy, relevant organizations and their interrelationships, professional roles, legal and regulatory issues, and payment systems. Must also address health reform initiatives in the U.S
The culture of healthcare	For individuals not familiar with health care, this course addresses job expectations in health care settings. It will discuss how care is organized inside a practice setting, privacy laws, and professional and ethical issues encountered in the workplace
Terminology in healthcare and public health settings	Explanation of specific terminology used by workers in health care and public health. Note that this is NOT a course in data representation or standards.
Intro to information and computer science	For students without an IT background, provides a basic overview of computer architecture; data organization, representation and structure; structure of programming languages; networking and data communication. Includes basic terminology of computing.
Installation and maintenance of HIT systems	Instruction in installation and maintenance of health IT systems, including testing prior to implementation. Introduction to principles underlying configuration. Materials must support hands-on experience in computer labs and on-site in health organizations
Fundamentals of health workflow	Fundamentals of health workflow process analysis and redesign as a necessary component of complete practice automation; includes topics of process validation and change management.

Configuring EHRs	A practical experience with a laboratory component, addressing approaches to assessing, selecting, and configuring EHRs to meet the specific needs of customers and end-users.
Public health IT	For individuals specifically contemplating careers in public health agencies, an overview of specialized public health applications such as registries, epidemiological databases, biosurveillance, and situational awareness and emergency response. Includes information exchange issues specific to public health.
Special topics course on vendor specific systems	Provides an overview of the most popular vendor systems highlighting the features of each as they would relate to practical deployments, and noting differences between the systems
Customer service in the health environment	Development of skills necessary to communicate effectively across the full range of roles that will be encountered in health care and public health settings
Working in teams	An experiential course that helps trainees become "team players" by understanding their roles, the importance of communication, and group cohesion
Planning, management and leadership for HIT	For those preparing for leadership roles, principles of leadership and effective management of teams. Emphasis on the leadership modes and styles best suited to IT deployment.
Intro to project management	An understanding of project management tools and techniques that results in the ability to create and follow a project management plan.
Training and instructional design	Overview of learning management systems, instructional design software tools, teaching techniques and strategies, evaluation of learner competencies, maintenance of training records, and measurement of training program effectiveness.

References

1. Ackoff RL. From data to wisdom. J Appl Syst Anal 1989;16:3-9
2. The DIKW Model of Innovation. www.spreadingscience.com (Accessed February 21 2010)
3. Bernstam EV, Smith JW, Johnson TR. What is biomedical informatics? Biomed Inform 2010;43(1):104-10
4. Shortliffe, E .What is medical informatics? Lecture. Stanford University, 1995.
5. MF Collen. Preliminary announcement for the *Third World Conference on Medical Informatics, MEDINFO 80*, Tokyo
6. UK Health Informatics Society http://www.bmis.org (Accessed September 5 2005)
7. Center for Toxicogenomics http://www.niehs.nih.gov/nct/glossary.htm (Accessed September 10 2005)
8. Biohealthmatics http://www.biohealthmatics.com/knowcenter.aspx (Accessed September 5 2008)
9. Hersh WR. A stimulus to define informatics and health information technology BMC Medical Informatics and Decision Making. 2009;9. www.biomedcentral.com/1472-6947/9/24 (Accessed November 4 2009)
10. The Joint Commission http://www.jcrinc.com/8250/ (Accessed March 18 2007)
11. Intel http://www.intel.com/technology/silicon/sp/glossary.htm (Accessed September 4 2005)

12. Computational Technology for Effective Health Care. Immediate Steps and Strategic Directions. 2009. National Academies Press. Stead WW and Li HS, editors http://books.nap.edu/openbook.php?record_id=12572&page=R1 (Accessed June 16 2010)

13. Balmer S. Keynote Address 2007 HIMSS Conference. February 26 2007

14. Nyweide DJ, Weeks WB, Gottlieb DJ et al. Relationship of Primary Care Physicians' Patient caseload With Measurement of Quality and Cost Performance. JAMA 2009;302(22):2444-2450

15. Data.Gov www.data.gov (Accessed June 5 2010)

16. Community Health Data Initiative www.hhs.gov/open/datasets/communityhealthdata.html (Accessed June 5 2010)

17. Community Health Data Initiative www.cdc.gov/nchs/data_access/chdi.htm (Accessed June 5 2010)

18. Health People. www.healthypeople.gov/hp2020/Objectives/TopicAreas.aspx (Accessed June 5 2010)

19. Sabbatini RME. Handbook of Biomedical Informatics. Wikipedia Books. Pedia-Press. 2009. Germany. http://en.wikipedia.org/wiki/Wikipedia:Books/BiomedicalInformatics

20. A history of computers http://www.maxmon.com/history.htm (Accessed September 30 2005)

21. Hersh WR. Informatics: Development and Evaluation of Information Technology in Medicine JAMA 1992;267:167-70

22. Laboratory of Computer Science. Massachusetts General Hospital www.lcs.mgh.harvard.edu (Accessed December 14 2009)

23. VUMC Dept. of Biomedical Informatics http://www.mc.vanderbilt.edu/dbmi/informatics.html (Accessed Oct 1 2005)

24. Health Informatics http://en.wikipedia.org/wiki/Medical_informatics (Accessed September 20 2005)

25. Howe, W. A Brief History of the Internet http://www.walthowe.com/navnet/history.html (Accessed September 24 2005)

26. Zakon, R. Hobbe's Internet Timeline v8.1 http://www.zakon.org/robert/internet/timeline (Accessed September 24 2005)

27. W3C http://www.w3.org/WWW/ (Accessed September 25 2005)

28. Advance for Healthcare Executives http://www.xwave.com/healthcare/cms/about_us/doc/industry_analysis_electronic_medical_record_system_3.doc (Accessed October 1 2005)

29. Koblentz, E. The Evolution of the PDA http://www.snarc.net/pda/pda-treatise.htm (Accessed Oct 3 2005)

30. Human Genome Project. US Dept of Energy http://www.ornl.gov/sci/techresources/Human_Genome/home.shtml (Accessed Oct 5 2005).

31. FAQ's about NHII. Dept. of Health and Human Services. http://aspe.hhs.gov/sp/nhii/FAQ.html (Accessed September 28 2005)

32. Crossing the Quality Chasm: A new health system for the 21st century (2001) The National Academies Press http://www.nap.edu/books/0309072808/html/ (Accessed October 5 2005)

33. Crossing the Chasm with Information Technology. Bridging the gap in healthcare. First Consulting Group July 2002 http://www.chcf.org/documents/ihealth/CrossingChasmIT.pdf (Accessed September 20 2005)

34. To Error is Human: Building a safer Healthcare System (1999) The National Academies Press http://www.nap.edu/catalog/9728.html (Accessed October 5 2005)

35. Association of American Medical Colleges. Better health 2010. http://www.aamc.org/programs/betterhealth/betterhealthbook.pdf (Accessed October 4 2005)

36. Bridges To Excellence http://www.bridgestoexcellence.org/bte/bte_overview.htm (Accessed June 10 2009)

37. E-health Initiative http://www.ehealthinitiative.org/default.mspx (Accessed June 10 2009)

38. Center for Information Technology Leadership www.citl.org (Accessed January 4 2008)

39. The Leapfrog Group http://www.leapfroggroup.org/ (Accessed October 5 2008)

40. National Alliance for Health Information Technology www.nahit.org (Accessed May 20 2009)

41. Connecting for Health www.connectingforhealth.org (Accessed June 15 2009)

42. National eHealth Collaborative. www.nationalehealth.org (Accessed September 30 2009)

43. Health Information Technology Standards Panel www.hitsp.org (Accessed May 10 2009)

44. Certification Commission for Health Information Technology www.cchit.org (Accessed June 18 2009)

45. Cost of New CCHIT EHR Certifications. EMR and HIPAA. www.emrandhipaa.com (Accessed October 1 2009)

46. National Committee on Vital and Health Statistics http://ncvhs.hhs.gov (Accessed May 22 2009)

47. Health Information Security and Privacy Collaborative. www.phdsc.org/hispc.htm (Accessed June 2 2008)

48. American Recovery and Reinvestment Act of 2009 Public Law 111 – 5. February 17 2009 http://en.wikisource.org/wiki/American_Recovery_and_Reinvestment_Act_of_2009 (Accessed March 2 2009)

49. Department of Health and Human Services. www.hhs.gov (Accessed May 12 2009)

50. Office of the National Coordinator for Health Information Technology. http://healthit.hhs.gov (Accessed June 3 2010)

51. Agency for Healthcare Research and Quality http://www.ahrq.gov/ (Accessed May 12 2009)

52. Centers for Medicare and Medicaid Services. Medicare Demonstrations http://www.cms.hhs.gov/DemoProjectsEvalRpts/MD/ (Accessed June 13 2009)

53. Centers for Medicare & Medicaid www.cms.hhs.gov (Accessed January 23 2010)

54. Centers for Disease Control and Prevention www.cdc.gov (Accessed January 24 2010)

55. Health Resources and Service Administration www.hrsa.gov (Accessed July 10 2009)

56. First Annual Report and Recommendations from the State Alliance for E-Health http://www.nga.org/Files/pdf/0809EHEALTHREPORT.PDF (Accessed June 12 2009)

57. Anderson GF et al. Health Care Spending and use of Information Technology in OECD countries. Health Affairs 2006;25:819-831

58. EHR and the Return on Investment. HIMSS 2003. www.himss.org/content/files/ehr-roi.pdf. Accessed December 1 2007

59. Hersh W .Health Care Information Technology JAMA 2004; 292 (18):2273-441

60. American Health Information Management Association site http://www.ahima.org (Accessed June 14 2009)

61. AMIA 10 x 10 Program. http://www.amia.org/10x10/partner.asp (Accessed June 14 2009)

62. Hilty DM, Benjamin S, Briscoe G et al. APA Summit on Medical Student Education Task Force on Informatics and Technology: Steps to Enhance the Use of Technology in Education Through Faculty Development, Funding and Change Management. Acad Psych 2006;30:444-450

63. Recommended Curriculum Guidelines for Family Medicine Residents http://www.aafp.org/online/etc/medialib/aafp_org/documents/about/rap/curriculum/medicalinformatics.Par.0001.File.tmp/Reprint288.pdf (Accessed June 12 2009)

64. Basch P et al .Electronic health records and the national health information network: affordable, adaptable and ready for prime time? Ann Intern Med 2005 143(3):165-73

65. Interview with Dr. Carolyn Clancy. Medscape June 2005. www.medscape.com (Accessed November 4 2005)

66. Machiavelli N, The Prince Chapter VI www.constitution.org/mac/prince06.htm (Accessed September 26 2008

67. Knoll, F. Medical Imaging in the Age of Informatics. Stanford University. November 15 2005

68. Rogers EM, Shoemaker FF. Communication of Innovation 1971 New York, The Free Press

69. Luddite www.wikipedia.org/luddite (Accessed November 1 2009)

70. Gartner hype cycle http://gsb.haifa.ac.il/~sheizaf/ecommerce/GartnerHypeCycle.html (Accessed November 21 2007)

71. Shcherbatykh I, Holbrook A, Thabane L et al. Methodologic Issues In Health Informatics Trials: The Complexities of Complex Interventions. JAMIA 2008; 15:575-580

72. Goldzweig CL, Towfligh A, Maglione M et al. Costs and Benefits of Health Information Technology: New Trends From the Literature. Health Affairs 2009 28 (2) w282-w293 www. content.healthaffairs.org/cgi/content/abstract/28/2/w282-w293 (Accessed February 4 2009)

73. Girosi, Federico, Robin Meili, and Richard Scoville. 2005. *Extrapolating Evidence of Health Information Technology Savings and Costs*. Santa Monica, Calif. RAND Corporation http://rand.org/pubs/research_briefs/RB9136/index1.html (Accessed May 20 2008)

74. Congressional Budget Office Paper: Evidence on the Costs and Benefits of Health Information Technology www.cbo.gov (Accessed May 20 2008)

75. Himmelstein DU, Wright A, Woolhandler S. Hospital Computing and the Costs and Quality of Care: A National Study. Am J Med 2009; 123(1):40-46

76. Laszewski R. Health IT Adoption and the Other myths of Health Care Reform. January 12 2009 www.ihealthbeat.org (Acessed January 12 2009)

77. Diamond CC, Shirky C. Health Information Technology: A Few Years of Magical Thinking? Health Affairs www.healthaffairs.org 2008;27 (5): w383-w390 (Accessed September 3 2008)

78. AMIA www.amia.org (Accessed June 10 2009)

79. Biohealthmatics   www.Biohealthmatics.com (Accessed October 15 2008)

80. International Medical Informatics Association www.imia.org (Accessed January 11 2010)

81. Health Information Management Systems Society www.himss.org (Accessed June 12 2009)

82. Alliance for Nursing Informatics http://www.allianceni.org/ (Accessed June 12 2009)

83. American Telemedicine Association www.atmeda.org (Accessed November 25 2009)

84. Health IT Job Spot. http://jobspot.healthcareitnews.com/home/5815_rec.cfm?site_id=5815 (Accessed June 16 2009)

85. Gardner RM, Overhage JM, Steen EB et al Core Content for the Subspecialty of Clinical Informatics JAMIA 2009;16 (2):153-157

86. Detmer DE, Munger BS, Lehmann CU. Clinical Informatics Board Certification: History, Current Status, and Predicted Impact on the Clinical Informatics Workforce. Appl Clin Inf 2009;1:11-18

87. Health Affairs. http://content.healthaffairs.org (Accessed June 14 2009)

88. Informatics Journals. www.dmoz.org/Health/Medicine/Informatics/Journals/ (Accessed June 18 2009)
89. Journals in Medical Informatics. Informatics Review www.informatics-review.com/journals/index.html (Accessed June 18 2009)
90. Health Informatics Journals and Publications www.hiww.org/jou.html (Accessed June 18 2009)
91. Online health informatics journals www.hi-europe.info/library/hi_journals.htm (Accessed June 18 2009)
92. Handbook of Biomedical Informatics.Wikipedia Books. http://pediapress.com (Accessed February 12 2010)
93. Guide to Health Informatics. Enrico Coiera. 2003. Arnold Publications
94. Biomedical Informatics: Computer Applications in Health Care and Biomedicine. EH Shortliffe and J Cimino. 2006 Springer.New York, NY
95. Sabbatini RME. Medical Informatics: Concepts, Methodologies, Tools and Applications. J Tan. Four Volumes. 2008. Information Science Reference. Hershey, PA
96. Journal of the American Medical Informatics Association http://www.amia.org/mbrcenter/pubs/jamia/ (Accessed June 10 2009)
97. International Journal of Medial Informatics http://www.sciencedirect.com/science/journal/13865056 (Accessed June 13 2009)
98. Journal of Biomedical Informatics http://www.elsevier.com/wps/find/journaldescription.cws_home/622857/description#description (Accessed February 10 2010)
99. Journal of AHIMA http://journal.ahima.org (Accessed February 14 2010)
100. CIN: Computers, Informatics, Nursing www.cinjournal.com (Accessed June 18 2009)
101. BMC Medical Informatics and Decision Making. www.biomedcentral.com/bmcmedinformdecismak/ (Accessed June 20 2009)
102. The Open Medical Informatics Journal www.bentham.org/open/tominfoj/ (Accessed June 21 2009)
103. The Journal of Medical Internet Research. http://www.jmir.org/ (Accessed June 21 2009)
104. Electronic Journal of Health Informatics http://ejhi.net (Accessed February 15 2010)
105. Applied Clinical Informatics. http://www.schattauer.de/index.php?id=558&L=1 (Accessed February 15 2010)
106. Perspectives in Health Information Management http://perspectives.ahima.org (Accessed February 15 2010)
107. Ihealthbeat www.ihealthbeat.org (Accessed June 1 2008)
108. eHealth SmartBrief http://www.smartbrief.com/news/EHEALTH/index.jsp?categoryid=7B651A9C-543B-43A9-909D-CC5F80F69335 (Accessed June 20 2009)
109. Health Data Management www.healthdatamanagement.com (Accessed June 16 2009)
110. University of West Florida. Introduction to Medical Informatics resource page www.uwf.edu/sahls/medicalinformatics/ (Accessed July 15 2009)
111. Agency for Healthcare Research and Quality. Knowledge Library. http://healthit.ahrq.gov/portal/server.pt?open=512&objID=653&parentname=CommunityPage&parentid=10&mode=2. (Accessed June 20 2009)
112. Family Medicine Digital Resources Library http://www.fmdrl.org/1503 (Accessed November 20 1009)
113. Health IT Buzz www.healthit.hhs.gov/blog/onc (Accessed February 20 2010)
114. Life as a Healthcare CIO http://geekdoctor.blogspot.com (Accessed March 1 2010)
115. The Health Care Blog www.thehealthcareblog.com (Accessed February 20 2010)

116. E-care management http://e-caremanagement.com (Accessed February 20 2010)

117. Health Informatics Forum www.healthinformaticsforum.com (Accessed February 20 2010)

118. Biological Informatics http://biologicalinformatics.blogspot.com (Accessed February 20 2010)

119. HealthTechTopia http://mastersinhealthinformatics.com/2009/top-50-health-informatics-blogs/ (Accessed February 20 2010)

120. Biomedexperts www.biomedexperts.com (Accessed February 20 1010)

121. EMR & HIPAA Blog. www.emrandhipaa.com (Accessed February 20 2010

2

Electronic Health Records

ROBERT E. HOYT
BRANDY G. ZIESEMER

Learning Objectives

After reading this chapter the reader should be able to:
- State the definition and history of electronic health records
- Describe the limitations of paper based health records
- Identify the benefits of electronic health records
- List the key components of an electronic health record
- Describe the ARRA-HITECH programs to support electronic health records
- Describe the benefits and challenges of computerized order entry and clinical decision support systems
- State the obstacles to purchasing and implementing an electronic health record
- Enumerate the steps to purchase an EHR

> *"If computers get too powerful, we can organize them into a committee.*
> *That will do them in"*
>
> Bradley's Bromide

There is no topic in Medical Informatics as important, yet controversial, as the electronic health record (EHR). In spite of their significance, the history of EHRs in the United States is relatively short. The Problem Oriented Medical Information System (PROMIS) was developed by The Medical Center Hospital of Vermont in collaboration with Dr. Lawrence Weed, the originator of the problem oriented record and SOAP formatted notes. Ironically, the inflexibility of the concept led to its demise.[1] In a similar time frame the American Rheumatism Association Medical Information System (ARAMIS) appeared. All findings were displayed as a flow sheet. The goal was to use the data to improve the care of rheumatologic conditions.[2] Other EHR systems began to appear throughout the US: the Regenstrief Medical Record System (RMRS) developed at Wishard Memorial Hospital, Indianapolis; the Summary Time Oriented Record (STOR) developed by the University of California, San Francisco; Health Evaluation Through Logical Processing (HELP) developed at the Latter Day Saints Hospital, Salt Lake City and The Medical Record developed at Duke University [3], the Computer Stored Ambulatory Record (COSTAR) developed by Octo Barnett at Harvard and the De-Centralized Hospital Computer Program (DHCP) developed by the Veterans Administration. [4]

In 1970 Schwartz optimistically predicted *"clinical computing would be common in the not too distant future".*[5] In 1991 the Institute of Medicine (IOM) recommended electronic health records as a solution for many of the problems facing modern medicine.[6] Since the IOM recommendation, little progress has been made during the last decade for multiple reasons. As Dr. Donald Simborg states, the slow acceptance of electronic health records is like the "*wave that never breaks*".[7]

The American Recovery and Reimbursement Act (ARRA) of 2009 was a major game changer for electronic health records, with reimbursement for the Meaningful Use of certified EHRs, as well as other programs that supported EHR education and health information exchange. Reimbursement details will be discussed in more detail later in this chapter.

We will primarily discuss outpatient (ambulatory) electronic health records. Inpatient EHRs share many similarities to ambulatory EHRs but the scope, price and complexity are different.

Electronic Health Record Adoption

Outpatient (Ambulatory) EHR Adoption: The adoption rate of ambulatory EHRs has been reported to be in the 10-20% range, depending on which study you read and what group is studied.[8] Many of the commonly quoted statistics come from surveys, with their obvious shortcomings. It is also important to realize that many outpatient practices may have EHRs but continue to run dual paper and electronic systems or may use only part of the EHR. It should also be noted that EHRs are being purchased largely by primary care practices, as opposed to surgical specialties, which may skew the statistics. Furthermore, a significant concern is that small and/or rural practices are more likely to lack the finances and information technology support to purchase and implement EHRs.

In 2008, an article in the New England Journal of Medicine (NEJM) reported the adoption rate of outpatient EHRs. In this study a sample of 5000 physicians was selected from the AMA master file. Osteopaths, residents and federal physicians were excluded. The return rate of the survey was just over 60%. The most significant finding was that only 4% of respondents reported using a comprehensive EHR (order entry capability and decision support), whereas 13% reported using a basic EHR system. As has been reported before, the adoption rate was higher for large medical groups or medical centers. Given the fact that most experts believe only comprehensive EHRs will impact patient safety and improve the quality of medical care, the 4% adoption rate is disturbing. Importantly, responding physicians did report multiple beneficial effects of using EHRs.[9] The National Ambulatory Medical Care Survey, conducted by the Centers for Disease Control on 2000 non-federal ambulatory physicians in 2009 reported that 44% of respondents had a full or partial EHR (that could include paper and electronic) but only 20% had a "basic system" and 6% had a "fully functioning "EHR.[10] In summary, very few practices (particularly small rural primary care) have a comprehensive EHR with robust order entry, clinical decision support and reporting capability.

Inpatient EHR Adoption: The American Hospital Association reported on the 2006 use of EHRs with more than 1,500 community hospitals responding. They noted that 68% of the hospitals surveyed had installed inpatient EHRs, but only 11% were fully implemented and these were mainly by large urban and/or teaching hospitals. In only 10% were physicians using computerized physician order entry (CPOE) to order medications, at least 50% of the time.[11]

In March 2009 an article about inpatient EHR adoption appeared in the New England Journal of Medicine by the same authors of the ambulatory NEJM study cited above. Of interest, both NEJM articles were co-authored by Dr. David Blumenthal, the National Coordinator for Health Information Technology. They surveyed all members of the American Hospital Association and had a return rate of 63% (3049 hospitals). Their results showed that 7.6% of the respondents reported a basic EHR system and only 1.5% reported a comprehensive EHR. Again, large urban and/or academic centers had the highest adoption rates. User satisfaction rates were not reported.[12]

A HIMSS Analytics study looked at data from over 5000 US hospitals to determine the actual level or degree of EHR adoption in 2008 and 2009. The scale they used rated hospitals from 0, meaning hospitals with an EHR with no functionality installed, to 7 indicating a fully functional paperless system that is interoperable and capable of advance reporting. As of March 2010, only seven hospital systems in the US had attained level 7 adoption (up from 2 in 2009).[13] (Table 2.1)

Table 2.1 EHR adoption statistics by stage in US (courtesy HIMSS Analytics)

Stage	Cumulative Capabilities	2008	2009
7	Paperless system. Able to generate Continuity of Care Document (CCD). Data warehousing in use	0.3%	0.7%
6	Physician documentation (structured templates), full clinical decision support systems (CDSS) and computerized physician order entry. Full picture archiving and communication systems (PACS)	0.5%	1.6%
5	Closed loop medication administration	2.5%	3.8%
4	CPOE and CDSS	2.5%	7.4%
3	Clinical documentation (flow sheets), CDSS, PACS available outside radiology	35.7%	50.9%
2	Clinical data repository (CDR), CDSS, document imaging	31.4%	16.9%
1	Lab, radiology and pharmacy modules installed	11.5%	7.2%
0	Lab, radiology and pharmacy modules not installed	15.6%	11.5%

One can only speculate why the medical profession has been willing to tolerate the lack of legible and accessible information for so many years. Many physicians believe that purchasing an EHR is not their responsibility and therefore someone else should pick up the tab. Others are concerned that they will purchase the wrong system and waste money and others are simply overwhelmed with the task of implementing and training for a completely different system. As a group, physicians are not noted for embracing innovation. In their defense, new technologies should be shown to improve patient care, save time or money, in order to be accepted.

There are approximately 275 EHR vendors but only about ten to twenty seem to be consistently successful in terms of a large client base. If the selection and purchase of EHRs was easy they would already be universal. As the reader will see later in this chapter, there are issues such as workflow, implementation and training that are just as important as the decision which EHR to purchase.

The United States is not the only country to face the challenge of trying to have a nationwide interoperable electronic health record. Canada planned for a universal EHR by 2009, Australia by 2010 and Great Britain by 2014, although delays have almost universally been reported.[14] According to a 2009 Study in Health Affairs, there is considerable difference in EHR and HIT adoption by developed countries. Based on their randomized surveys of primary care physicians the following EHR adoption rates were reported: Netherlands 99%, New Zealand 97%, Norway 97%, United Kingdom 96%, Australia 95%, Sweden 94%, Italy 94%, Germany 72%, France 68%, United States 46% and Canada 37%. Clearly, only a few countries had a high percent of fully functional EHRs such as Australia, Netherlands and New Zealand.[15] Importantly, because many of these other countries are smaller, more homogeneous, have regulated health care and may use only one EHR, adoption and implementation of EHRs is easier.

The HITECH Act and EHR Reimbursement

Arguably, the most significant EHR-related initiative occurred in 2009 as part of the American Recovery and Reinvestment Act (ARRA). Two major parts of ARRA, Title IV and Title XIII are known as the Health Information Technology for Economic and Clinical Health or HITECH Act. Approximately $20 billion was dedicated for Medicare and Medicaid reimbursement for EHRs to clinicians and hospitals. In this chapter we will primarily focus on reimbursement to eligible professionals (EPs) and not hospitals or Medicare Advantage organizations, even though they are also potentially reimbursable. In June 2010 the Centers for Medicare and Medicaid Services (CMS) revealed a new web site that explains the EHR Incentive Programs at www.cms.gov/EHRIncentivePrograms that we will summarize in the following sections.

In order for a clinician to participate in this program they must be: 1. Eligible 2. Use a certified EHR 3. Demonstrate Meaningful Use, 4. Register for reimbursement and 5. Receive reimbursement.

1. **Eligible Professionals (EPs).**
   Medicare: According to the Notice of Proposed Rule Making (NPRM) released on January 13 2010, Medicare defined EPs as doctors of medicine or osteopathy, doctors of dental surgery or dental medicine, doctors of podiatric medicine, doctors of optometry and chiropractors. Hospital based physicians such as pathologists and emergency room physicians are not eligible for reimbursement. Hospital-based is defined as providing 90% or more of care in a hospital setting. The exception is if more than 50% of a physician's total patient encounters in a six-month period occur in a federally qualified health center or rural health clinic. Also, in April 2010 the law was modified to include hospital-based outpatient physicians. Physicians may select reimbursement by Medicare or Medicaid, but not both. Also, they cannot receive Medicare EHR reimbursement and reimbursement by other Medicare programs such as e-prescribing or Physicians Quality Reporting Initiative (PQRI).
   Medicaid. Medicaid EPs are defined as physicians, nurse practitioners, certified nurse midwives, dentists and physician assistants (physician assistants must provide services in a federally qualified health center or rural health clinic that is led by a physician assistant). Medicaid physicians must have at least 30% Medicaid volume (20% for pediatricians). If a clinician practices in a federally qualified health center (FQHC) or rural health clinic (RHC), 30% of patients must be "needy individuals". The Medicaid program will be administered by the states and physicians can receive a one-time incentive payment for 85% of the allowable purchase and implementation cost of a certified EHR in the first year, even before Meaningful Use is demonstrated. [16]
2. **Certified EHRs:** An EHR will have to be certified by a recognized certifying organization. An Interim Final Rule (IFR) published January 13, 2010 listed the initial standards, implementation specifications and EHR certification criteria with a final rule published in July 2010. The final certification criteria are posted at http://www.ofr.gov/OFRUpload/OFRData/2010-17210_PI.pdf . ONC made every effort to have these standards match the Meaningful Use criteria. The following is a listing of standards (these will be covered in more detail in the chapter on data standards):
   a. Transport standards: SOAP version 1.2 and REST were removed from final rule
   b. Content exchange and vocabulary standards:

    i. Patient summary: Continuity of Care Document (CCD) or Continuity of Care Record (CCR), using HITSP C32 version 2.5
    ii. Problem summary list: ICD-9 (ICD-10) or SNOMED CT
    iii. Drug formulary benefits: NCPDP formulary and benefits standard 1.0
    iv. E-prescribing: RxNorm and NCPDP version 8.1 and 10.6
    v. Lab reporting: HL7 2.3.1 or 2.5.1, LOINC and UCUM
    vi. Public health or immunization reports: HL7 2.3.1 or 2.5.1
    vii. Quality reporting: CMS PQRI 2008 Registry XML Specification
    viii. Procedures: CPT-4 and ICD-9 (ICD-10 in 2013) [17]

It is unclear how many certifying organizations there will be in addition to The Certification Commission for Health Information Technology (CCHIT). In March 2010 ONC published its NPRM on EHR certification programs:

Temporary certification program: organizations can submit an application to ONC to become a certifying and testing organization. This was done to try to have more organizations ready by 2011. The final rule was published on June 18 2010 and outlined how organizations can become ONC-Authorized Testing and Certification Bodies. Certification by one of these bodies will show that the EHR technology is capable of supporting Meaningful Use.

Permanent certification program: many of the administrative processes assumed by ONC will move to the public sector and the permanent program will replace the temporary one. The National Institute of Standards and Technology (NIST) will be responsible for accrediting testing laboratories.[18] They will post draft testing criteria at http://healthcare.nist.gov .

CCHIT has already published criteria for 2011 EHR certification and approximately 20 EHRs have been certified. Ironically, although most certified products meet or exceed federal standards they will need to be re-tested to be certain they meet Meaningful Use criteria. For further details about CCHIT we refer you to chapter 1.

3. **Meaningful Use (MU):** The goals of MU are the same as the national goals for HIT: (a) improve quality, safety, efficiency and reduce health disparities; (b) engage patients and families; (c) improve care coordination; (d) ensure adequate privacy and security of personal health information; (e) improve population and public health. Three processes stressed by ARRA to accomplish this are: e-prescribing, health information exchange and the production of quality reports.

Meaningful Use will occur in three stages:

Stage 1 (2011): As previously noted, on January 13 2010 a NPRM was published to outline the MU criteria and those measurements that would have to be submitted to CMS. Six months later the final rule was released by HHS and included fewer criteria. The final rule divided objectives into a *core set* (15 objectives for EPs) and a *menu set*. To be a meaningful stage 1 user, participants must meet all of the core objectives (Appendix A, end of chapter) and select 5 out of 10 menu objectives (Appendix B, end of chapter). However, they must choose at least one population and public health measure. The appendices delineate criteria and measures for EPs, not hospitals.

In 2011 the results of all objectives and measures, to include clinical quality measures will be reported by clinicians and hospitals to CMS and Medicaid clinicians will report to states by attestation. Quality measures are derived from the Physician Quality Reporting Initiative (PQRI) and the National Quality Forum (NQF). Each EP will submit information on three core quality measures in 2011 and 2012 (tobacco use, blood pressure measurement and adult weight screening). They must also choose three other measures that are ready for incorporation into EHRs.

Stage 2 (2013): Will likely include all menu objectives and expand on stage 1 criteria.

Stage 3 (2015 or later): Focus on improving patient safety, quality, efficiency, decision support for core measurements, patient access to self-management tools , access to comprehensive patient data and improving population health outcomes. The following table demonstrates the fact that the longer a clinician waits to implement an EHR, the more complex the requirements to meet Meaningful Use will be. The implementation schedule, however, is approximate and not fixed. [19]

Table 2.3 Stage of Meaningful Use criteria by payment year. TBD = to be determined

First Payment Year	2011	2012	2013	2014	2015
2011	Stage 1	Stage 1	Stage 2	Stage 2	TBD
2012		Stage 1	Stage 1	Stage 2	TBD
2013			Stage 1	Stage 2	TBD
2014				Stage 2	TBD

4. **Registration.** Registration will begin in January 2011. It is known that Medicare physicians must have a National Provider Identifier (NPI) and be enrolled in the CMS Provider Enrollment, Chain and Ownership System (PECOS) to participate.

5. **Reimbursement:** It is not known how the federal government determined the reimbursement amount. In their NPRM they estimated that the cost to adopt/implement/purchase a certified EHR would be $54,000 per full time physician and approximately $10,000 annually per full time physician for maintenance and support. Medicare payments will begin in mid-May 2011 for EPs. Tables 2.2-2.3 list the Medicare and Medicaid reimbursement levels for EHRs. Medicare physicians may earn an additional 10% if they practice in a healthcare professional shortage area (HPSA). It is important to note that no monies are paid upfront and contrary to what is published by EHR vendors and others, the amount listed yearly in table 2.2 is a maximum. Physicians will be reimbursed 75% of allowable Part B charges or up to, for example, $18,000 in the first year. It is anticipated that clinicians will be paid in a single annual payment and it will have to be for at least 90 days of continuous EHR use in the first year and the entire year thereafter.

Medicare physicians who do not use a certified EHR nor demonstrate "Meaningful Use" will receive penalties of 1% in 2015, 2% in 2016 and 3% in 2017 when they bill Medicare. Penalties could reach 5% in 2018 and beyond if fewer than 75% of physicians are using EHRs at that point. In addition, late adoption will mean that more complex Meaningful Use (Stage 2 or 3) will be required, likely to make purchase and implementation more difficult.

Table 2.4 Maximum **Medicare** reimbursement for EHR adoption

Year	2011 (year 1)	2012 (year 1)	2013 (year 1)	2014 (year 1)	2015 (year 1)
2011	$18,000				
2012	$12,000	$18,000			
2013	$8,000	$12,000	$15,000		
2014	$4,000	$8,000	$12,000	$15,000	
2015	$2,000	$4,000	$8,000	$12,000	$0
2016	$0	$2,000	$4,000	$8,000	$0
Total	**$44,000**	**$44,000**	**$42,000**	**$35,000**	**$0**

Medicaid is administered by states and will use the same Meaningful Use criteria. In addition to the states being given the reimbursement money by the federal government to give to clinicians and hospitals, they will also receive 90% reimbursement for the cost of administering the program. Medicaid EPs and hospital-based physicians are not subject to possible payment reductions. Unlike Medicare, Medicaid physicians can be paid the first year just to adopt or upgrade an EHR and not yet meaningfully use the EHR. Medicaid EPs must demonstrate Meaningful Use in years 2-6. Medicaid physicians are not eligible for the 10% HPSA bonus but can receive the e-prescribing and PQRI bonuses.

Table 2.5 Maximum **Medicaid** reimbursement for EHR adoption

Eligible Clinician	Base Year: Max 85% of EHR cost	Year 1	Year 2	Year 3	Year 4	Year 5	Year 6	Total
Physician	$21,250	$8500	$8500	$8500	$8500	$8500	$8500	$63,750
Dentist	$21,250	$8500	$8500	$8500	$8500	$8500	$8500	$63,750
Nurse mid-wife	$21,250	$8500	$8500	$8500	$8500	$8500	$8500	$63,750
Physician assistant	$21,250	$8500	$8500	$8500	$8500	$8500	$8500	$63,750
Nurse practitioner	$21,250	$8500	$8500	$8500	$8500	$8500	$8500	$63,750
Pediatrician	$14,167	$5667	$5667	$5667	$5667	$5667	$5667	$42,500

Hospitals can also be reimbursed for the purchase of EHRs and can share this technology with the known limits of the "safe harbor act" discussed later in this chapter. Hospitals will start at a base of $2 million annually with decreasing amounts over five years, plus an additional amount dependent on patient volume.[15] Hospitals may receive reimbursement from both Medicare and Medicaid. [16, 20-21]

Electronic Health Record Definition

There is no universally accepted definition of an EHR. As more functionality is added the definition will need to be broadened. Importantly, EHRs are also known as electronic medical records (EMRs), computerized medical records (CMRs), electronic clinical information systems (ECIS) and computerized patient records (CPRs). Throughout this book we will use electronic health record as the more accepted and inclusive term, but either term is acceptable.

Figure 2.1 demonstrates the relationship between EHRs, EMRs and personal health records (PHRs).[22] As indicated in the diagram, PHRs can be part of the EMR/EHR system which may cause confusion.

Figure 2.1 Relationship between EHR (electronic health record), PHR (personal health record) and EMR (electronic medical record)

In May 2008 the National Alliance for Health Information Technology released the following definitions in an effort to standardize terms used in HIT:

Electronic Medical Record: "*An electronic record of health-related information on an individual that can be created, gathered, managed and consulted by authorized clinicians and staff within one healthcare organization*".

Electronic Health Record: "*An electronic record of health-related information on an individual that conforms to nationally recognized interoperability standards and that can be created, managed and consulted by authorized clinicians and staff across more than one healthcare organization*".

Personal Health Record: "*An electronic record of health-related information on an individual that conforms to nationally recognized interoperability standards and that can be drawn from multiple sources while being managed, shared and controlled by the individual*".[23]

Why do we need Electronic Health Records?

The following are the most significant reasons why our healthcare system would benefit from the widespread transition from paper to electronic health records:

- **The paper record is severely limited**. Much of what can be said about handwritten prescriptions can also be said about handwritten office notes. Figure 2.2 illustrates the problems with a paper record. In spite of the fact that this clinician used a template, the handwriting is illegible and the document cannot be electronically shared or stored. It is not structured data that is computable and hence sharable with other computers and systems. Other shortcomings of paper: expensive to copy, transport and store; easy to destroy; difficult to analyze and determine who has seen it; and the negative impact on the environment. Electronic patient encounters represent a quantum leap forward in legibility and the ability to rapidly retrieve information. Almost every industry is now computerized and digitized for rapid data retrieval and trend analysis. Look at the stock market or companies like Walmart or Federal Express. Why not the field of medicine?

  With the relatively recent advent of "pay for performance" there is a new reason to embrace technology in order to receive more reimbursement. It is much easier to retrieve and track patient data using EHRs and patient registries than to use labor intensive paper chart reviews. EHRs are much better organized than paper charts, allowing for faster retrieval of lab or x-ray results. It is also likely that EHRs will have an electronic problem summary list that outlines a patient's major illnesses, surgeries, allergies and medications. How many times does a physician open a large paper chart, only to have loose lab results fall out? How many times does a physician re-order a test because the results or the chart is missing? It is important to note that paper charts are missing as much as 25% of the time, according to one study.[24] Even if the chart is available; specifics are missing in 13.6% of patient encounters, according to another study. [25]

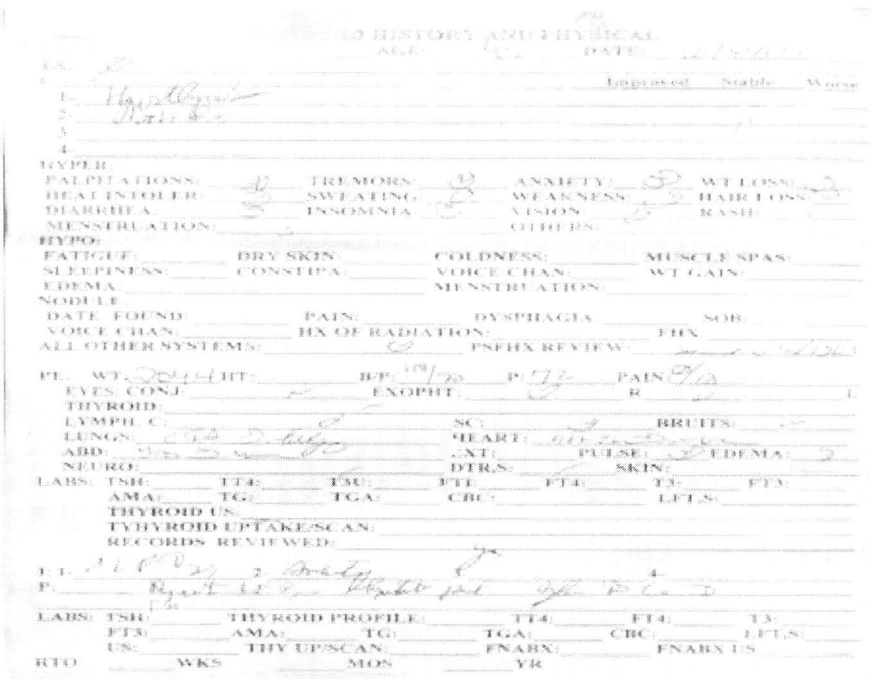

Figure 2.2 Outpatient paper-based patient encounter form

Table 2.6 shows the types of missing information and its frequency.[19] According to the President's Information Technology Advisory Committee, 20% of laboratory tests are re-ordered because previous studies are not accessible.[26] This statistic has great patient safety, productivity and financial implications.

Table 2.6 Types and frequencies of missing information

Information Missing During Patient Visits	% Visits
Lab results	45%
Letters/dictations	39%
Radiology results	28%
History and physical exams	27%
Pathology results	15%

EHRs allow easy navigation through the entire medical history of a patient. Instead of asking to "pull paper chart volume 1 of 3" to search for a lab result, it is simply a matter of a few mouse clicks. Another important advantage is the fact that the record is available 24 hours a day, 7 days a week and doesn't require an employee to pull the chart, nor extra space to store it. Adoption of electronic health records has saved money by decreasing full time equivalents (FTEs) and converting records rooms into more productive space, such as exam rooms. Importantly, electronic health records are accessible to multiple healthcare workers at the same time, at multiple locations. While a billing clerk is looking at the electronic chart, the primary care physician and a specialist can be analyzing clinical information simultaneously. Moreover, patient information should be available to physicians on call so they can review records on patients who are not in their panel. Furthermore, it is believed that electronic health records improve the level of coding. Do clinicians routinely submit a lower level of care for billing purposes because they know that handwritten patient notes are short and incomplete? Templates may help remind clinicians to add more history or details of the physical exam, thus justifying a higher level of coding (templates are disease specific electronic forms that essentially allow you to point and click a history and physical exam). A study of the impact of an EHR on the completeness of clinical histories in a labor and delivery unit demonstrated improved documentation, compared to prior paper-based histories.[27] Lastly, EHRs provide clinical decision support such as alerts and reminders, which we will cover later in this chapter

- **The need for improved efficiency and productivity.** The goal is to have patient information available to anyone who needs it, when they need it and where they need it. With an EHR, lab results can be retrieved much more rapidly, thus saving time and money. It should be pointed out however, that reducing duplicated tests benefits the payers and patients and not clinicians so there is a misalignment of incentives. Moreover, a study in 1987 using computerized order entry showed that simply displaying past results reduced duplication and the cost of testing by only 13%. [28] If lab or x-ray results are frequently missing, the implication is that they need to be repeated which adds to this country's staggering healthcare bill. The same could be said for duplicate prescriptions. It is estimated that 31% of the United State's $2.3 trillion dollar healthcare bill is for administration.[29] EHRs are more efficient because they reduce redundant paperwork and have the capability of interfacing with a billing program that submits claims electronically. Consider what it takes to simply get the results of a lab test back to a patient using the old system. This might involve a front office clerk, a nurse and a physician. The end result is frequently placing the patient on hold or playing "telephone tag". With an EHR, lab results can be forwarded via secure messaging. Electronic health records can help with productivity if templates are used judiciously. As noted, they allow for point and click histories and physical exams, thus saving time. Embedded educational content for clinicians is one of the newest features of a comprehensive EHR. Clinical practice guidelines, linked educational content and patient handouts can be part of the EHR. This may permit finding the answer to a medical question while the patient is still in the exam room. Several EHR companies also offer a centralized area for all physician approvals and signatures of lab work, prescriptions, etc. This should improve work flow by avoiding the need to pull multiple charts or enter multiple EHR modules

- **Quality of care and patient safety.** As we have previously suggested, an EHR should improve patient safety through many mechanisms: (a) Improved legibility of clinical notes (b) Improved access anytime and anywhere (c) Reduced duplication (d) Reminders that tests or preventive services are overdue (e) Clinical decision support that reminds us of patient allergies, the correct dosage of drugs, etc. (f) Electronic problem summary lists provide diagnoses, allergies and surgeries at a glance. In spite of the before mentioned benefits, a study by Garrido of quality process measures before and after implementation of a widespread EHR in the Kaiser Permanente system, failed to show improvement.[30] To date there has only been one study published that suggested use of an EHR decreased mortality. This particular EHR had a disease management module designed specifically for renal dialysis patients that could provide more specific medical guidelines and better data mining to potentially improve medical care. The study suggested that mortality was lower compared to a pre-implementation period and compared to a national renal dialysis registry. Unfortunately, the lead author was also the founder of the company that sells the software.[31] As pointed out by Wears and Berg, "*systems cannot be adequately evaluated by their developers*", not to mention the conflict of interest.[32]

  It is likely that we are only starting to see the impact of EHRs on quality. Based on internal data Kaiser Permanente determined that the drug Vioxx had an increased risk of cardiovascular events before that information was published based on its own internal data.[33] Similarly, within 90 minutes of learning of the withdrawal of Vioxx from the market, the Cleveland Clinic queried its EHR to see which patients were on the drug. Within 7 hours they deactivated prescriptions and notified clinicians via e-mail.[34]

Quality reports are far easier to generate with an EHR compared to a paper chart that requires a chart review. Quality reports can also be generated from a data warehouse or health information organization that receives data from an EHR and other sources.[35]Quality reports are the backbone for pay for performance which we will discuss further in another chapter

- **Public expectations**. According to a 2006 Harris Interactive Poll for the Wall Street Journal Online, 55% of adults thought an EHR would decrease medical errors; 60% thought an EHR would reduce healthcare costs and 54% thought that the use of an EHR would influence their decision about selecting a personal physician.[36] The Center for Health Information Technology would argue that EHR adoption results in better customer satisfaction through fewer lost charts, faster refills and improved delivery of patient educational material [37]

- **Governmental expectations**. EHRs are considered by the federal government to be transformational and integral to healthcare reform. As an example, EHR reimbursement is a major focal point of the HITECH Act. It is the goal of the US Government to have an interoperable electronic health record by 2014. In addition to federal government support, states and payers have initiatives to encourage EHR adoption. Many organizations state that we need to move from the "*cow path*" to the "*information highway*". CMS is acutely aware of the potential benefits of EHRs to help coordinate and improve disease management in older patients

- **Financial savings**. The Center for Information Technology Leadership (CITL) has suggested that ambulatory EHRs would save $44 billion yearly and eliminate more than $10 in rejected claims per patient per outpatient visit. This organization concludes that not only would there be savings from eliminated chart rooms and record clerks; there would be a reduction in the need for transcription. There would also be fewer callbacks from pharmacists with electronic prescribing. It is likely that copying, faxing and mail expenses, chart pulls and labor costs would be reduced with EHRs, thus saving full time equivalents (FTEs). More rapid retrieval of lab and x-ray reports results in time/labor saving as does the use of templates. It appears that part of the savings is from improved coding. More efficient patient encounters mean more patients could be seen each day. Improved savings to payers from medication management is possible with reminders to use the "*drug of choice*" and generics.[38]

  EHRs should reduce the cost of transcription if clinicians switch to speech recognition and/ or template use. Because of structured documentation with templates, they may also improve the coding and billing of claims.

  It is not known if EHR adoption will decrease malpractice, hence saving physician and hospital costs. A 2007 Survey by the Medical Records Institute of 115 practices involving 27 specialties showed that 20% of malpractice carriers offered a discount for having an EHR in place. Of those physicians who had a malpractice case in which documentation was based on an EHR, 55% said the EHR was helpful [39]

- **Technological Advances**. The timing seems to be right for electronic records partly because the technology has evolved. The Internet and World Wide Web make the application service provider (ASP) concept for an electronic health record possible. An ASP option means that the EHR software and patient data reside on a remote web server that you access via the Internet from the office, hospital or home. Computer speed, memory and bandwidth have advanced such that digital imaging is also a reality, so images can be part of an EHR system. Standard PCs, laptops and tablet PCs continue to add features and improve speed and memory while purchase costs drop. Wireless and mobile technologies permit access to the hospital information system, the electronic health record and the Internet using a personal digital assistant, smartphone or laptop computer. The chapter on health information exchange will point out that health information organizations can link EHRs together, in order to share information and services

- **The need for aggregated data**. In order to make evidence based decisions, we need high quality data that should derive from multiple sources: inpatient and outpatient care, acute and chronic care settings, urban and rural care and populations at risk. This can only be accomplished with electronic health records and discrete structured data. Moreover, we need to combine or aggregate data to achieve statistical significance. Although most primary care is delivered by small practices, it is difficult to study because of relatively small patient populations, making aggregation necessary. [40]

- **The EHR as a transformational tool**. It is widely agreed that US Healthcare needs reform in multiple areas. To modernize its infrastructure we would need to have widespread adoption of EHRs. Large organizations such as the Veterans Health Administration and Kaiser Permanente use robust EHRs (VistA and Epic) that generate enough data to change the practice of medicine. In 2009 Kaiser Permanente reported two studies, one pertaining to the management of bone disease (osteoporosis) and the other chronic kidney disease. They were able to show that with their EHR they could focus on patients at risk and use all of the tools available to improve disease management and population health. [41-42] In a study reported in 2009 Kaiser-Permanente reported that electronic visits that are part of the electronic health record system were likely responsible for a 26.2% decrease in office visits over a four year period. The posited that this was good news for a system that aligns incentives with quality, regardless whether the visit was virtual or face-to-face. [43] Other fee-for-service organizations might find this alarming if office visits decreased and e-visits were not reimbursed

- **Older and more complicated patients require more coordinated care**. According to a Gallup poll it is very common for older patients to have more than one physician: 0 physicians (3%), 1 physician (16%), 2 physicians (26%), 3 physicians (23%), 4 physicians (15%), 5 physicians (6%) and 6+ physicians (11%). [44]
  Having more than one physician mandates good communication between the primary care physician, the specialist and the patient. This becomes even more of an issue when different healthcare systems are involved. A 2000 *Harris Interactive* survey reported that physicians understand that adverse outcomes result from poor care coordination with chronically ill patients. [45] Nevertheless, a survey of patients with chronic conditions showed that 18% of the population received duplicate tests or procedures, 17% received conflicting information from other clinicians and 14% received different diagnoses from different physicians. [46] O'Malley et al. surveyed 12 medical practices and found that in-office coordination was improved by EHRs but the technology was not mature enough to improve coordination of care with external physicians. [47] Electronic health records are being integrated with health information organizations so that inpatient and outpatient patient-related information can be accessed and shared, thus improving communication between disparate healthcare entities. Home monitoring (telehomecare) can transmit patient data from home to an office's EHR also assisting in the coordination of care

Institute of Medicine's Vision for EHRs

The history and significance of the Institute of Medicine (IOM) is detailed in chapter 1. They have published multiple books and monographs on the direction US Medicine should take, including *The Computer-Based Patient Record: An Essential Technology for Health Care.* This visionary work was originally published in 1991 and was revised in 1997 and 2000.[48] In this book and their most recent work *Key Capabilities of an Electronic Health Record System: Letter Report* (2003) they outline eight core functions all EHRs should have:

1. **Health information and data**: In order for the medical profession to make evidence-based decisions, you need a lot of accurate data and this is accomplished much better with EHRs than paper charts. "If you can't measure it, you can't manage it."
2. **Result management**: Physicians should not have to search for lab, xray and consult results. Quick access saves time and money and prevents redundancy and improves care coordination.
3. **Order management**: CPOE should reduce order errors from illegibility for medications, lab tests and ancillary services and standardize care.
4. **Decision support**: Should improve overall medical care quality by providing alerts and reminders.
5. **Electronic communication and connectivity**: Communication among disparate partners is essential and should include all tools such as secure messaging, text messaging, web portals, health information exchange, etc.
6. **Patient support**: Recognizes the growing role of the Internet for patient education as well as home telemonitoring.
7. **Administrative processes and reporting:** Electronic scheduling, electronic claims submission, eligibility verification, automated drug recall messages, automated identification of patients for research and artificial intelligence can speed administrative processes.
8. **Reporting and population health**: We need to move from paper based reporting of immunization status and biosurveillance data to an electronic format to improve speed and accuracy. [49]

Electronic Health Record Key Components

The following components are desirable in any EHR system. The reality is that many EHRs do not currently have all of these functions.
- Clinical decision support systems (CDSS) to include alerts, reminders and clinical practice guidelines. CDSS is associated with computerized physician order entry (CPOE). This will be discussed in more detail in this chapter and the patient safety chapter
- Secure messaging (e-mail) for communication between patients and office staff and among office staff. Telephone triage capability is important
- An interface with practice management software, scheduling software and patient portal (if present). This feature will handle billing and benefits determination. We will discuss this further in the chapter on practice management systems
- Managed care module for physician and site profiling. This includes the ability to track Health plan Employer Data and Information Set (HEDIS) or similar measurements and basic cost analyses

- Referral management feature
- Retrieval of lab and x-ray reports electronically
- Retrieval of prior encounters and medication history
- Computerized Physician Order Entry (CPOE). Primarily used for inpatient order entry but ambulatory CPOE also important. This will be discussed in more detail later in this chapter
- Electronic patient encounter. One of the most attractive features is the ability to create and store a patient encounter electronically. In seconds you can view the last encounter and determine what treatment was rendered
- Multiple ways to input information into the encounter should be available: free text (typing), dictation, voice recognition and templates
- The ability to input or access information via a PDA, smartphone or tablet PC
- Remote access from the office, hospital or home
- Electronic prescribing
- Integration with a picture archiving and communication system (PACS), discussed in a separate chapter
- Knowledge resources for physician and patient, embedded or linked
- Public health reporting and tracking
- Ability to generate quality reports for reimbursement, discussed in the chapter on pay for performance
- Problem summary list that is customizable and includes the major aspects of care: diagnoses, allergies, surgeries and medications. Also, the ability to label the problems as acute or chronic, active or inactive. Information should be coded with ICD-9 or SNOMED CT so it is structured data
- Ability to scan in text or use optical character recognition (OCR)
- Ability to perform evaluation and management (E & M) determination for billing
- Ability to create graphs or flow sheets of lab results or vital signs
- Ability to create electronic patient lists and disease registries
- Preventive medicine tracking that links to clinical practice guidelines
- Security and privacy compliance with HIPAA standards
- Robust backup systems
- Ability to generate a Continuity of Care Document (CCD) or Continuity of Care Record (CCR), discussed in the data standards chapter
- Support for client server and/or application service provider (ASP) option [50]

Computerized Physician Order Entry (CPOE)

CPOE is an EHR feature that processes orders for medications, lab tests, x-rays, consults and other diagnostic tests. The majority of articles written about CPOE have discussed medication ordering only, possibly giving readers the impression that CPOE is the same as electronic prescribing. The reality is that CPOE has a great deal more functionality as we will later point out, in this and other chapters. Many organizations such as the Institute of Medicine and Leapfrog see CPOE as a powerful instrument of change. There is limited evidence that CPOE will:

1. **Reduce medication errors**. CPOE has the potential to reduce medication errors through a variety of mechanisms.[51] Because the process is electronic, you can embed rules (clinical decision support) that check for allergies, contraindications and other alerts.

a) **Potential advantages.** Koppel et al.[52] lists the following advantages of CPOE compared to paper-based systems for patient safety:

    i.   CPOE overcomes the issue of illegibility
    ii.   Fewer errors associated with ordering drugs with similar names
   iii. More easily integrated with decision support systems than paper
   iv. CPOE is easily linked to drug-drug interaction warnings
    v.   More likely to identify the prescribing physician
   vi. Able to link to adverse drug event (ADE) reporting systems
  vii. Able to avoid medication errors like trailing zeros
 viii. CPOE will create data that is available for analysis
   ix. CPOE can point out treatment and drugs of choice
    x.   Has the potential to reduce under and over-prescribing
   xi. Prescriptions reach the pharmacy quicker

b) **Inpatient CPOE.** This functionality was recommended by the IOM in 1991. Most studies so far have looked primarily at inpatient CPOE and not ambulatory CPOE. A 1998 study by David Bates in JAMA showed that CPOE can decrease serious inpatient medication errors by 55% (relative risk reduction).[53] Many of the studies showing reductions in medication errors by the use of technology were reported out of the same institution. Other hospital systems are unlikely to experience the same optimistic results. A 2008 systematic review of CPOE with CDSS by Wolfstadt et al. only found 10 studies of high quality and those dealt primarily with inpatients. Only half of the studies were able to show a statistically significant decrease in medication errors and none of the studies were randomized. Of the ten studies, seven evaluated homegrown (not commercial) CPOE/CDSS systems so results are difficult to generalize.[54]

    With the inception of CPOE we are, in fact, seeing evidence of new errors that result from technology. A December 2005 article in the medical journal *Pediatrics* suggested that the mortality rate increased from 2.8% to 6.5% after implementing Cerner's EHR at Children's Hospital of Pittsburgh. They point out however, with the new system they could not use it "*until after the patient had physically arrived*" and registered in the system. This may have led to delays in diagnosis and treatment. The situation was corrected and we will have to see if mortality rates drop back down to baseline.[55] In another article in the July 2006 issue of *Pediatrics* from Children's Hospital and Regional Medical Center in Seattle, they implemented the same EHR and found no increase in mortality. It appears that this was due to better planning and implementation. In that article Dr. Del Beccaro stated that the CPOE system eliminated handwriting errors, improved medication turnaround time and helped standardize care.[56] An article by Nebeker demonstrated that substantial ADEs continued at a VA hospital following the adoption of CPOE that lacked full decision support, such as medication alerts.[57] A study of pediatric inpatients admitted to hospitals with the Eclipsys EHR showed a reduction in preventable adverse drug events (46 vs. 26) and potential adverse drug events (94 vs. 35) compared to pre-EHR statistics.[58] Suffice it to say, clinicians and staff must be properly trained in CPOE; otherwise errors will likely increase, at least in the short term.

c) **Outpatient CPOE.** Americans made 906.5 million outpatient visits in the year 2000. By sheer numbers there is more of a chance for a medication error written for outpatients. According to an optimistic report by the Center for Information

Technology Leadership, adoption of an ambulatory CPOE system (ACPOE) will likely eliminate about 2.1 million ADEs per year in the USA. This would prevent 1.3 million ADE-related visits, 190,000 hospitalizations and more than 136,000 life threatening ADEs. They estimated ambulatory CPOE could save as much as $44 billion/year.[59] A 2007 systematic review of outpatient CPOE by Eslami was not as optimistic. They concluded that only 1 of 4 studies demonstrated reduced ADEs; 3 of 5 studies showed decreased medical costs; most showed improved guideline compliance, but it took longer to electronically prescribe and there was a high frequency of ignored alerts.[60] Kuo et al. reported medication errors from primary care settings. 70% of medication errors were related to prescribing, 10% were administration errors, 10% were documentation errors, 7% dispensing errors and 3% were monitoring errors. ADEs resulted from 16% of medication errors with 3% hospitalizations and no deaths. In their judgment, 57% of errors might have been prevented by electronic prescribing.[61]

2. **Reduce costs**. Several studies have shown reduced length of stay and overall costs in addition to decreased medication costs with the use of CPOE.[62] Tierney was able to show in 1993 an average savings of $887 per admission when orders were written using guidelines and reminders, compared to paper based ordering that was not associated with clinical decision support.[63]

3. **Reduce variation of care**. One study showed excellent compliance by the medical staff when the drug of choice was changed using decision support reminders.[64] Study conclusions should be interpreted with some note of caution. Many of the studies were conducted at medical centers with well established medical informatics programs where the acceptance level of new technology is unusually high. Several of these institutions such as Brigham and Women's Hospital developed their own EHR and CPOE software. Compare this experience with that of a rural hospital trying CPOE for the first time with potentially inadequate IT, financial and leadership support. It is likely that smaller and more rural hospitals and offices will have a steep learning curve.

On the surface CPOE seems easy, just replace paper orders with an electronic format. The reality is that CPOE represents a significant change in work flow and not just new technology. An often repeated phrase is "*it's not about the software, dummy*", meaning, regardless which software program is purchased, it requires change in work flow and extensive training.

Adoption of CPOE has been slow, partly because of cost and partly because work flow is slower than scribbling on paper.[65] Although physicians have been upset by new changes that do not shorten their work day, many authorities feel EHRs greatly improve numerous hospital functions. There has been less resistance traditionally in teaching hospitals with a track record of good informatics support. Also, young house staff who work in teaching hospitals and who write the majority of orders are more likely to be tech savvy and amenable to change. It does require great forethought, leadership, planning, training and the use of physician champions in order for CPOE to work. According to some, CPOE should be the last module of an EHR to be turned on and alerts should be phased in to bring about change more gradually. Others have recognized nurses as more accepting of change and willing to teach docs "*one-on-one*" on the wards.

For more information on CPOE we refer you to a monograph "*A Primer on Physician Order Entry*" and an article "*CPOE: benefits, costs and issues*".[66-67]

Clinical Decision Support Systems (CDSS)

Traditionally, CDSS meant computerized drug alerts and reminders to perform preventive tests as part of computerized physician order entry (CPOE) applications. Most of the studies in the literature evaluated those two functions. However, according to Hunt, CDSS is "*any software designed to directly aid in clinical decision making in which characteristics of individual patients are matched to a computerized knowledge base for the purpose of generating patient specific assessments or recommendations that are then presented to clinicians for consideration*".[68] Therefore, CDSS should have a broader definition than just alerts and reminders.

Two papers published in 2005 addressed the effects of CDSS on clinical care. Garg et al. performed a systematic review of the literature looking at how CDSS would affect practitioner performance and patient outcomes. He concluded that overall, CDSS improved performance in 64% of the 97 studies analyzed. This included CDSS involved with diagnostic, reminder, disease management and drug-dosing support. They were able to show, however, that only 13% of the 52 studies analyzed reported improvement in patient outcomes.[69] The other paper looked at those factors that contributed to the success of CDSS: automatic CDSS that was part of clinician work flow; recommendations and not just assessments; provision of CDSS at the point of care and computer based CDSS (not paper based). When these four features were present, CDSS improved clinical care about 94% of the time.[70]

According to a 2009 article, clinical decision support by nine commercial EHRs was extremely variable and tended not to offer choices.[71] Clearly, the most sophisticated CDSS are developed at medical centers with home grown EHRs and a long record of extensive HIT adoption. With Meaningful Use criteria, certified EHRs will have to conform to CDSS standards which may reduce variability.

Sheridan and Thompson have discussed various levels of CDSS: (level 1) all decisions by humans, (level 2) computer offers many alternatives, (level 3) computer restricts alternatives, (level 4) computer offers only one alternative, (level 5) computer executes the alternative if the human approves, (level 6) human has a time line before computer executes, (level 7) computer executes automatically, then notifies human, (level 8) computer informs human only if requested, (level 9) computer informs human but is up to computer and (level 10) computer makes all decisions.[72] Most EHR systems may offer alternatives and provide reminders but make no decisions on their own.

Table 2.7 outlines some of the clinical decision support available today. Calculators, knowledge bases and differential diagnoses programs are primarily standalone programs but they are slowly being integrated into EHR systems.

1. **Knowledge support**. Numerous digital medical resources are being integrated with EHRs. As an example, the American College of Physician's PIER resource is integrated into Allscript's Touch Chart.[73] UpToDate is now available in General Electric's Centricity EHR and eClinicalWorks.[74] iConsult (offered by Elsevier) is a primary care information database available for integration into EHRs. Diagnostic (ICD-9) codes can be hyperlinked to further information or you can use *infobuttons*. Other products such as Dynamed, discussed in the chapter on online medical resources are available as *infobuttons*. Figure 2.3 shows an example of iConsult integrated with the Epic EHR.[75] Another interesting integrated knowledge program is the Theradoc Antibiotic Assistant. The program integrates with an inpatient EHR's lab, pharmacy and radiology sections to make suggestions as to the antibiotic of choice with multiple alerts. Clinicians can be alerted via cell phones, pagers or e-mail. Other modules include Adverse Drug Event (ADE) Assistant, Infection Control Assistant and Clinical Alerts Assistant.[76] A study in the New England Journal of Medicine (NEJM) using this product showed considerable

improvement in the prescription of appropriate antibiotics resulting in cost saving, reduced length of stay and fewer adverse drug events.[77]

Table 2.7 Clinical decision support

Type of CDSS	Examples
Knowledge	iConsult®, Theradoc®
Calculators	Medcalc 3000®
Trending/ Patient tracking	Flow sheets graphs
Medications	CPOE and drug alerts
Order sets/protocols	CPGs and order sets
Reminders	Mammogram due
Differential diagnosis	Dxplain®
Radiology CDSS	What imaging studies to order?
Laboratory CDSS	What lab tests to order
Public health alerts	Infection disease alerts

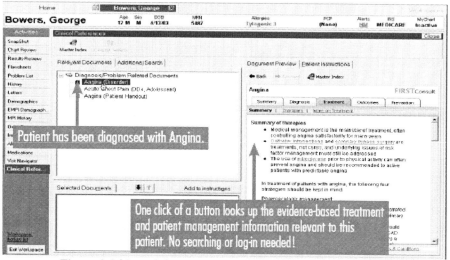

Figure 2.3 iConsult integrated with Epic EHR (courtesy iConsult)

2. **Calculators**. It is likely with time that more calculators will be embedded into the EHR, particularly in the medication and lab ordering sections. Below is the standalone online and Pocket PC based program Medcalc3000 with over 100 calculations available. They now offer a "Connect" option that will integrate with EHRs by linking calculators, clinical criteria tools, labs and decision trees.[78] Important calculations, such as kidney function (creatinine clearance) should be calculated on all patients.

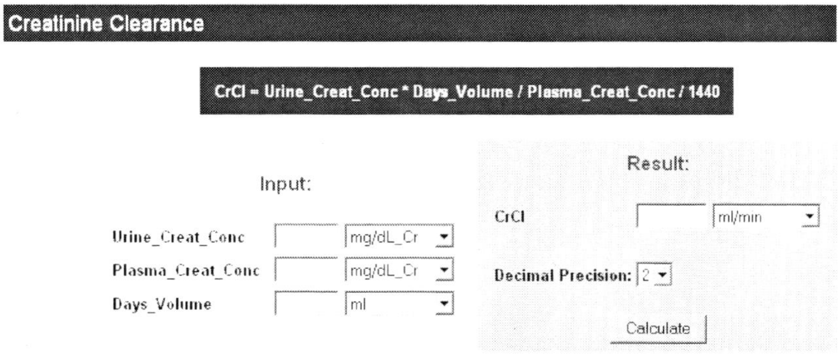

Figure 2.4 Medcalc3000 (courtesy Medcalc3000)

3. **Flow sheets, graphs, patient lists and registries**. The ability to track and trend lab results and vital signs, for example, in diabetic patients will greatly assist in their care. Furthermore, the ability to use a patient list to contact every patient taking a recalled drug will improve patient safety. Registries will be covered in more detail in the disease management chapter.

4. **Medication ordering support**. Decision support as part of CPOE possesses several rules engines to detect known allergies, drug-drug interactions, drug-condition and drug-food allergies, as well as excessive dosages. As EHRs and CPOE mature, they will factor in age, gender, weight, kidney (renal) and liver (hepatic) function of the patient, known contraindications based on known diagnoses, as well as the pregnancy and lactation status. Incorporation of these more robust features is complicated and best implemented at medical centers with an established track record of CDSS and CPOE development. As has been pointed out, there are programs that improve antibiotic ordering based on data residing in the EHR.[79] Computerized drug alerts have obvious potential in decreasing medication errors but have not been universally successful to date. According to a systematic review by Kawamoto et al. successful alerts need to be automatic, integrated with CPOE, require a physician response and make a recommendation.[80] In an interesting study of all alert overrides for 3 months at Brigham and Women's Hospital they noted that 80% of alerts were over-ridden because: 55% of respondents stated they were "*aware of the issue*"; 33% stated "*patient doesn't have this allergy*" and 10% stated "*patient already taking the medicine*". Only six percent of patients experienced ADEs from alert overrides, but half were serious. Their conclusion was that alert overrides are common but don't usually result in serious errors.[81] In a newer study of outpatient alerts at the same institution, they found better acceptance when alerts were interruptive only for critical situations. Sixty seven percent of interruptive alerts were accepted which represents an improvement. Many alerts were still incorrect so further improvements are needed.[82] In a third study from Brigham and Women's Hospital, critical lab alerts were automated to call a physician's cell phone. This strategy led to 11% quicker treatment and reduced the duration of a dangerous condition by 29%.[83] Non-drug related alerts for inpatients have a variable success record. A computer alert program at the Brigham and Women's Hospital alerted physicians about the risk of blood clots in legs and was able to show a substantial improvement in the preventive measures used, as well as a decrease in the actual number of blood clots reported in the legs and lungs compared to a group who did not receive alerts.[84] WakeMed Health and Hospitals created a robust library on their Intranet as decision support for IV drugs. They reasoned that making clinical content more accessible at the point of care would decrease pharmacy questions and promote evidence based nursing. They posted local

hospital guidelines and standard drug information that used infobuttons to quickly link to content. A survey showed that 87% of nurses used the information and 95% believed the content improved productivity and efficiency.[85] An excellent review by Kuperman et al. describes basic and advanced medication-related CDSS.[86] Further information about alerts is included in the chapters on patient safety and e-prescribing.

5. **Reminders**. Computerized reminders that are part of the EHR assist in tracking the yearly preventive health screening measures, such as mammograms. Shea performed a meta-analysis of 16 randomized controlled trials that looked at the role of computerized reminders and preventive care and concluded that there was clear benefit for vaccinations, breast cancer and colorectal screening, but not cervical cancer screening.[87] A well designed system should allow for some customization of the reminders as national recommendations change. Reminders are not always heeded by busy clinicians who may choose to ignore them. As a possible solution, preventive reminders could be reviewed by the office nurse and overdue tests ordered prior to the visit with the physician.

6. **Order sets and protocols**. Order sets are groups of pre-established inpatient orders that are related to a symptom or diagnosis. For instance, you can create an order set for pneumonia that might include the antibiotic of choice, oxygen, repeat chest x-ray, etc that saves keystrokes and time. Order sets can also reflect best practices (clinical practice guidelines), thus offering better and less expensive care. Over one hundred clinical practice guidelines are incorporated into the electronic health record at Vanderbilt Medical Center.[88] An excellent 2007 review of order sets as part of CDSS was reported in the Journal of the American Medical Informatics Association.[89]

7. **Differential Diagnoses**. Dxplain is a differential diagnosis program developed at Massachusetts General Hospital. When you input the patient's symptoms it generates a differential diagnosis (the diagnostic possibilities). The program has been in development since 1984 and is currently web based. A licensing fee is required to use this program. At this time it cannot be integrated into an EHR.[90] In spite of the potential benefit, an extensive 2005 review of CDSSs revealed that only 40% of the 10 diagnostic systems studied showed benefit, in terms of improved clinician performance.[91]

8. **Radiology CDSS**. Physicians, particularly those in training, may order imaging studies that are either incorrect or unnecessary. For that reason, several institutions have implemented clinical decision support to try to improve ordering. Appropriateness criteria have been established by the American College of Radiologists. Massachusetts General Hospital has had radiology order entry since 2001 and studied the addition of decision support. They noted a decline in low utility exams from 6% down to 2% as a result of decision support.[92]

9. **Laboratory CDSS**. It should be no surprise that clinicians occasionally order inappropriate lab tests, for a variety of reasons. It would be helpful if clinical decision support would alert them to the indications for a test, as well as the price. A Dutch study of primary care demonstrated that 20% fewer lab tests were ordered when clinicians were alerted to lab clinical guidelines.[93]

10. **Public Health Alerts.** The New York Department of Health and Mental Hygiene used Epic EHR's "Best Practice Advisory" to alert New York physicians about several infectious disease issues. The EHR-based alert also hyperlinked to disease specific order sets for educational tips, lab and medication orders.[94]

How well clinicians use CDSS programs such as those discussed, remains to be seen. They will have to be intelligently designed and rigorously tested in order to be accepted. Bates et al. offers the Ten Commandments for effective clinical decision support:

- Speed is everything: provide the information rapidly
- Anticipate needs and deliver in real time: make it convenient and provide at the point of care
- Fit into the users workflow: provide information in a screen that is logical and not in a separate standalone application
- Little things make a big difference: subtle details may make or break CDSS
- Recognize that physicians will strongly resist stopping: because of this some monitoring of clinicians who ignore important guidelines may be necessary
- Changing direction is easier than stopping: a helpful reminder to make a minor change (like lower dose of med) is better accepted than a complete change
- Simple interventions work best: guidelines should fit on a single screen
- Ask for additional information only when you really need it: clinicians will object to having to find information that is not readily available
- Monitor impact, get feedback and respond: suffice it to say that no system will work without getting feedback from the end user and tweaking it as needed
- Manage and maintain your knowledge based systems: there is a need to update recommendations as new evidence appears [95]

For more information on CDSS, we refer you to these excellent resources.[96-99]

Electronic Health Record Examples

There are approximately 275 EHRs available in the United States that vary in price from free to about $40,000 with features that vary from basic to comprehensive. We will start by mentioning a few less expensive choices that offer a moderate amount of functionality.

Open Source and Free EHRs.

Wikipedia lists more than 10 national and international open source EHRs.[100] While most open source EHRs are inexpensive, the term "free and open source" or FOSS means freedom to copy or distribute and not without cost.[101] In spite of the fact that there are no license fees, there are usually charges to install and support the system. There are also charges to obtain ICD-9 and CPT codes and possible charges to interface with labs, etc. With proprietary EHRs there are license fees that tend to be expensive and part of "vendor lock in" which means that you are tied to one vendor and often pay when software is upgraded. Many open source and "free" EHRs will struggle to meet Meaningful Use criteria for reimbursement because they may lack the finances to market their product and to make major programming changes to be CCHIT certified. Open source software in healthcare is discussed in more detail in the last chapter on emerging trends.

- **Open Source**
  - **OpenVista** is an open source initiative based on the popular VistA program the VA system uses and is available for download without charge [102]
  - **WorldVistA** is basically the same EHR the Veteran's Affairs (VA) system uses without the inpatient function. It was released in October 2005 as a joint venture between the VA and the Centers for Medicare and Medicaid Services to promote a more affordable EHR for the US and third world countries (VistA-Office Project).

The software was written with the old MUMPS programming language. Version VOE/1 was approved by CCHIT in 2006 [103]

- o **Medsphere OpenVista** is a vendor that will offer an open source version on a subscription basis that covers upgrades and support. Their program is offered in two configurations: enterprise for an organization and clinic configuration for smaller clinics and multi-specialty groups. The software will operate on a Linux or Windows operating system [104]

- o **vxVistA** is an open source Freedom of Information Act (FOIA) EHR, released by the vendor DSS in 2009 that targets moderate-sized hospitals, long term care facilities and ambulatory practices. VistA has been modified to upgrade its existing 125 modules and includes add-on products such as dental electronic records and document imaging. Software is free to download but does not cover CPT code or Cache™ licenses [105]

- o **OpenEMR** is an open source GPL license EHR with practice management, electronic billing, the ability to generate electronic prescriptions and HL7 support. It is intended for small ambulatory practices and available as a client-server or ASP model. It supports multiple languages [106]

- **"Free" EHRs**

  - o **PracticeFusion** is mentioned because it is a free, easy to use ASP model EHR funded by small advertisements and the sale of de-identified data. They claim 28,000 users (number of physicians or unique practices unknown) as of early-2010. The program includes scheduling, patient encounters, knowledge resources, growth charts, advanced directives, customizable templates, personal health record, e-prescribing, the ability to upload documents and a fee-based billing interface and fee-based document copying service. One of the more useful features is templates that can not only be shared and customized "on the fly", different sections of a SOAP (subjective, objective, assessment and plan) note can be coordinated. In other words, for a patient who presents with bronchitis, the history, physical exam, assessment and treatment will already contain information related to bronchitis. A typical SOAP note is demonstrated in figure 2.5. This vendor has gone on record guaranteeing compliance with Meaningful Use [106]

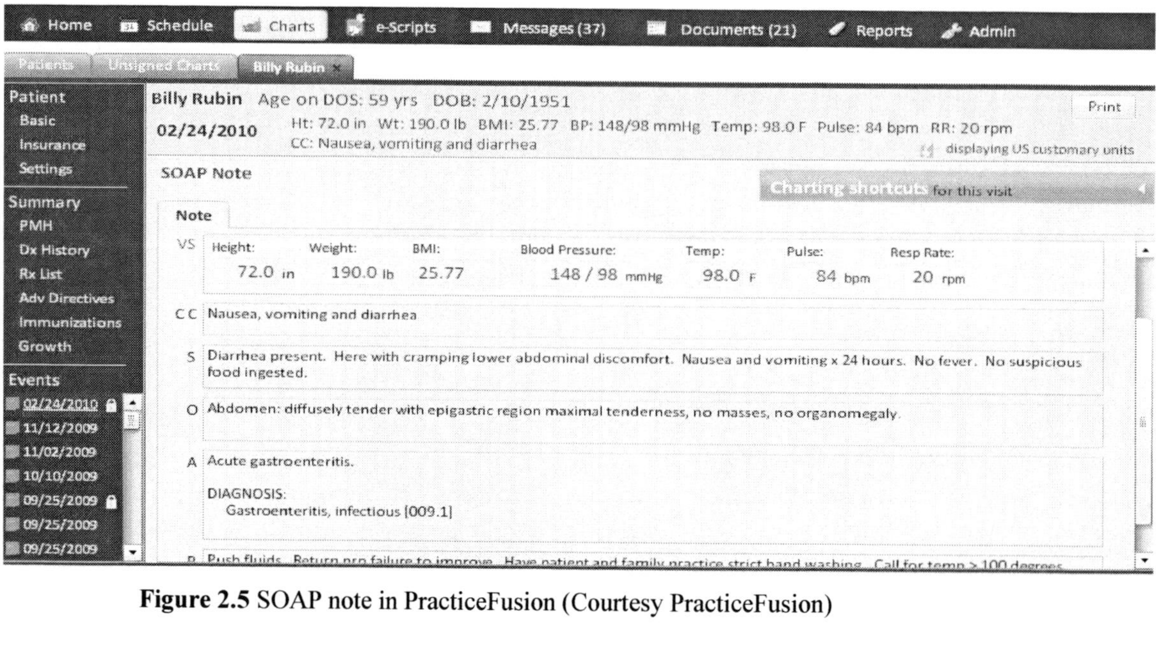

Figure 2.5 SOAP note in PracticeFusion (Courtesy PracticeFusion)

- Other "Free" EHRs
  - **MDBug** www.mdbug.com
  - **alloFactor** www.allofactor.com
  - **EMDfix** http://www.emdfix.com/
  - **Mitochon eEMR** http://mitochonsystems.com/

Low Cost EHR Programs

Medical Office Online. This web-based (ASP) EHR is priced at $300/physician per month, plus a user license of $15 per user per month and an initial set up and training fee of $695 to import data. They are not CCHIT certified as of early 2010. [107]

Amazing Charts. The vendor offers this CCHIT certified (2008) client-server based EHR for a single payment of $995/clinician as well as a free 3 month trial. There is a charge of $500/physician/year for updates and a charge of $500/physician/year for maintenance, support and e-prescribing. Interfaces are extra. Package includes scheduling, internal secure messaging, charting (template driven), billing (superbills) and ad hoc reporting. While functionality is limited (but expanding) this platform continues to be ranked very high in terms of user satisfaction on multiple surveys. [108]

SoapWare. This program is popular among small primary care practices but also covers more than 60 specialties and exists in more than 30+ countries. Patient encounter data can be inputted by templates or voice recognition. They are CCHIT certified (2007) and claim to be qualified for Medicare/Medicaid reimbursement. Training and implementation are extra. As of mid-2010 they offer three packages:
- SOAPwareRX: $995 for basic EHR with e-prescribing. No charge for staff access
- SOAPware Standard: $1995 for CCHIT certified EHR
- SOAPware Professional: $3995 with functionality like standard model plus data mining and reports, E&M coding, alerts and interfaces

Training and implementation are extra. The support package is an annual cost and depends on which package you purchase but the first year is included in the purchase price. The ASP professional model starts at $250 per month. Optional features include: HL7 interface, billing, knowledge resources, fax capability and back up services. [109]

Other similar low cost EHR programs:

- InteGreat (2006 CCHIT certified) www.igreat.com
- Praxis EMR (2006 CCHIT certified) www.praxisemr.com
- E-MDS 2008 CCHIT certified www.e-mds.com
- SpringCharts (2006 CCHIT certified) www.springcharts.com
- ComChart www.comchart.com

More Comprehensive EHR Examples

Armed Forces Health Longitudinal Technology Application (AHLTA). This is the program (previously known as CHCS II) that has been deployed worldwide for every Department of Defense (DOD) medical facility (9.2 million patients). It began as CHCS I in about 1992 and was used primarily for outpatient order entry and the ability to retrieve lab, x-ray and pharmacy results. AHLTA is more robust than CHCS I with the ability to use templates to input information into a patient encounter. A unique feature of this EHR is the use of the MEDCIN terminology engine. MEDCIN is a collection of 200,000+ terms you can select to construct a history and physical exam. This was selected in order to create a structured note, so each data field can be saved and retrieved and the information can build an Evaluation and Management (E & M) code (MEDCIN is also discussed in the data standards chapter). The limitation is that it is very time consuming to learn to use this technique and then create your own list of customized profiles. Specialists are used to dictating a very comprehensive note, whereas templates tend to generate very cryptic notes. Needless to say, templates and MEDCIN have their advocates and their skeptics. Figure 2.6 shows the MEDCIN tree and the navigation menu on the left. Other limitations include the lack of interoperability with the VA system, lack of an inpatient module and the inability to scan documents into the EHR. Excellent demos can be found on a training web site, customized for providers, nurses, support techs and clerks.[110-111]

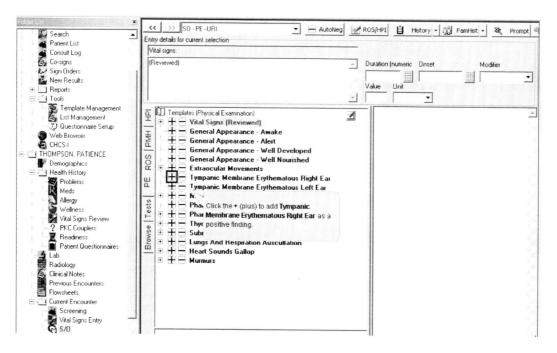

Figure 2.6 AHLTA and MEDCIN terminology engine
(adapted from the Naval Medical Center Portsmouth web site)

Veterans Health Administration VistA. In 1996, the VA introduced VISTA (Veterans Health Information Systems and Technology Architecture). Computerized Patient Record System (CPRS) is the graphic user interface (GUI) for the Veteran's EHR system that serves approximately 7.5 million enrollees (Figure 2.7). CPOE accounts for 93% of all prescriptions and the VA system processes approximately 860,000 orders daily. Image archiving and bar coded medication administration is part of VistA. This EHR is currently used in outpatient, inpatient, mental health, intensive care unit (ICU), emergency department, clinic, homecare, nursing home and other settings. The Indian Health Service has modified Vista to include women's and children's health functions and is known as Resource and Patient Management System (RPMS). One disadvantage of VistA is that it was built on the MUMPS programming language and the plan is to eventually migrate to a more modern architecture and will be known as "HealtheVet". The newer architecture will likely be an ASP model and will allow for improved interoperability with other VA facilities and information exchanges. In spite of the older programming language this EHR has been well received by clinicians and considered one of the major transformational factors in VA healthcare.

The program has a new patient portal, MyHealtheVet that allows patients to enter lab and vital signs and create a personal health record. [112-115]

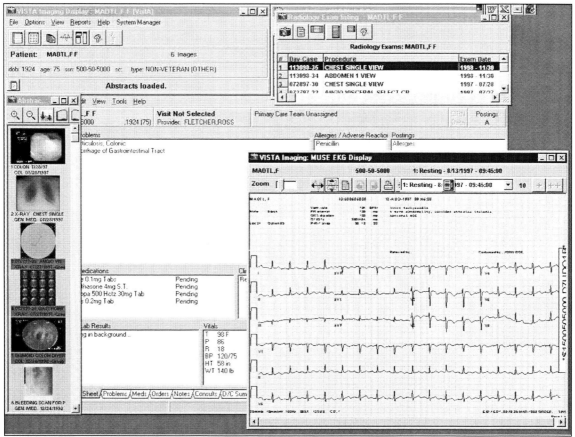

Figure 2.7 VistA CPRS (courtesy Veteran's Affairs)

Kaiser Permanente HealthConnect™. The largest roll-out of a civilian EHR in the world (8.6 million members) was begun by Kaiser-Permanente in 2003 and completed by 2008. This four billion dollar project used the Epic Systems EHR with modules from other vendors. The goal of this non-ASP model EHR is to integrate all electronic aspects of patient care into one system: billing, inpatient and outpatient records, data repository, registration, scheduling, lab, x-ray and pharmacy records. They offer a patient portal, integrated with the EHR known as My Health Manager with the following features: online appointments, online refills, access to lab results, eligibility and benefits, access to immunization records and secure messaging with physicians. [116-117]

eClinicalWorks. This EHR was selected by the Massachusetts Medical Society because it is multi-featured and directed at small, medium and large practices. In 2010 they claimed 40,000 users and they have CCHIT 2011 certification. It is also one of the few that lists its pricing schedule on their web site. Available features include:
- "Do Lists" are always posted at the top of the screen and are colored coded as to urgency
- Multiple means of inputting data such as templates, handwriting recognition, voice recognition and free text
- Sticky note feature for reminders
- Separate window to let physician know if patient owes the practice money
- Tab to access the resource UpToDate
- Patient can pre-register ahead of time on the patient portal eClinicalWeb

- True e-prescribing transmission of scripts to a pharmacy and not just a printed script or fax
- Continuity of Care Record (electronic patient summary) available
- The ability to create customizable digital practice forms
- Patient education resource
- Comprehensive patient/disease registries with customizable alerts
- Referral letters can be automatically generated
- A summary of the visit can be printed and given to the patient
- E-visit capability
- eClinicalMobile hosts the EHR on an iPhone
- Mid-2010 pricing:
  - Client-server purchase price of $10,000 for first clinician, then $5,000 for additional clinicians. Maintenance fee is 0.18% of total price and support fees of $600 per month
  - Client server lease at $4800 for first clinician, $4320 for second clinician
  - Remote hosting (ASP) at charge of $400 per month per clinician
  - EHR ASP model was offered along with Dell Computers as a packaged deal by select Sam's Club stores in 2009 [118]

GE Centricity Practice for Community Health Centers. Most EHRs are designed for general practices but some may be more appropriate for small versus very large practices and some offer unique features. GE offers an EHR specifically for community health centers (CHCs) that have unique practice and billing requirements. They are able to connect to the Patient Electronic Care System (PECSYS)™ disease registry discussed in the chapter on disease management. In addition, they are able to share best practices with other CHCs using this same EHR. [119]

Specialty EHRs. A customized EHR for sub-specialties makes sense because their needs are narrower and different than those of a primary care physician. The EHR vendor NexGen EHR offers 24 sub-specialty patient encounter modules. An excellent example is their ophthalmology module that organizes the EHR into a very logical order for eye physicians. Their product is available as a client-server or ASP option. The history and physical exam can be specialty specific and images can be stored in each patient's record. Coding is automatically generated, as are letters back to the referring physician. A PDA version is also available. They offer practice management, document scanning, a patient portal and a community health module (central data repository). [120-121]

Barriers to Electronic Health Record Adoption

Many of the same barriers to HIT adoption covered in chapter 1 also pertain to EHR adoption. According to Shortliffe there are four historic constraints to EHR adoption: 1) the need for standardized clinical terminology; 2) privacy, confidentiality and security concerns; 3) challenges to data entry by physicians and 4) difficulties with integrating with other systems. In addition, other barriers exist to include: [122]

1. **Financial Barriers.** Although there are models that suggest significant savings after the implementation of ambulatory EHRs, the reality is that it is expensive. Surveys by the Medical Records Institute, MGMA and HIMSS report that lack of funding is the number one barrier to EHR adoption, cited by about 50% of respondents. [123] In a 2005 study published in *Health Affairs*, initial EHR costs averaged $44 K (range $14-$63,000) per FTE (full time equivalent) and ongoing annual costs of $8.5 K per FTE. These costs included the purchase of new hardware, etc. Financial benefits averaged about $33,000 per FTE provider per year. Importantly, more than half of

the benefit derived was from improved coding. This study looked at fourteen primary care practices using two well known EHRs. The average practice showed a return on investment in only 2.5 years.[124] This is not a surprise given the fact that studies have shown that physicians often "*under code*" for fear of punishment or lack of understanding what it takes to code to a certain level.[125] A 2008 survey on EHR purchase price conducted by Physician's Practice reported about 1/3 of physicians paid between $500-$3000 per clinician, 1/3 paid between $3001-$6000 and about 1/3 paid more than $6000. [126]

It is important to consider that integration with other disparate systems such as practice management systems can be very expensive and hard to factor into a cost-benefit analysis. The web based application service provider (ASP) option is less expensive in the short term and perhaps in the long term, when you factor in the expenses to maintain and upgrade an office client-server network. According to many studies, including a survey by the Commonwealth Fund, adoption of EHR and CPOE was far higher in large physician practices that could afford the initial high cost.[127] Lack of cost transparency is a serious EHR issue. Only a few EHRs vendors such as Amazing Charts, SoapWare and eClinicalWorks post their pricing models on their web sites. Most require a formal reason for proposal (RFP) to be submitted before they will provide an estimate. It will be challenging for regional extension centers (RECs) to advise physicians about EHRs without transparent cost comparison.

2. **Physician resistance**. In a monograph by Dr. David Brailer, lack of support by medical staff is consistently the second most commonly perceived obstacle to adoption, behind lack of resources.[128] They have to be shown a new technology makes money, saves time or is good for their patients. None of these can be proven for certain for every practice. Although you should not expect to implement CPOE or go paperless from the beginning, at some point it can no longer be optional. It seems clear that CPOE does take longer than written orders but offers multiple advantages over paper as pointed out previously. One systematic review suggests that documentation time for CPOE was far greater for physicians than nurses.[129] Many studies that looked at this issue were based on surveys or theoretical models and not high quality research. Implementation will not fix old work flow issues and will not work if several physicians in a group are opposed to going electronic. We now know that some practices have opted to change or discontinue their use of an EHR. A 2007 survey by the Medical Records Institute demonstrated that fewer than 20% of respondents had uninstalled their EHR in an effort to step down to a less expensive alternative and 8% had returned to paper.[130]

   EHRs are not the only issue on the table for most physicians and healthcare organizations. They have to face increases in overhead while reimbursement wanes along with ICD-10, HIPAA 5010, new healthcare reform and Red Flag rules, just to mention several looming challenges.

3. **Loss of productivity**. It is likely physicians will have to work at reduced capacity for several months with gradual improvement depending on training, aptitude, etc. This is a period when physician champions can help maintain morale and momentum with a positive attitude.

4. **Work flow changes**. Everyone in the office will have to change the way they route information compared to the old paper system. If planning was well done in advance you should know how your work flow will change. As an example, many offices

place the patients chart in the exam room door to indicate that the patient is ready to be seen. How will you do that with an electronic system? Initially, you will have to maintain a dual system of paper and electronic records. Work flow analysis will also determine where you will place computer terminals. Importantly, clinicians will have to maintain eye contact as often as possible and learn to incorporate the EHR into the average patient visit. Use of a movable monitor or tablet PC may help diminish the time the clinician spends not looking at the patient.

5. **Usability Issues**. Usability has been defined as the "effectiveness, efficiency and satisfaction with which specific users can achieve a specific set of tasks in a particular environment".[131] Is the software well organized and intuitive such that the user can find what they are looking for with a minimal number of mouse clicks? This is more complicated than what one would expect because there are multiple sub-specialties with unique needs, as well as multiple clinicians who are used to working in a set sequence. Based on several surveys included in this chapter, usability does not necessarily correlate with the amount of money paid for the software. HIMSS has an EHR usability task force and is collecting survey information that should prove to be interesting.[132]

6. **Integration with other systems**. Hopefully, integration with other systems like the practice management software was already solved prior to implementation. Be prepared to pay significantly for programmers to integrate a new EHR with an old legacy system. An average cost is about $3-$15,000 per interface.[133] Most office and hospitals have multiple old legacy systems that do not talk to each other. Systems are often purchased from different vendors and written in different programming languages. If either the EHR or practice management sytem's software is upgraded, then interfaces need to be checked and possibly changed. It is now popular to purchase an EHR already integrated with practice management, billing and scheduling software programs. It is worth noting that the open source Mirth Connect application is a HL7 integration engine that should decrease the cost to interface with outside laboratories, etc.[134]

7. **Lack of standards**. Data standards and medical vocabularies are necessary for interoperability. The initial standards have been proposed by ONC and will be covered in more detail in another chapter. Reimbursement for "Meaningful Use" will mandate that EHRs demonstrate the ability to exchange information. Although we have numerous standards already accepted (separate chapter) they will likely need to be updated and new standards added based on use cases.

8. **Adverse legislation**. There is concern that previously passed legislation will make it difficult for hospitals and physicians to combine forces and create information networks. The Stark Law prohibited a physician from referring Medicare patients to an entity if he/she had a financial relationship with the entity. The Anti-kickback Act made it illegal for an individual or entity to offer remuneration of any kind to another individual or entity for referring a patient. It is illegal to have to purchase or lease any covered item or service. In 2006 "*safe harbor*" exceptions became law and allowed hospitals to provide electronic health records to physicians. However, the physician or office must pay 15% of the donor's cost for the technology.[135]

9. **Legal Aspects**. Because EHRs are new, many legal aspects are still being discussed and debated. As an example, the paper medical record resides in one area and is well

defined. The EHR, on the other hand, can be composed of images, text and data from multiple areas that can be dynamic. Information can be shared with personal health records and health information organizations. Therefore, the concept of the "legal EHR" is complex and will require healthcare organizations to develop governance and policies as part of health information management. [136]

10. **Inadequate proof of benefit**. Although there is no shortage of hype regarding the benefits of EHRs, the reality is we need better research. A systematic review by Hunt showed that the effects of clinical decision support systems, as an example, have not been adequately studied. [68] Moreover, successful CDSS programs at a medical center where they have been in use for an extended period of time does not mean they will be successful at another medical center with no such track record. The systematic review by Chaudry et al. is often cited as proof of the benefits of HIT, but in his conclusion he states "four benchmark institutions have demonstrated the efficacy of health information technologies in improving quality and efficiency. Whether and how other institutions can achieve similar benefits and what costs, are unclear". [137] Furthermore, research is hindered by the fact that most hospitals don't have EHR systems with robust CDSS so we can't adequate evaluate comprehensive EHRs.

In 2009 three interesting studies about EHRs became available. In the first observational study by Yu et al. they evaluated 3364 hospitals (2004 HIMSS Analytics database) that participated in the CMS Hospital Quality Alliance. They concluded that there was a positive association between certain medication and non-medication quality indicators and CPOE implementation, but the association was not strong. [138] In the second study (unpublished) by Jha and DesRoches they compared 3000 hospitals with different levels of EHR adoption. Specifically, they looked at standard quality measures, as well as length of hospital stays. They concluded that there were only marginal improvements in cost and quality related to EHR adoption. [139] The third study was the *European Commission EHR IMPACT study* that looked at EHRs and e-prescribing in eleven large practices in Europe, the USA and Israel. They concluded that there were definite benefits to interoperable EHRs but that the benefits often took 4-9 years to realize a return on investment and "net cash injections" were necessary. [140]

There were several studies in 2005 that have shown increased errors as a result of implementing CPOE [86, 141-143] Most of the studies have been criticized for one reason or another but it should come as no surprise that any new technology will create new possibilities for errors. Weiner coined the term "*e-iatrogenesis*" to mean "*patient harm caused at least in part by the application of health information technology*".[144] Ash et al. wrote an excellent 2004 review on the unintended consequences of information technology in health care.[145] Berger and Kichak wrote an insightful but negative appraisal of CPOE entitled "*Computerized Physician Order Entry: Helpful or Harmful*" in which they disputed findings by groups such as the IOM and Leapfrog.[146] Eventually, with better training or re-design some of the technology-related errors are likely to be overcome.

The Center for Medicare and Medicaid Services (CMS) outlined a 5 year $150 million pilot program that began in 2008 to test the effects of EHR implementation using 1200 recruited physicians, covering 12 states and cities.[147] We anticipate that this study will provide some answers about the benefits and impact of EHR implementation. Once EHR adoption is more widespread and achieves HIMSS level

6 or 7 in a variety (location and size) of clinical settings, we will be better able to evaluate their true impact.

More research is needed to obtain a balanced opinion of the impact of EHRs on quality of care, patient safety and productivity. Furthermore, we will need to study the impact on all healthcare workers and not just physicians. A survey of 600 + nurses was reported in 2010 and showed that 49% felt EHRs had a positive benefit, while 47% felt they had a negative effect or none at all.[148]

Selecting an Electronic Health Record: The Logical Steps

There is a tendency to pick a well know EHR vendor and hope for the best, much like picking an automobile. Unfortunately, the process is far more complex and less dependent on the vendor selected. More important are the specific needs of the group, careful implementation, adequate training, integration with other systems and buy-in by all staff. The following are steps that should be taken to plan for the purchase and implementation of an EHR.

1. **Develop your office strategy**. List priorities for the practice. Are you trying to save time and/or money or do you just want to go paperless? Are you looking to be more competitive by offering patient satisfaction-related features like secure messaging, virtual visits, a portal and connectivity with your medical community? Do you need remote computing and remote access for the clinicians? Are you seeking improved workflow to expedite chart pulls and provide easier refills? Do you need more reporting capability than what you currently have? Do you need better integration with your practice management system? Are you trying to integrate disparate programs? Now is the time to study work flow and see how it will change your practice. This is when frequent conferences with your front office staff will be critical to get their input about the processes that need to improve. Make sure physicians are committed to using the EHR. Look for at least one physician champion and be sure your staff is onboard. Do not proceed if there are hold-outs. Factor in your future requirements. Do you plan to add more partners or offices or specialties? Plan for initial decreased productivity.

2. **Research the EHR topic**.
   a. Take a short course
   b. Utilize expertise from regional extension centers (RECs) (see chapter 1)
   c. Read a text book. Examples: Gartee, R. *Electronic Health Records: Understanding and Using Computerized Medical Records*. 2006; Carter, J. *Electronic Health Records*, Second Edition. 2008; Amatayakul, MK. *Electronic Health Records*, Fourth Edition. 2009; Hamilton, B. Electronic Health Records. 2010.
   d. Read important articles, monographs and surveys
      i. EHR articles reside on the University of West Florida HIT Resource Site http://uwf.edu/sahls/medicalinformatics/EHR.cfm
      ii. *Electronic Medical Record, Implementation Guide. The Link to a Better Future*, 2nd edition. Texas Medical Association provides this free 104 page monograph. Also available for CME (3 AMA category I credits) for a $25 fee.[149]
      iii. 2009 EHR User Satisfaction survey was the third EHR survey reported by Family Practice Management. They received 2012 responses from Family physicians, reporting on 142 EHR systems. Only 22 systems were used by the majority of respondents (87%). A chart correlates the top 22 EHRs by

practice size and the number of respondents using the specific EHR. Twenty percent of respondents were from practices of 50 or more clinicians and 16 percent were from solo practices, so there was a mix of large and small practices. Respondents were then asked 13 questions about EHR usability, features, training and support in order to rank them. Clearly, cost did not correlate with user satisfaction. [150]

      iv. An EHR survey was performed by Medscape in 2009 with similar results. [151]

e. Consultants

      i. Consulting firms such as AC Group provide consulting for EHR purchase. In addition they have several fee-based monographs on the subject.[152]

      ii. KLAS is an independent HIT rating service that vendors pay to join and end-users pay to receive reviews. Their reviews cover EHRs and components based on practice size and include letter grades on "implementation, service and product". Their input usually comes from office managers or IT specialists and not necessarily end-users. Physicians can evaluate survey data on individual vendors free if they are willing to complete an online questionnaire. [153-154]

3. **List the EHR features you need**. Review the key components section of this chapter. Choose the method of inputting: keyboard, mouse, stylus, touch-screen or voice recognition? Don't forget backup systems, e.g. "dual failover".

4. **Analyze and re-engineer workflow**. Processes such as prescriptions, telephone triage, lab ordering, appointments, etc will change with the use of an electronic health record. Healthcare must embrace business process engineering (BPR) and business process automation (BPA) to create a digital office. It is wise to map the various processes to see what changes must occur and where you might add computer terminals to execute the process electronically. Some choose to use workflow software to map office workflow. HIMSS offers a toolkit "Workflow redesign in Support of the Use of Information Technology within Healthcare" for its members. [20]

5. **Use Project Management Tools.** A variety of tools exist that improve organizational skills. Consider using standard matrices that are glorified checklists and timelines that organize your efforts:

    a. EHR Implementation Checklist [149]
    b. DOQ-IT EHR implementation Roadmap [155]
    c. "Build your EHR Timeline" [156]
    d. DOQ-IT EHR Implementation checklist [157]

6. **Decide on client-server or the application service provider (ASP) option**. One early decision that must be made is whether you want to purchase a standard client-server EHR package which means having the software on your own computers. The other choice is an ASP model which uses a remote server that hosts the EHR software and your patient data. Each has its merits and shortcomings. Features of an ASP Model: [50]

    a. Vendor charges monthly fees to provide access to patient data on a remote server. Fees will usually include maintenance, software upgrades, data backups and help desk support.
    b. Lease agreement commitments range from 1-5 years.
    c. ASP may charge a fixed amount or charge for the number of users.

    d.   ASP can be completely web based or can require a small software program (thin client) to help share processing tasks.

    e.   Pros: Lower start-up costs; ASP maintains and updates software; requires very little local tech support, thus saves money. Often a better choice for small practices with less IT support. Enables remote log-ons, for example, from home or satellite offices

    f.   Cons: If your ISP is not working you aren't either; concerns about security and HIPAA; concerns about who owns the data and cost of monthly cable fees. Speed may be a little slower compared to the client-server model. Must have fast Internet connection; should be cable modem, DSL or T1 line.

    g.   An informal 2010 survey, using an EHR search engine revealed that 67% of the vendors listed an ASP alternative.[158] Some larger practices and hospitals are hosting EHR ASP services for smaller practices to make it affordable.[159] An excellent review of ASPs was published in October 2006 by the California HealthCare Foundation. [160]

7. **Decide on an inputting strategy.** Different types of inputting are necessary because clinicians have different specialties, personal preferences and document requirements.

    a.   Dictation: in spite of the desire by most people who purchase an EHR to avoid dictation, many physicians will not want to give this up because it is part of their routine or they practice in a specialty where the historical narrative is best told with a dictation. Besides cost (10-20 cents per line), the disadvantages are the fact it is non-structured data, the physician must proof read and someone must "cut and paste" the narrative into the EHR, thus causing some delay.

    b.   Speech recognition: is an attractive alternative to standard dictation for many but not all physicians. The cost to purchase, example Dragon Naturally Speaking (DNS) 10 ®, is approximately $1600 per physician (on-site training not included) and includes a choice of multiple medical specialty vocabularies. DNS is available for the iPhone and wireless platforms. [161] There is preliminary evidence to suggest speech recognition improves the patient narrative and has a reasonable return on investment. [162] While it is true that speech recognition has improved dramatically in the last few years, it will not be satisfactory for all users. In 2010 Hoyt and Yoshihashi reported a failure rate of 31% in a large scale implementation of voice recognition into an EHR in a military treatment facility. [163]

    c.   Handwriting recognition. A few EHRs utilizing the tablet PC platform will allow a clinician to write on the tablet and have the information converted to text.

    d.   Templates: most EHRs offer a template or "point and click" option to facilitate inputting history and physical exam data into the EHR. In addition to saving time, templates input data as structured data so it is machine and human readable. Practices can create templates ahead of time before going live and thereby, try to standardize care within a practice. Multiple template designs are available. With MEDCIN every phrase must be located and selected for inputting. Others "document by exception" which means there is standard language for most exams; if verbiage does not pertain to a patient, it can be deleted. Most templates can be customized (some on the fly) and shared. Many are disease specific such as "low back pain" or "headache" templates. One concern with templates, besides a potential "robotic note", is the over use of options such as "auto-negative" where the review of systems can be performed rapidly with the potential for false documentation. Clicking history or physical exam choices that the clinician did not ask or examine is considered fraud. Conversely, submitting an overly detailed history or physical exam that is not justified by the diagnosis could be considered abuse.

e. Typing: a minority of physicians will be happy to input their data by typing, particularly if they are tech savvy and excellent typists. Most physicians, however, will complain that typing notes is not why they went to medical school.

f. Scribes: emergency rooms were the first hospital area to hire scribes that "shadowed physicians" and in addition to multiple duties were responsible for inputting information into the EHR by typing, templates, dictation or transcription. [164]

g. A blended approach: Medical practices would be wise to offer multiple means to input patient data. As an example, for simple patient encounters for flu, templates may be adequate. For more complex visits dictation or voice recognition may be necessary. Organizations will have to balance the need for productivity by finding better ways to input into an EHR with the needs to have discrete or structured data. As an example, hospitals rated as stage 6 by HIMSS used templates 35%, dictation/transcription 62% and speech recognition 4% for inputting into EHRs. Newer software, using natural language processing, will extract discrete data known as "narradata" from dictations that can be used secondarily for decision support, reporting and billing. This approach is known as discrete reportable transcription (DRT) and may be important for Meaningful Use of EHRs.[165]

8. **Discuss mobility.** Will clinicians need to be wireless? Will they benefit from access of the EHR remotely using a smart phone? Multiple vendors, like Epic, offer their software on, example an iPhone. They offer a free app (known as Haiku) that permits access to schedules, patient lists, health summaries, test results and notes.[166]

9. **Do you need a combined EHR and Practice Management System?** Decide early on if you plan to purchase a combined EHR and practice management system or do you need an interface to be created?

10. **Survey hardware and network needs.** How many more computers will you need to buy? Do you need to hardwire a network and/or are you going wireless? Are you going to need an in-house server with its dedicated closet, air conditioning and backup? Do you need a network switch and commercial grade firewall? Do you need to hire short term or long term IT staff? You will need a data back up and disaster plan. Plan for a commercial grade uninterruptible power supply. Also, plan for a service level agreement if you opt for the ASP model.

11. **What interfaces do you need?** Will you need interfaces to external laboratory, pharmacy and radiology services or is that part of the package purchased?

12. **Do you need third party software?** As an example: patient education material, ICD-9 codes, CPT codes, HCPCS database, SNOMED, drug database, voice recognition, etc. Ask if that is part of the purchased package.

13. **Develop your vendor strategy.**

a. Write a simple "request for proposal" (RFP) or request for information (RFI). This will cause you to put on paper all of your requirements and will provide the vendor with all of the important details regarding your practice. This formal request will standardize your responses from vendors as they will need to respond in writing how they plan to address your requirements. Exact pricing should be part of the RFP. Sample RFPs are available on the Web.[149, 167]

b. There is a free web-based search engine that will search most EHR vendors with the filters for practice specialty, practice size, ASP versus client server, EMR or EMR plus PMS and whether the EHR is CCHIT certified.[158] In mid-2010 a new free EHR search engine became available through the auspices of the American College of Physicians. The site offers EHR readiness assessments, vendor comparisons, vendor profiles, EHR top 10 ratings (11 categories) and eventually will offer a web based RFI submission process. [168]

c. Obtain several references from each vendor and visit each practice if possible. Be sure to select similar practices to yours. Questions to ask your references are available. [148]

d. The following excellent reference provides an EHR demonstration rating form, questions to ask vendors, EHR references and a vendor rating tool. [169] Create a scoring matrix to compare vendors. Examples of questions to ask:
   i.   How many licenses have been sold overall?
   i.   Number of years in business of selling EHRs?
   ii. Number of employees and salesmen?
   iii. Does the company focus primarily on a certain size practice?
   iv. Is there a problem with multiple log-ons at the same time?
   v.   Does the EHR interface with other electronic systems?
   vi. Does the maintenance fee cover travel and does it cover nights, weekends and holidays?
   vii. How much for software upgrades?
   viii. Interface costs?
   ix. Is e-prescribing included?
   x.   Confirm that the EHRs qualifies for Medicare/Medicaid reimbursement
   xi. What is the training time required to become truly operational?
   xii. Training cost per user or practice?
   xiii. On-site training available?
   xiv. Are you willing to put terms in a sales agreement?
   xv. Do you plan to stay up to date with CCHIT certification?
   xvi. Hardware and software requirements? [50, 149]

e. The following reference also has a scoring sheet with sections for vendor software, interfaces, third party software, conversion services, implementation services, training services, data recovery services, annual support and maintenance, financing alternatives and terms. It also includes red flags and FAQ's. This reference is intended to compare costs and not EHR functionality between candidate vendors. [170]

f. Obtain in writing commitments for implementation and technical support, including data conversion from paper records; interfacing with practice management (PM) software; exact schedule and time line for training.

14. Look for funding

a. The most obvious choice is Medicare or Medicaid reimbursement under the HITECH Act.

b. As noted before, hospitals can donate EHR systems to physician offices under the "safe harbor" with physicians having to pay 15% of costs.

c. Physician Quality Reporting Initiative (PQRI) will reward physicians for quality reports that can be generated by an EHR. We will cover this more in the chapter on pay for performance. [171]

d. There are incentive programs by states, payers and hospitals. [172-173] Maryland has passed a law to provide incentives for clinicians who adopt EHRs that can connect to the state's HIO. [174]

15. **Select a vendor and develop a contract**

    a. Most practices will need to create a contract with legal help. This will ensure the vendor meets their obligations and will define the contract period, duties and obligations, license stipulations, scope of license, payment schedules, termination clauses, upgrades, support, warranties, liabilities, downtime clauses, etc. [149]

16. **Decide on a strategy to convert paper encounters to electronic format.** Most experts advise that key information (medications, allergies, major illnesses, immunizations, lab results, etc) be keyed in by staff on active patients several months before going live. Decide what documents such as prior encounters, consultations, discharge summaries, etc you need to upload into the EHR. Several resources will help you develop a strategy. [175-177] Digital Island posts an approximate charge of 15 cents per page for less than 30,000 pages to scan in paper forms. As an example, for 5000 pages this would amount to a charge of $825. [178]

17. **Training.** It can be said that you cannot train too much. Determine if your vendor has an electronic training database clinicians and staff can use before going live. Assess IT competencies of the clinicians and staff and train for gaps in knowledge.

18. **Implementation.** Consider a phased in approach where clinicians and staff begin with processes such as e-prescribing, internal messages and laboratory retrieval before tackling patient encounters. Develop a "go live" plan to determine reduced schedules and frequent debriefs. [149]

An interesting article written by Trachtenbarg in 2007 reviews the myths and truths about EHR (Table 2.8). [179]

Table 2.8 Myths and Truths about EHRs

Myths	Truths
"A new EHR will fix everything"	The reality is that it will only solve a few problems and probably create a few new ones
"Brand A is the best"	No single vendor has an ideal product and you must determine which features you need as each practice is unique
"Our software needs to work the way we currently work"	Expect to make some changes in your workflow and analyze that ahead of time
"Software will eliminate errors"	Expect a reduction, not elimination of medication errors, with a few new errors appearing
"Discrete data is always best"	In spite of the fact that structured (computable) data has advantages, it is clumsy and reads poorly
"The more templates the better"	Don't make any more templates than you actually need
"Mobile is best"	Having your EHR on a PDA phone or tablet PC sounds cool but a substantial number of clinicians won't use them
"You must have a detailed plan and stick to it"	You must remain flexible as needs, equipment and reimbursement change over time
"You can stop planning"	Planning never stops because better systems will be available in the future

For a recent real-world study of EHR implementation by a group of four internists reported in the literature, see reference.[180] According to one author:

> "Despite the difficulties and expense of implementing the electronic health record, none of us would go back to paper"

Key Points

- Electronic health records are central to creating health information organizations and a nationwide health information network

- The current paper-based system is fraught with multiple shortcomings

- It is likely that reimbursement for e-prescribing and electronic health records by the federal government will promote adoption

- In spite of the potential benefits of electronic health records, obstacles and controversies persist

- Computerized physician order entry and clinical decision support are still in their infancy and will likely improve in the future with artificial intelligence

- Advance planning and training is mandatory for successful implementation of electronic health records

Conclusion

In spite of the slow acceptance of EHRs by clinicians and healthcare organizations, they continue to proliferate and improve over time. Electronic health records have been transformational for large organizations like the VA, Kaiser-Permanente and the Cleveland Clinic, but the reality is that most medicine in this country is practiced by small medical groups, with limited finances and IT support. As a new trend, we are seeing outpatient clinicians opt to re-engineer their business model based on an EHR. Their goal is to reduce overhead by having fewer support staff and to concentrate on seeing fewer patients per day but with more time spent per patient. When this is combined with secure messaging, e-visits and e-prescribing the goal of the "e-office" is achievable.[181]

Buyers have a wide choice of features and cost to choose from. At this time cost is a major obstacle as well as the lack of high quality economic studies demonstrating reasonable return on investment. As more studies show cost savings, medical groups that have been sitting on the fence will make the financial commitment.

Without doubt, Medicare and Medicaid reimbursement for EHRs and e-prescribing is the most significant impetus to jump start EHR adoption we have seen. We anticipate a flurry of activity in preparation for reimbursement. Optimistically, we will see a variety of EHRs qualify for reimbursement. It is too early to know how well received stage 1 Meaningful Use objectives and measures will be received. Clearly, the statement "one size does not fit all" applies to EHRs. For those practices that can afford and need complexity, multiple high-end vendors exist. For smaller, rural, primary care practices, simpler alternatives exist. It is also worth noting that purchasing EHRs is only one of multiple difficult challenges facing clinicians and their staff. According to a mid-2009 Medical Group Management Association (MGMA) survey implementing an EHR was ranked third in difficulty preceded by rising operating costs and maintaining clinician salaries in the face of decreasing reimbursement. [182]

Appendix A. Meaningful Use Core Objectives and Measures

Stage 1 objectives	Stage 1 measures
Use CPOE for medication orders	Measure more than 30% of patients on medications
Implement drug-drug and drug-allergy interaction checks	Demonstrate functionality is enabled during entire period
Maintain an up to date problem list of current and active diagnoses	More than 80% have at least one entry as structured data
Generate and transmit electronic prescriptions	More than 40% are transmitted using certified EHRs
Maintain active medication list	At least 80% of patients have at least one entry as structured data
Maintain active medication allergy list	More than 80% of patients have at least one entry as structured data
Record demographics: sex, race, ethnicity, date of birth and preferred language	More than 50% of data recorded as structured data
Record and chart vital signs: height, weight, blood pressure, BMI, growth charts for children	More than 50% of patients 2 yrs of age or older have vitals recorded as structured data
Record smoking status in patients age 13 older	More than 50% of patients 13 years of age or older have smoking status as structured data
Report quality measures to CMS or states	For 2011, provide aggregate numerator and denominator through attestation; for 2012 submit reports electronically
Implement 1 clinical decision support rule and track compliance	One clinical decision support rule implemented
Provide patients with a copy of their record upon request: diagnostic tests, problem list, medication lists, allergies	More than 50% of requesting patients receive copy within 3 days
Provide clinical summaries for patients for each office visit	Summaries provided for more than 50% of visits within 3 business days
Capability to exchange electronically key clinical information among clinicians or patient-approved entities	Perform at least 1 test of EHR's capability to exchange information
Implement systems to protect privacy and security of patient data in the EHR	Conduct or review a security risk analysis, implement security updates as necessary and correct identified security deficiencies

Appendix B. Meaningful Use Menu Objectives and Measures

Stage 1 objectives	Stage 1 measures
Implement drug-formulary checks	EP has enabled this function for entire reporting period
Incorporate lab-test results into EHR as structured data	At least 40% of lab results entered
Generate patient lists by specific conditions	Generate at least one report with specific condition
Send reminders to patients as per their preference	Send to at least 20% of patients 65 or older or 5 years or younger
Provide patients with timely electronic access to their lab results, problem list, medication list, allergy list within 4 business days	At least 10% are provided electronic access within 4 business days
Use EHR to identify and provide patient specific education resources	At least 10% are able to receive electronic educational resources
Perform medication reconciliation following transitions of care	Perform medication reconciliation at least 50% of the transitions of care
Provide summary of care for any transition of care	Provide summary for more than 50% of transitions of care
Capability to submit electronic data to immunization registry or information system	Perform at least one test of EHR's capability to submit data electronically to immunization registry
Capability to submit electronic syndromic surveillance to public health agencies	Perform at least one test of EHR's capability to submit syndromic data to public health agencies

References

1. Weed LL. Medical Records that guide and teach. NEJM 1968;278:593-600
2. Fries JF, McShane DJ. ARAMIS: a proto-typical national chronic disease data bank. West J Med 1986;145:798-804
3. Atkinson, JC, Zeller, GG, Shah C. Electronic Patient Records for Dental School Clinics: More than Paperless Systems. J of Dental Ed. 2002;66 (5): 634-642
4. National Institute of Health Electronic Health Records Overview. April 2006. MITRE Corp. http://www.ncrr.nih.gov/publications/informatics/EHR.pdf (Accessed October 5 2009)
5. Schwartz WB. Medicine and the computer. The promise and problems of change NEJM 1970;283:1257-64
6. The Computer-Based Patient Record: An Essential Technology for Health Care, Revised Edition (1997) Institute of Medicine. The National Academies Press www.nap.edu/books/0309055326 (Accessed October 15 2005)
7. Berner ES, Detmer DE, Simborg D. Will the wave finally break? A brief view of the adoption of electronic medical records in the United States JAIMA 2004;12
8. Hing, ES et al. Electronic Medical Record Use by Office-Based Physicians and Their Practices: United States, 2006. Advance Data from Vital and Health Statistics. October 26 2007. No.393. www.cdc.gov/nchs. (Accessed December 3 2007)
9. DesRoches, CM et al. Electronic Health Records in Ambulatory Care—A National Survey of Physicians. NEJM 2008;359:50-60
10. Hsiao J, Beatty PC. Electronic Medical Record/Electronic Health Record Use By Office Based Physicians: United States, 2008 and Preliminary 2009. http://www.cdc.gov/nchs/data/hestat/emr_ehr/emr_ehr.pdf (Accessed November 1 2009)
11. Continued Progress: Hospital Use of Information Technology. American Hospital Association 2007. www.aha.org. (Accessed August 18 2007)
12. Jha AK, DesRoches CM, Campbell EG, et al. Use of Electronic Health Records in US Hospitals NEJM 2009;360:1-11
13. HIMSS Analytics www.himssanalytics.org (Accessed April 10 2009)
14. England's NHS experiencing delays with National IT Program www.ihealthbeat.org April 28 2004 (Accessed January 29 2006)
15. Schoen C, Osborn R, Doty M et al. A Survey of Primary Care Physicans in Eleven Countries, 2009: Perspectives on Care, Cost and Experiences. Health Affairs 2009 Online 11/5/2009. 10.1377/hlthaff.28.6.w1171 (Accessed October 5 2009)
16. Notice of Proposed Rule Making. Medicare and Medicaid Programs; EHR Incentive Program. January 13 2010. http://www.cms.hhs.gov/Recovery/Downloads/CMS-2009-0117-0002.pdf (Accessed January 20 2010)
17. Interim Final Rule. Initial set of standards, implementation specifications and certification criteria. January 13 2010. http://edocket.access.gpo.gov/2010/E9-31216.htm (Accessed January 13 2010)
18. NPRM. Proposed establishment of certification programs for HIT. http://edocket.access.gpo.gov/2010/2010-4991.htm (Accessed March 15 2010)
19. Final Rule. Federal Register. July 28 2010. http://edocket.access.gpo.gov/2010/pdf/2010-17207.pdf (Accessed July 28 2010)
20. Health Information Management Systems Society. http://www.himss.org/EconomicStimulus/ (Accessed January 20 2010)
21. Journal of the American Health Information Association. Meaningful Use Monograph Series of Whitepapers. http://journal.ahima.org (Accessed March 25 2010)

22. Stead WW, Kelly BJ, Kolodner RM. Achievable steps toward building a National Health Information Infrastructure in the United States JAMIA 2005;12:113-120

23. Defining Key Health Information Technology Terms April 28 2008 www.nahit.org (Accessed May 20 2008)

24. Tang PC et al. Measuring the Effects of Reminders for Outpatient influenza Immunizations at the point of clinical opportunity. JAMIA 1999;6:115-121

25. Smith PC et al. Missing Clinical Information During Primary Care Visits JAMA 2005;293:565-571

26. The President's Information Technology Advisory Committee (PITAC) http://www.nitrd.gov/pubs/pitac/ (Accessed January 28 2006)

27. Eden KB, Messina R, Li H et al. Examining the value of electronic health records on labor and delivery. Am J Obstet Gynecol 2008;199;307.e1-307.e9

28. Tierney WM. Computerized Display of Past Test Results. Annals of Int Medicine 1987;107:569-574

29. Lohr S Building a Medical Data Network. The New York Times. November 22 2004. (Accessed December 20 2005)

30. Garrido T, Jamieson L, Zhou Y, Wiesenthal A, Liang L. Effect of electronic health records in ambulatory care: retrospective, serial, cross sectional study. BMJ 2005;330:1313-1316

31. Pollak, VE, Lorch JA. Effect of Electronic Patient Record Use on Mortality in End Stage Renal Disease, a Model Chronic Disease: A Retrospective Analysis of 9 years of Prospectively Collected Data. Biomed Central http://www.biomedcentral.com/1472-6947/7/38. (Accessed December 20 2007)

32. Wears, RL, Berg M. Computer Technology and Clinical Work. Still Waiting for Godot. JAMA 2005;293:1261-1263

33. US Food and Drug Administration http://www.fda.gov/ola/2004/vioxx1118.html (Accessed Aug 15 2006)

34. Badgett R, Mulrow C. Using Information Technology to transfer knowledge: A medical institution steps up to the plate [editorial] Ann of Int Med 2005;142;220-221

35. Housman D. Quality Reporting Through a Data Warehouse Pat Safety & Qual Healthcare Jan/Feb 2009;26-31

36. Wall Street Journal Online/Harris Interactive Health-Care Poll www.wsj.com/health (Accessed October 24 2006)

37. Potential benefits of an EHR. AAFP's Center for Health Information Technology. www.centerforhit.org/x1117.xml (Accessed January 23 2006)

38. Center for Information Technology Leadership. CPOE in Ambulatory Care. www.citl.org/research/ACPOE.htm (Accessed November 9 2005)

39. Patient Safety & Quality Healthcare. New Survey Addresses Relationship of EMRs to Malpractice Risk. August 22 2007. www.psqh.com (Accessed August 23 2007)

40. Nweide DJ, Weeks WB, Gottlieb DJ et al. Relationship of Primary Care Physicians' Patient Caseload with Measurement of Quality and Cost Performance. JAMA 2009;302(22):2444-2450

41. Dell RM, Greene D, Anderson D et al. Osteoporosis Disease Management: What Every Orthopedic Surgeon Should Know. J Bone Joint Surg 2009;91:79-86

42. Lee BJ, Forbes K. The role of specialists in managing the health of populations with chronic disease: the example of chronic kidney disease. BMJ. 2009;339:b2395

43. Chen C, Garrido T, Chock D et al.The Kaiser Permanente Electronic Health Record: Transforming and Streamlining Modalities of Care. Health Affairs 2009;28(2):323-333

44. Jacobe D. Worried about…the Financial Impact of Serious Illness. Gallup Serious Chronic Illness Survey 2002 http://poll.gallup.com/content/default.aspx?ci=6325&pg=1 (Accessed January 29 2006)

45. *Chronic Illness and Caregiving*, a survey conducted by Harris Interactive, Inc., 2000. www.harrisinteractive.com (Accessed January 29 2006)

46. National Public Engagement Campaign on Chronic Illness–Physician Survey conducted by Mathematica Policy Research, Inc., 2001. (Accessed October 15 2005)

47. O'Malley AS, Grossman JM, Cohen GR et al. Are Electronic Medical Records Helpful for Care Coordination? Experiencces of Physician Practices. J Gen Int Med 2009 DOI: 10.1007/s11606-009-1195-2 22 December 2009

48. The Computer Based Record: An Essential Technology for Health Care. www.iom.edu/Reports (Accessed January 12 2010)

49. *Key Capabilities of an Electronic Health Record System: Letter Report.*Committee on Data Standards for Patient Safety. http://www.iom.edu/Reports/2003/Key-Capabilities-of-an-Electronic-Health-Record-System.aspx (Accessed January 12 2010)

50. Carter J Selecting an Electronic Medical Records System, second edition, 2008 Practice Management Center. American College of Physicians. www.acponline.org/pmc (Accessed January 10 2009)

51. Bates DM, Teich JM, Lee J et al. The impact of Computerized Physician Order Entry on Medication Error Prevention JAMIA 1999;6:313-321

52. Koppel R et al. Role of Computerized Physician Order Entry Systems in Facilitating Medication Errors. JAMA 2005;293:1197-1203

53. Bates DW et al. Effect of computerized physician order entry and a team intervention on prevention of serious medication errors JAMA 1998;280:1311-1316

54. Wolfstadt JI, Gurwitz JH, Field TS et al. The Effect of Computerized Physician Order Entry with Clinical Decision Support on the Rates of Adverse Drug Events: A Systematic Review. J Gen Int Med 2008;23(4):451-458

55. Han YY et al. Unexpected increased mortality after implementation of a commercially sold computerized physician order entry system Pediatrics 2005;116:1506-1512

56. Del Beccaro MA et al. Computerized Provider Order Entry Implementation: No Association with Increased Mortality Rate in An Intensive Care Unit Pediatrics 2006;118:290-295

57. Nebeker J et al. High Rates of Adverse Drug Events in a highly computerized hospital Arch Int Med 2005;165:1111-16

58. Holdsworth MT et al. Impact of Computerized Prescriber Order Entry on the Incidence of Adverse Drug Events in Pediatric Inpatients. Pediatrics. 2007;120:1058-1066

59. Center for Information Technology Leadership. CPOE in Ambulatory Care http://www.citl.org/research/ACPOE.htm (Accessed April 5 2006)

60. Eslami S, Abu-Hanna, A, de Keizer, NF. Evaluation of Outpatient Computerized Physician Medication Order Entry Systems: A Systematic Review. JAMIA 2007;14:400-406

61. Kuo GM, Phillips RL, Graham D. et al. Medication errors reported by US family physicians and their office staff. Quality and Safety In Health Care 2008;17(4):286-290

62. Mekhjian HS. Immediate Benefits Realized Following Implementation of Physician Order Entry at an Academic Medical Center JAMIA 2002;9:529-539

63. Tierney, WM et al. Physician Inpatient Order Writing on Microcomputer Workstations: Effects on Resource Utilization. JAMA 1993;269(3):379-383

64. Teich JM et al. Toward Cost-Effective, Quality Care: The Brigham Integrated Computing System. Pp19-55. Elaine Steen [ed] The Second Annual Nicholas E. Davis Award: Proceedings of the CPR Recognition symposium. McGraw-Hill 1996

65. Ashish KJ et al. How common are electronic health records in the US? A summary of the evidence. Health Affairs 2006;25:496-507

66. A Primer on Physician Order Entry. California HealthCare Foundation. First Consulting Group September 2000. www.chcf.org (Accessed September 20 2006)

67. Kuperman GJ, Gibson RF. Computer Physician Order Entry: Benefits, Costs and Issues Ann Intern Med 2003;139:31-39

68. Hunt DL et al. Effects of Computer-Based Clinical Decision Support Systems on Physician Performance and Patient Outcomes: A systematic review JAMA 1998;280(15);1339-1346

69. Garg AX et al. Effects of Computerized Clinical Decision Support Systems on Practitioner Performance and Patient Outcomes. JAMA 2005;293(10):1223-1238

70. Kawamoto, K, Houlihan, CA, Balas, EA, Loback DF. Improving clinical practice using clinical decision support systems: a systematic review of trials to identify features critical to success. BMJ. March 2005. http://bmj.com/cgi/content/full/bmj.38398.500764.8F/DC1. (Accessed June 13 2007)

71. Wright A, Sittig DF, Ash JS et al. Clinical Decision support Capabilities of Commerically Available Clinical Information Systems. JAMIA 2009;16:637-644.

72. Sheridan TB, Thompson JM. People versus computers in medicine. In: Bogner MS, (ed). Human Error in Medicine. Hillsdale, NJ: Lawrence Erlbaum Associates, 1994, pp141-59

73. American College of Physicians Physician Information and Educational Resource (ACP Pier) http://pier.acponline.org/index.html?hp (Accessed January 28 2006)

74. Hoffheinz F. UpToDate. Personal Communication November 20 2006

75. FirstConsult integration into EHR using iConsult http://iconsult.elsevier.com/demo.html (Accessed December 15 2007)

76. Theradoc Antibiotic Assistant www.theradoc.com (Accessed December 24 2007)

77. Evans RS et al. A Computer-Assisted management Program for Antibiotics and Other Anti-infective agents NEJM 1998;338 (4):232-238

78. MEDCALC 3000 www.medcalc3000.com (Accessed December 24 2007)

79. Bates DW et al. A Proposal for Electronic Medical Records in US Primary Care JAMIA 2003;10:1-10

80. Kawamoto K et al. Improving clinical practice using clinical decision support systems: a systematic review of trials to identify features critical to success. BMJ 2005 330: 765-772

81. Hsiegh TC et al. Characteristics and Consequences of Drug Allergy Alert Overrides in a Computerized Physician Order Entry System JAIMA 2004;11:482-491

82. Shah NR et al. Improving Acceptance of Computerized Prescribing Alerts in Ambulatory Care JAMIA 2006;13:5-11

83. Bates DW, Gawande AA. Patient Safety: Improving Safety with Information Technology NEJM 2003;348:2526-2534

84. Kucher N et al. Electronic Alerts to Prevent Venous Thromboembolism among Hospitalized Patients. NEJM. 2005;352-969-977

85. Smith, A. Online support for IV drug administration. Patient Safety & Quality Healthcare March/April 2007:50-54

86. Kuperman GJ, Bobb A, Payne TH et al. Medication-related Clinical Decision Support in Computerized Provider Order Entry Systems: A Review. J Am Med Inform Assoc 2007;14:29-40

87. Shea, S, DuMouchel, W, Bahamonde, L. A Meta-Analysis of 15 Randomized Controlled Trials to Evaluate Computer-Based Clinical Reminder Systems for Preventive Care in the Ambulatory Setting. JAMIA 1996;3:399-409

88. Giuse NB et al. Evolution of a Mature Clinical Informationist Model JAMIA 2005;12:249-255

89. Bobb AM, Payne TH, Gross PA. Viewpoint: Controversies surrounding use of order sets for clinical decision support in computerized order entry. JAMIA 2007;14:41-47

90. DxPlain. http://www.lcs.mgh.harvard.edu/projects/dxplain.html . (Accessed January 1 2008)

91. Osheroff JA et al. Improving Outcomes with Clinical Decision Support: An Implementer's Guide. HIMSS Publication 2005. www.HIMSS.org

92. Rosenthal DI, Wilburg JB, Schultz T et al. Radiology Order Entry With Decision Support: Initial Clinical Experience. J Am Coll Radiol 2006;3:799-806

93. Van wijk MAM, Van der lei, J, MOssveld, M, et al. Assessment of Decision Support for Blood Test Ordering in Primary Care. Ann Int Med 2001;134:274-281

94. Lurio J, Morrison FP, Pichardo M et al. Using electronic healthr record alerts to provide public health situational awareness to clinicians. JAMIA 2010;17:217-219

95. Bates DM et al. Ten Commandments for Effective Clinical Decision Support: Making the Practice of Evidence-based Medicine a Reality JAMIA 2003;10:523-530

96. Briggs B. Decision Support Matures. Health Data Management August 15 2005. www.healthdatamanagement.com/html/current/CurrentIssueStory.cfm?Post ID=19990 (Accessed August 20 2005)

97. M.J. Ball. Clinical Decision Support Systems: Theory and Practice. Springer. 1998

98. Clinical Decision Support Systems in Informatics Review. www.informatics-review.com/decision-support. (Accessed January 23 2006)

99. Improving Outcome with Clinical Decision Support: An Implementer's Guide. Osheroff, JA, Pifer EA, Teich JM, Sittig DF, Jenders RA. HIMSS 2005 Chicago, Il

100. List of Open Source healthcare software. Wikipedia. www.wikipedia.org (Accessed April 10 2009)

101. Open-Source EHR Systems for Ambulatory Care: A Market Assessment. California HealthCare Foundation. January 2008 (Accessed April 1 2009)

102. Open Vista. http://sourceforge.net/projects/openvista (Accessed April 16 2009)

103. VistA-Office EHR http://worldvista.org (Accessed April 16 2009)

104. Medshere. www.medsphere.com (Accessed April 9 2009)

105. OpenEMR www.oemr.org (Accessed July 10 2009)

106. Practice Fusion www.practicefusion.com (Accessed March 20 2010)

107. MDBug https://www.mdbug.com (Accessed July 3 2009)

108. Medical Office Online. www.medicalofficeonline.com. (Accessed April 9 2009)

109. Soapware EMR www.docs.com (Accessed March 20 2010)

110. CHCSII/ALHTA site. **http://tiny.cc/0dUbM** (Accessed January 25 2009)

111. AHLTA-Electronic Health Records www.ha.osd.mil/AHLTA/ (Accessed December 2 2005)

112. Murff HJ, Kannry J. Physician satisfaction with two order entry systems. JAMIA 2001;8:499-509

113. The VA electronic health record http://www.hhs.gov/healthit/attachment_2/iii.html (Accessed January 28 2006)

114. Veterans Health Information System and Technology Architecture (VISTA)http://www1.va.gov/vha_oi/docs/What_is_VistA.pdf (Accessed January 28 2006)

115. From VistA to HealtheVet-Vista. GAO Training week. Kolodner R. http://www1.va.gov/vha_oi/docs/GAO_Education_Week_November_2004.ppt (Accessed January 28 2006)

116. Kaiser Offers No-Cost Personal Health Records to Members Nationwide. November 7 2007. www.ihealthbeat.org (Accessed November 9 2007)

117. Kaiser Permanente. In depth focus: Kaiser Permanente HealthConnect. November 2007.http://ckp.kp.org/kpindepth/archive/indepth_faq_all.html (Accessed December 8 2007)

118. Eclinicalworks. www.eclinicalworks.com (Accessed May 1 2010)
119. GE Centricity https://www2.gehealthcare.com (Accessed June 28 2010)
120. NexGen web site www.nexgen.com (Accessed April 9 2009)
121. Vision Associates. www.visionassociates.net (Accessed January 28 2006)
122. Shortliffe E. The Evolution of electronic medical records Acad Med 1999;74:414-419
123. Wang SJ et al. A Cost-Benefit Analysis of Electronic Medical Records in Primary Care Amer J of Med 2003;114:397-403
124. Miller RH et al. The Value of Electronic Health Records In solo or small group practices Health Affairs 2005;24:1127-1137
125. King MS, Sharp L, Lipsky M. Accuracy of CPT evaluation and management coding by Family Physicians. J Am Board Fam Pract 2001;14(3):184-192
126. Moore P. Tech Survey: Navigating the Tech Maze. September 2008. Physicians Practice www.physicianspractice.com
127. 2003 Commonwealth Fund National Survey of Physicians and Quality of Care. http://www.cmwf.org/surveys/surveys_show.htm?doc_id=278869 (Accessed December 7 2004)
128. Brailer DJ, Terasawa EL. Use and Adoption of Computer Based Patient Records California Healthcare Foundation 2003 www.chcf.org (Accessed February 10 2005)
129. Poissant L, Pereira J, Tamblyn R et al. The impact of electronic health records on time efficiency of physicians and nurses: a systematic review. JAMIA 2005;12:505-516
130. Medical Records Institute's Ninth Annual EHR Survey of Electronic Medical Records, Usage and Trends 2007. http://www.medrecinst.com/MRI/emrsurvey.html. (Accessed December 4 2007)
131. Boone, E. EMR Usability: Bridging the Gap Beween the Nurse and Computer. Nursing Management 2010;41(3): 14-16
132. HIMSS EHR Usability Task Force. June 2009 www.himss.org (Accessed March 15 2010)
133. Sujansky WV, Overhage JM, Chang S et al. The Development of a Highly Constrained Health Level 7 Implementation Guide to Facilitate Electronic Laboratory Reporting to Ambulatory Electronic Health Record Systems. JAMIA 2009;16:285-290
134. Mirth Connect http://www.mirthcorp.com/products/mirth-connect (Accessed July 31 2009)
135. Physician Self-Referral Exceptions For Electronic Prescribing and Electronic Health Records Technology. Centers for Medicare and Medicaid Services. www.cms.hhs.gov/apps/media/press/release.asp?Counter=1920 (Accessed November 24 2006)
136. Dougherty M, Washington L. Still Seeking The Legal Record. Journal of AHIMA February 10; 42-45
137. Chaudry B, Wang J, Wu S et al. Systematic Review: Impact of Health Information Technology on Quality, Efficiency and Costs of Medical Care. Ann of Int Med 2006;144:E12-22.
138. Yu FB, Menachemi N, Berner ES et al. Full Implementation of Computerized Physician Order Entry and Medication-Related Quality Outcomes: A Study of 3364 Hospitals. Am J of Qual Meas 2009; E1-9doi:10.1177/106286060933626
139. Study: EHR Adoption Results in Marginal Performance Gains. November 16 2009. www.ihealthbeat.org (Accessed November 23 2009)
140. EHR IMPACT. The socio-economic impact of interoperable electronic health record (EHR) and ePrescribing systems in Europe and beyond. Final study report. October 2009. www.ehr-impact.eu (Accessed March 1 2010)

141. Han YY et al. Unexpected Increased Mortality After Implementation of a commercially sold computerized physician order entry system Pediatrics 2005; 116:1506-1512

142. Nebecker JR et al. High Rates of Adverse Drug Events in a Highly Computerized Hospital. Arch Int Med 2005; 165:1111-1116

143. Bates DW. Computerized physician order entry and medication errors: Finding a balance. J of Bioinform 2005;38:259-261

144. Weiner, JP et al. e-Iatrogenesis: The Most Critical Unintended Consequence of CPOE and other HIT. JAMIA. 2007;14(3):387-388

145. Ash JS, Berg M, Coiera E. Some Unintended Consequences of Information Technology in Health Care: The Nature of Patient Care Information System-Related Errors. JAMIA. 2004;11:104-112

146. Berger RG, Kichak, JP. Computerized Physician Order Entry: Helpful or Harmful JAMIA 2004;11:100-103

147. Ackerman, K. Details Sparse, But Federal EHR Program Draws Praise, Optimism. November 15th 2007. www.ihealthbeat.org. (Accessed November 15th 2007)

148. Nursing Survey. April 21 2010. iHealthbeat. www.ihealthbeat.org (Accessed April 23 2010)

149. Electronic Medical Record Implementation Guide 2009. www.physiciansfoundation.org (Accessed June 10 2010)

150. Edsall RL, Adler KG. The 2009 EHR User Satisfaction Survey. November/December http://www.amazingcharts.com/images/printmedia/10the2.pdf (Accessed January 15 2010)

151. Medscape EHR Survery www.medscape.com/viewarticle/709856 (Accessed December 3 2009)

152. AC Group www.acgroup.org (Accessed March 20 2010)

153. Lowes R. Report Names Top Electronic Health Records for Physician Practices. December 2009. www.medscape.com (Accessed March 15 2010)

154. KLAS Research www.klasresearch.com (Accessed March 15 2010)

155. DOC-IT Roadmap www.healthinsight.org (Accessed March 16 2010)

156. Build your EHR Timeline http://www.physiciansehr.org/index.asp?PageAction=Custom&ID=48 (Accessed March 16 2010)

157. DOC-IT EHR Implementation Checklist www.archive.healthit.ahrq.gov/ (Accessed March 3 2010)

158. EHR Scope www.ehrscope.com (Accessed March 21 2010)

159. Havenstein H Emerging ASP model targets health records. May 9 2005. Computerworld http://tiny.cc/DL9zD (Accessed January 29 2006)

160. Physician Practices: Are Application Service Providers Right for You? October 2006 www.chcf.org (Accessed October 20 2006)

161. Nuance www.nuance.com

162. Speech Recognition Improves EMR ROI. Health Management Technology October 2009. www.healthmgttech.com (Accessed February 20 2010)

163. Hoyt R, Yoshihashi. Lessons Learned from Implementation of Voice Recognition for Documentation. Perspectives in Health Information Management. Winter 2010. http://perspectives.ahima.org (Accessed February 1 2010)

164. Conn J. Docs using scribes to ease EHR transition. Modern Healthcare. February 8 2010. www.modernhealthcare.com (Accessed February 20 2010)

165. Cannon J, Lucci S. Transcription and EHRs, benefits of a blended approach. Journal of AHIMA February 10 2010:36-40

166. Epic EHR. www.epic.com (Accessed March 15 2010)

167. HRSA. www.hrsa.gov/healthit/ehrguidelines.htm (Accessed July 20 2009)

168. AmericanEHR Partners. www.americanehr.com (Accessed July 13 2010)

169. Adler KG. How to Select an Electronic Health Record System. Family Practice Management. February 2005 www. Aafp.org/fpm (Accessed January 20 2006)

170. eHealth Initiative EHR Master Quotation Guide. eHealth Initiative. http://tiny.cc/7RJZg (Accessed January 30 2006)

171. Physician Quality Reporting Initiative (PQRI). www.cms.hhs.gov/pqri (Accessed April 15 2009)

172. Levick, D. "Sure I want an EMR, but I don't want to pay for it". HIMSS presentation. April 4 2009. Chicago, Il.

173. The CCHIT Incentive Index. April 15 2009. Certification Commission for Healthcare Information Technology. www.cchit.org (Accessed April 24 2009)

174. Lewis, M. Maryland doctors to see double bonus for EHR adoption. July 10 2009. Modern Medicine. http://www.modernmedicine.com/modernmedicine/article/articleDetail.jsp?id=607959 (Accessed July 23 2009)

175. California Healthcare Foundation. Chart Abstraction: EHR Deployment Techniques Feb 2010 www.chcf.org (Accessed March 10 2010)

176. Terry K. Technology: EMR Success in 8 Easy Steps. www.physicianpractice.com (Accessed March 10 2010)

177. Nelson R. Getting Data into an EHR. Medpage Today March 3 2008. www.printhis.clickability.com (Accessed October 9 2009)

178. Digital Island http://www.digitisle.com (Accessed March 13 2010)

179. Trachtenbarg DE. EHRs fix everything-and nine other myths. Family Practice Management March 2007 www.aafp.org/fpm/20070300/26ehrs.html (Accessed September 19 2008)

180. Baron RJ, Fabens EL, Schiffman M. Electronic Health Records: Just around the corner? Or over the cliff? Annals of Int Med 2005;143:222-226

181. Diamond J, Fera B. Implementing an EHR. 2007 HIMSS Conference February 25-March 1.New Orleans

182. MGMA. Medical Practice Today, 2009. www.mgma.com (Accessed August 12 2009)

3

Practice Management Systems

BRANDY G. ZIESEMER
ROBERT E. HOYT

Learning Objectives

After reading this chapter the reader should be able to:

- Document the workflow in a medical office that utilizes a practice management (PM) system integrated with an electronic health record system
- Compare the functionality of a standalone PM system with fully integrated PM software as part of a robust EHR system
- Identify the features of an integrated practice management system
- List the most common integrated PM software packages currently available
- Identify the key advantages and obstacles in converting from paper records and a standalone PM system to an integrated PM with EHR system
- Discuss emerging trends in practice management

"In an environment where quality is quantitatively and qualitatively measured, an administrator will be judged to be productive if his or her work leads to an organization performing at a higher level in the metrics of cost, quality, and service."

Rick Madison and Ken Clarke, Group Practice Journal

Most medical offices have had computerized practice management (PM) systems for many years, regardless of whether that office maintains paper medical records, electronic health records (EHRs) or a hybrid of these two. As we will point out there are many reasons why PM systems have become so prevalent but one of the main reasons is for more rapid claims submission and adjudication. Without an electronic system, time and money would be lost on faxes, phone calls and snail mail. The American Medical Association estimated that inefficient claims submission systems lead to about $210 billion annually in unnecessary costs.[1] A PM system is designed to capture all of the data from a patient encounter necessary to obtain reimbursement for the services provided. This data is then used to:

- Generate claims to seek reimbursement from healthcare payers
- Apply payments and denials
- Generate patient statements for any balance that is the patient's responsibility
- Generate business correspondence
- Build databases for practice and referring physicians, payers, patient demographics and patient encounter transactions (i.e., date, diagnosis codes, procedure codes, amount charged, amount paid, date paid, billing messages, place and type of service codes, etc)

Additionally, a PM system provides routine and ad hoc reports so that an administrator can analyze the trends for a given practice and implement performance improvement strategies based on the findings. For example, a medical office administrator is able to use the PM system to compare and contrast different payers with regards to the amount reimbursed for each given service or the turnaround time between claims submission and payment. The results lead to deciding which managed care plans the practice will participate in versus those plans that the practice may want to consider not accepting in the future. Another example is to analyze all payers for a given service performed in the practice to determine if that service is a good use of the practice's clinical time. This analysis provides one aspect of whether or not the practice should consider continuing to offer a certain service such as case management of a patient who is receiving a home health service. Of course, the administrator has to weigh services that aren't profitable against any negative impact on overall patient satisfaction but the PM system provides a means of analyzing payment performance.

Most PM systems also offer patient scheduling software that further increases the efficiency of the business aspects of a medical practice. Finally, some PM systems offer an encoder to assist the coder in selecting and sequencing the correct diagnosis (International Classification of Diseases, 9th revision, clinically modified for use in the US or ICD-9-CM) and procedure (Current Procedural Terminology, fourth edition or CPT-4® and Healthcare Common Procedure Coding System or HCPCS) codes. Even when a physician determines the appropriate codes using a "*superbill*," (a list of the common codes used in that practice along with the amount charged for each procedure), there are times when a diagnosis or procedure is not listed on the superbill and an encoder makes it efficient to do a search based on the main terms and select the best code. Furthermore, some encoders are packaged with tools such as a subscription to "*CPT Assistant*" that help the practice comply with correct coding initiatives which in turn optimize the reimbursement to which the practice is legally and ethically entitled and avoids fraud or abuse fines for improper coding.

Clinical and Administrative Workflow in a Medical Office

Several steps are common to almost any medical practice with regards to treating patients and getting reimbursed properly for the services provided. The steps are subdivided based on whether or not the patient has been to this practice previously for any type of service. The first step is to get the patient registered. This can be accomplished via a practice website or by the patient calling the office to schedule an appointment. Figure 3.1 demonstrates typical outpatient office workflow.

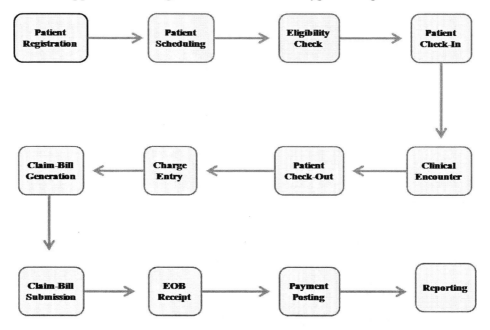

Figure 3.1 Typical Outpatient office workflow (EOB = explanation of benefits)

Patient Registration. This step includes obtaining demographic information, including any healthcare plan or plans the patient has and establishing which member of the patient's household is financially responsible for any balances due either at the time of the visit or after claims adjudication by any healthcare payer(s) the practice agrees to bill for the patient.

Patient Scheduling. The patient is then scheduled for an appointment. If the patient had a previous encounter with the physician, the office receptionist simply has to update any changes to the patient information already on file.

Eligibility Check. For a new patient the insurance information must be verified to ensure that the patient is currently covered by a plan accepted by the practice and the planned services are a covered benefit. If not, the patient must be notified in advance of the visit to determine if they are willing to accept full financial responsibility for the services (i.e. full payment then attempt to get reimbursement from their healthcare plan on their own) or cancel the appointment and find a participating physician. If a practice offers web-based patient registration, there are some choices ranging from designing the website and all applicable online forms internally to contracting with a forms services company. Based on the amount of money the practice is willing to spend, a forms company offers basic forms design for use on the practice's own website. Alternately, they can subcontract to use the company's server and website for forms design, updating, processing and transmitting information to the practice's EHR or PM system. See Medical Web Office services for a sample range of forms and communications services available for medical practices.[2]

Patient Check-In. The patient checks in for the scheduled visit. If already established with the practice the receptionist simply verifies/updates the patient information. If the patient is new, and the data gathered to schedule an appointment was obtained via telephone, the patient is asked to complete a registration form and provide a copy of his or her insurance card(s). Any information not previously obtained is keyed into the computer system for use by the PM system and the source document is added to the paper medical record, if applicable. Scanning the information is an option with an EHR. Most practices that have a PM system that is integrated with an EHR can scan the documents (including bubble sheets completed by the patient at time of registration) into the system once and the information is posted to the appropriate places in both the EHR and the PM system. Sometimes the data that is used by both the EHR and the PM software, such as patient name, is saved to a common database in an integrated system. At other times, however, the shared data is communicated electronically between the EHR and the PM system even though the databases are separate. It is important to know that when the systems have a shared database, this database only contains the part of the clinical record that is used to obtain reimbursement such as the patient demographics, diagnoses and procedures, dates of service, etc. However, the purely financial information is only found in the PM system – such as amount billed and amount paid or information about health plans. This is because it is not advisable to combine the business aspects of health information with clinical aspects. What procedure is done on a given date and the diagnosis that justifies the medical necessity of a procedure is both clinical and financial but how much the procedure costs and how much the patient paid out-of-pocket, etc is purely financial.

Clinical Encounter. The patient is generally first seen by a nurse or medical assistant, to have vitals taken, collect blood and urine samples, if needed, and update the patient's subjective history. The patient is then examined by the physician who takes additional history and completes the objective physical exam and updates the clinical notes in SOAP order – Subjective, Objective, Assessment and Plan. In a paper system, the physician dictates either during the visit or as soon afterward as possible and a transcriptionist creates a paper copy of the notes. Alternately, some physicians use voice recognition technology to dictate directly into a laptop or other device then print out the report generated by the software to file in the paper record. For a sample of a full range of voice recognition software that can be used as a standalone product for creating a paper document or that interfaces with electronic health records, visit Nuance's website.[3]

As discussed previously, in an EHR system clinicians have several options for inputting patient information into the clinical record. They can use voice recognition software, standard dictation or templates. This can be accomplished on a PC or in a wireless mode with a tablet PC or PDA. For example, an EHR is formatted with physician workflow in mind and then, customizable by each individual physician to optimize efficiency based on specialty and personal preferences. The customization "initially takes time and patience but is well worth the effort in a practice that sees a lot of patients daily with the same symptoms such as an ear infection for a pediatrician."[4] Therefore, when the physician is face-to-face with the patient, the EHR would have already been started for that encounter by a nurse or other physician extender who would have entered the patient's chief complaint, vital signs and possibly any updates to the patient's subjective history (the subjective portion of the SOAP note).

The physician will continue building the encounter notes by using a series of drop-down menus to indicate body systems examined, tests performed, tests or prescriptions ordered, (the objective portion of the SOAP note), the assessment and the plan. Each selection made by the physician adds to the clinical notes. Clinical notes are a good example of data that is maintained in the EHR but not shared with a PM system. However, EHRs that use computer assisted coding technology can convert the standardized notes into codes and the codes are used by both the EHR and the PM system. For example, many EHRs can run the office notes through logic to assign CPT evaluation and management (E&M) codes based on either the 1995 or 1997 guidelines. The EHR system can pass these codes plus many ICD-9-CM codes over to integrated (same vendor) or interfaced (different vendors) PM system when the systems are compatible. The physician concludes the clinical aspects of the encounter by giving the patient discharge/follow-up instructions and patient education literature. Any lab samples are sent to the lab and, if the patient needs a prescription and the practice uses e-prescribing, a prescription is sent from the EHR to the pharmacy electronically or via Fax. If not, the patient is given a paper copy of the prescription.

Patient Check-Out. The patient is discharged after a receptionist collects any money due and schedules any follow-up visits. If the practice has chosen this feature, the EHR can interface with the PM system scheduler so the physician can schedule a follow-up visit and the patient can take home a printout of the office notes, any education material, the next appointment, plus a paper copy of physician orders or prescriptions for facilities not linked with the EHR.

Charge Entry Claims-Bill Generation. In a standalone PM system, the charges are entered, often from a superbill but sometimes the services are coded from the information in the medical record. In an integrated PM with EHR system, the information needed is sent directly from the EHR to the PM system and a claim is built as described above. However, a person responsible for correct coding and billing must still verify that all applicable codes were brought over to the PM system, add any codes that the system did not assign automatically and scrub codes which means to link the diagnoses to the correct procedures that justify medical necessity and check for obvious errors in order to get them ready to submit as claims to payers.

Claims-Bill Submission. The claims are sent electronically in all but rare cases but they are sent in cycles so once the PM system is updated, the claim is in queue waiting for transmission to a clearinghouse or directly to the payer, such as Medicare.

Remittance Advice (RA) and Explanation of Benefits (EOB) Receipt. Once the claim is sent, the payer electronically (again, there are some exceptions in which the practice will actually get a paper check in the mail) sends a remittance advice (RA) containing the details for each charge paid or denied in that cycle. The RA contains an EOB (payments, denials, denial reason, reduced payments and reasons, patient responsibility, whether or not the claim was sent automatically to a secondary payer, etc.) for each charge by patient.

Payment Posting. The money is electronically deposited into the practice's account. The payer generally mails a paper copy of each individual explanation of benefits to the patient. Billing personnel also have to follow-up when a person has more than one payer, to determine that the claim was transmitted to the appropriate secondary payer. If there is still a balance after the biller has applied payments and written off any charges in excess of the allowed amount for a particular payer, the system moves the balance into a queue to await patient billing. The biller is also responsible for tracking claims and initiating the collections process if a balance due by the patient is not paid in a timely fashion.

Reporting. Daily reports are run and verified to ensure deposits match, all patients who were seen that day have charges in the system, etc. There are both routine reports (daily, weekly, monthly and end-of-year) and ad hoc reports used by the practice.

Telephone calls in a Medical Office. Calls to the practice may be for various reasons from cancelling an appointment to asking if the doctor can see a patient who does not have an appointment. The calls should be prioritized into categories for emergency (in which the practice should advise the patient to hang-up and dial 911 and then confirm the patient is capable of doing that before disconnecting), urgent or non-urgent. Many EHR systems enable the message-taker to route the call directly to the intended recipient instead of having to take a paper message. In this case, protocols exist in which the person answering the phone can take certain actions or make some decisions. For example, the receptionist may be able to determine if outside lab results have been received by the practice or not. If they have, the receptionist can route a message directly to the appropriate clinician requesting the patient be called regarding the results. In more urgent cases, a patient may have a non-emergency, but urgent condition and request to be seen that day. The receptionist may be able to schedule that person in and tell them when to arrive. If the receptionist is uncertain, he or she may route an urgent message to the most appropriate clinician for a decision about whether the patient can be seen that day. The messages go directly to the recipient's attention and may be color-coded to highlight urgent versus non-urgent.[5-6]

Practice Management Systems and EHRs

When the administrators of a medical practice commit to the conversion of paper-based or hybrid records to an electronic health record system, they should strongly consider converting their existing practice management system to one that is integrated with the EHR they chose. Although this strategy means a higher initial investment of both time and money many practices report that part of their overall success with implementing an EHR with a PM system is due to the increased efficiency and accuracy of the billing process when the systems are integrated. One alternative to discontinuing an existing PM system – especially one that works very well and that everyone in the office who uses it is comfortable with, is to find a reputable EHR vendor that offers interface capabilities with your existing PM system vendor. For example, Eclipsys (formerly MediNotes), has a list of PM systems with which their EHR interfaces. One potential setback is anytime a vendor upgrades its software, the interfaces have to be tested and both vendors may need to get involved.

As mentioned in the Electronic Health Records chapter, formal studies of return on investment (ROI) for an EHR in medical practices are very limited at this time and although some of these studies mention how many of the practices surveyed implemented an EHR in conjunction with an integrated PMS, they don't differentiate the results as to the impact on ROI of an integrated EHR/PM system versus a standalone EHR project.

There are some testimonials and customer information pages available on many of the vendor websites that discuss success stories of EHR implementation in which some of the cost savings were realized by the integration of the EHR with a PM system. Some of these websites also list the ROI based on an integrated EHR/PM system. [7-8]

In March of 2008, the Medical Society of New York reported on the experience of a solo, internal medicine practice in identifying, purchasing, implementing and maintaining an EHR with a PM system. Some of the highlights from this article include:

- Chose in-office server over application service provider (ASP) because it was more affordable since the practice already had old computers that could be used as workstations and they did not already have business-grade broadband Internet service provider necessary for an ASP arrangement

- Chose to pay a monthly subscription fee to use the EHR/PM software because it was less expensive than purchasing software and hardware required. Lease agreement was approximately $400 per month per physician for software and off-site HIT support

- HIT vendor provided most of the on-site technical support during the first three months

- Chose the EHR from same vendor as PM because, "*to run an efficient office, the two systems should be totally compatible and ideally share the same database*"

- The criteria most important to Dr. Volpe included: perform the same functions as a paper record, only better; affordability; navigability for all users between both the EHR and the PM systems; adaptability or the records and templates for information were easily modified and customized as patient information and practice needs changed; track record of vendor; ease of data entry; tight security and, CCHIT certification

- The transition from paper was easy (by comparison) for the doctor and his staff because they had already used an e-prescribing system and a patient web portal. They transitioned from the old billing system to the new PM system gradually over 30-90 days by using the old system for any patient billing that was in-process or occurred before the staff training was complete and then used the new system for all encounters occurring after the staff was trained

- The overall assessment considering the improvement in efficiency, improved patient care and improved claims coding and submission is: "*within one year, Dr. Volpe had paid off the cost of his new hardware, his office had returned to full productivity, and he was earning over $30,000 more than he had the year before, due primarily to reduced overhead costs (space for paper records has been converted to an additional exam room, for example). He, his family, his staff and his patients all love the new system*" [9]

According to a recent newsletter offered by the Health Information Management Systems Society (HIMSS), "*A tool was recently launched to measure the healthcare industry's move from paper-based processes to electronic business applications. Called the US Healthcare Efficiency Index, this tool is intended to raise awareness about the cost savings that can be achieved by increasing the adoption of electronic business processes. One third of respondents believed that the largest cost savings would come in the area of claims submission. Nearly half of respondents that work for a healthcare provider organization reported that they submitted claims payments electronically. A similar percent also reported that their organization submitted claims eligibility and claim remittance advice transactions electronically. Another potential area of cost savings would be to have claims payments issued via direct deposit.*" [10]

Practice Management System Examples

There are more than 200 practice management systems on the market with a variety of PM features to include integration with an EHR and the availability as both a client server model and/or an ASP model. In Table 3.1 we have included a list of the better known vendors who offer a combined EHR-PM that is CCHIT certified and available in the ASP model. In the resource section we will direct you to methods to search all available PM systems.

Resources

Although there is very little written about the merits and limitations of practice management systems, we can direct you to several helpful resources:

- A 2009 monograph by the California HealthCare Foundation "Practice Management Systems for Safety-Net Clinics and Small Group Practices: A Primer" discusses how important PM systems are for safety net clinics but also provides an excellent overview on the subject. [11]
- Capterra is a web site that includes a search engine with filters for operating systems (platform), number of users, PM features, inclusion with EHRs, location (example USA) and annual revenue. Over 200 products are included with hyperlinks to the individual web sites, demos and tours [12]
- EHRScope is a search engine for EHRs but includes the ability to filter the search for EHRs that include PM systems, as well as differentiate between systems that are web-based or client-server based. Without filtering, 342 EHRs are included [13]
- "Selecting a Practice Management System" is a monograph by the American College of Physicians. Members can access this resource that focuses how to go about selecting a PM system, in terms of the steps that are necessary prior to purchase [14]
- "Medical Practice Management Buyer's Guide" is a May 2008 web-based resource that includes pricing and tips before purchase [15]
- The Online Consultant is a fee based generator of "requests for proposal" (RFP). In the case of PM systems they generate detailed questions about price and functional requirements. Once the RFP is complete they offer the ability to graph and create comparison reports between vendors. Charge is $695 for PM RFP [16]
- Wikipedia offers standard information on PM as well as vendor comparison chart and operating system compatibility chart [17]

Table 3.1. EHR-PM integrated systems that are CCHIT approved and available as an ASP model

EHR-PM System	Web site	Features
ABELMed v8	www.abelsoft.com	ASP available
Allscripts MyWay PM Allscripts Professional PM	www.allscripts.com	ASP available. Small-medium practices
Athena clinicals	www.athenahealth.com	Only ASP. Can outsource the billing service
Cerner PowerWorks	www.cerner.com	ASP available. Small-large practices
e-MDs Solution Series	www.e-mds.com	ASP available
eClinicalWorks	www.eclinicalworks.com	ASP available. Small-large practices
Eclipsys PeakPractice	www.medinotes.com	ASP available
GE Centricity Practice Solution	www.gehealth.com	ASP available. Medium-large practices
iMedica Prima	www.imedica.com	ASP available.
NextGen Healthcare	www.nextgen.com	ASP available.
Purkinje Care Series Plus	www.purkinje.com	ASP available.
Sage Intergy V 5.5	www.sagehealth.com	ASP available
Waiting Room Solutions	www.waitingroomsolutions.com	ASP available. 26 specialties

Key Points

- Many medical practices are struggling with implementation of EHRs and how to determine the return on investment (ROI) for a new integrated PM system

- As reimbursement methodologies become increasingly more complex and tied to quality measures, it is of utmost importance to ensure a medical office has the tools to obtain full payment to which the practice is legally and ethically entitled by collecting all of the appropriate data that justifies medical necessity and compliance with quality guidelines

- Practitioners not only have to provide high quality and safe patient care but they must do so as efficiently and effectively as to remain competitive

Conclusion

Ambulatory care practices have many options to consider when converting to a robust electronic health record system that is fully integrated with or interfaced with a comprehensive practice management system. The chapter on electronic health records discusses the steps a practice can take to identify the best overall electronic health record and practice management system based on factors such as size and type of practice, degree to which each physician supports the transition, information technology preferences (servers, technical support, purchase versus lease etc.), priority of various features, projected return on investment (ROI) and other considerations. The practice management piece fits in nicely with the overall EHR selection, implementation, training and maintenance process but should be included from the start rather than starting the selection and implementation piece after the EHR system is already in place. The combination of functionality between the clinical and business aspects of medical practice is considered the ideal future direction by many physicians and administrators who have had their combined systems long enough to enjoy an excellent return on their investment, improved efficiency, improved quality of care documentation and reimbursement.

References:

1. Pulley J. The Claims Scrubbers. Government Health IT. November 2008 pp10-14
2. Medical Web Office http://www.medicalweboffice.com (Accessed February 24 2009)
3. Nuance Communications, Inc http://www.nuance.com (Accessed February 24 2009)
4. Gartee, R. (2007). Electronic Health Records: Understanding and Using Computerized Medical Records. Upper Saddle River. Pearson/Prentice Hall.
5. Gartee, R. (2007). Electronic Health Records: Understanding and Using Computerized Medical Records. Upper Saddle River. Pearson/Prentice Hall.
6. Torpey, D. Physician Office Workflows. Lecture. UWF 2009.
7. E-MDs, inc. http://www.e-mds.com/education/articles/roi.html (Accessed March 29 2009)
8. eClinicalWorks http://www.eclinicalworks.com/casestudy6.php (Accessed March 29 2009)
9. HIT Taken to the Next Level: NYS Practices that have Successfully Adopted EMR & PMS. March 2008. *Medical Society of the State of New York: News of New York*, pp. 5-6.
10. Electronic Claims. *Health Information Management Systems Society: Vantage Point*. February 2009.

http://www.himss.org/content/files/vantagepoint/pdf/VantagePoint_200902.pdf (Accessed February 26 2009)

11. Sujansky W, Sterling R, Swafford R. Practice Management Systems for Safety-Net Clinics and Small Group Practices: A Primer www.chcf.org (Accessed October 1 2009)

12. Medical Practice Management Software. Capterra. http://www.capterra.com/medical-practice-management-software?srchid=135420&pos=1 (Accessed October 5 2009)

13. EHRScope http://www.ehrscope.com/emr-comparison/ (Accessed October 6 2009)

14. Selecting a Practice Management System. (2007) ACP Online. http://www.acponline.org/running_practice/technology/pms/ (Accessed October 6 2009)

15. Medical Practice Mangement Buyers Guide. May 2008. BuyerZone. http://www.buyerzone.com/software/mpm/buyers_guide1.html (Accessed October 6 2009)

16. Online Consultant. Selecting a new Physician Practice System? http://www.olcsoft.com/physician_practice_management_software_requirements.htm (Accessed October 6 2009)

17. Practice Management Software. Wikipedia http://en.wikipedia.org/wiki/Practice_management (Accessed March 21 2010)

4

Health Information Exchange

ROBERT E. HOYT
ROBERT W. CRUZ
STEPHAN L. STEFFENSEN

Learning Objectives

After reading this chapter the reader should be able to:
- Identify the need for and benefits of health information exchange and interoperability
- Describe the concept of health information organizations (HIOs) and how they fit into the Nationwide Health Information Network (NHIN)
- Compare and contrast the differences between NHIN Direct and NHIN Exchange
- Enumerate the basic and advanced features offered by HIOs
- Detail the obstacles facing HIOs

Health information exchange (HIE) is a critical element of Meaningful Use and integral to the future success of healthcare reform at the local, regional and national level. Exchange of health-related data is important to all healthcare organizations, particularly federal programs such as Medicare or Medicaid for several reasons. The federal government determined that HIE is essential to improve: the disability process, continuity of medical care issues, bio-surveillance, research and natural disaster responses. [1] As a result, the federal government has been a major promoter of HIE and the development of data standards to achieve interoperability. Electronic transmission of data results in faster and less expensive transactions, when compared to standard mail and faxes. If the goal of the federal government was only to promote electronic health records, then the end result would be electronic, instead of paper, silos of information. Instead, they have created a comprehensive game plan to share health information among disparate partners. Chapter 1 discusses multiple HITECH programs that support HIE and interoperability.

Table 4.1 lists some of the common types of health related data that are important to exchange among the many healthcare partners.

Table 4.1 Common types of health-related data exchanged

Data	Examples
Clinical results	Lab, pathology, medication data, microbiology reports
Images	Radiology reports or actual images
Documents	Office notes, discharge notes, emergency room notes
Clinical Summaries	Continuity of Care Documents (CCDs)
Financial information	Claims data, eligibility checks
Performance data	Quality measures such as cholesterol levels
Public health data	H1N1 outbreak data

In this chapter we will begin with important HIE-related definitions and then chronicle the evolution of local, state and national organizations created for HIE.

Definitions

In 2008 the National Alliance for Health Information Technology (funded by ONC) released a new set of definitions that would help clarify the ambiguity of several important terms related to HIE:

Health Information Exchange (HIE) is the *"electronic movement of health-related information among organizations according to nationally recognized standards"*.

Health Information Organization (HIO) is *"an organization that oversees and governs the exchange of health-related information among organizations according to nationally recognized standards"*.

Regional Health Information Organization (RHIO) is *"a health information organization that brings together health care stakeholders within a defined geographic area and governs health information exchange among them for the purpose of improving health and care in that community"*.[2]

Note that the term RHIO is inexact because HIOs do not have to be regional; they can include only one city or an entire state. Furthermore, HIOs are being created to exchange health information solely for Medicaid patients or focus on uninsured populations. In keeping with these new definitions we will use the acronym HIO when addressing health information organizations and RHIO when addressing specific defined regional HIOs. We will use HIE to describe the movement or exchange of health information.

History of the Nationwide Health Information Network

In April 2004 President Bush signed Executive Order 13335 creating the Office of the National Coordinator for Health Information Technology (ONC) and at the same time calling for interoperable electronic health records within the next decade.[3] How that would be accomplished was not stated nor was it known at the time of the executive order. In November 2004 ONC sent out a *"Request for Information"* (RFI) as to how the Nationwide Health Information Network (NHIN) should be established. In particular, they wanted to know how the NHIN would be governed, financed, operated and maintained. A 2005 report concluded that the NHIN should *"be a decentralized architecture built using the Internet linked by uniform communications and a software framework of open standards and policies"and a"network of networks"*.[4] That meant that there would not be a single centralized data repository of patient health information. Creation of the NHIN would require hundreds of HIOs to be interoperable with thousands of individual healthcare entities. It is important to point out that the NHIN is not a separate network; it is instead a set of standards, services and policies that direct the secure exchange of health information over the Internet. (Fig. 4.1)

Executive Order 13410 (August 2006) required federal agencies that dealt with health information to select systems capable of meeting national interoperability standards. This provided new guidance regarding interoperability for federal agencies and contributed one more building block towards the creation and deployment of the NHIN.[5]

In 2005 ONC awarded contracts to develop prototype architectures of the NHIN to four contractors (Accenture, Computer Sciences Corporation, IBM and Northrop Grumman). The results reported in January 2007 discussed the important issues of security, data standards and technology. The contractors had to support the *"use cases"* of EHR-lab use, consumer empowerment and biosurveillance. ONC required that the NHIN interface with electronic health records, personal health records, health information organizations and other organizations that dealt with secondary use of data like public health and research. Clearly, those that chose to participate in the NHIN

Nationwide Health Information Network (NHIN)

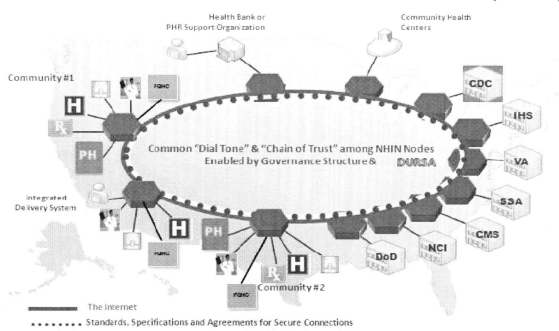

Figure 4.1 NHIN Model   (Courtesy ONC)

would have to accept uniform standards, services and requirements.[6] Specifics of the four different NHIN architectures can be found in an extensive monograph published by Gartner in May 2007.[7]

In June 2007 the Department of Health and Human Services released a request for proposal (RFP) to participate in phase 2 known as *"Trial Implementation"*. Contracts totaling $22.5 million were awarded to nine HIOs in October 2007 as part of the NHIN Cooperative: CareSpark, Delaware Health Information Network, Indiana University, Long Beach Network for Health, Lovelace Clinic Foundation, MedVirginia, New York eHealth Collaborative, North Carolina HealthCare Information and Communication Alliance, Inc and West Virginia Health Information Network. In addition, the CDC awarded contracts to study the use of HIOs to support public health information exchange and biosurveillance.[8]

In February 2008 ONC announced that 20 federal agencies would connect to the NHIN, as the *"tenth partner"*. Individual participants in the trial implementation were referred to as Nationwide Health Information Exchanges or NHIEs. This overall effort was known as the NHIN-Connect Gateway (previously referred to as NHIN-C). The Department of Defense and Veterans Administration jointly represent the largest NHIN participants, in terms of patient populations. The other government agencies involved are the National Cancer Institute and the Indian Health Service. Additional organizations that were added in the 2008 time frame include the Cleveland Clinic, HealthLINC, Community Health Information Collaborative, HealthBridge, Kaiser Permanente and Wright State University.

In late 2008, HHS hosted a national demonstration of phase 2 of the NHIN, wherein the aforementioned participants exchanged live health information (using test patient data). Specifically, participants tested the ability for a health entity to query a record, compile a patient summary record and send that information back to the person or entity that requested it. The standard used for interoperability by the NHIN was the C32 specification for Continuity of Care Documents (see chapter on data standards), that included patient registration and medication information.[9-10] In summary, the NHIN strategy was to establish cross-agency collaboration, develop gateway tools and participate in trial implementations.

One impediment to rapid NHIN progress has been the need for a common data use and reciprocal sharing/support agreement (DURSA). This is important because the NHIN will have civilian partners from multiple states as well as federal partners. The process of achieving a common trust on how data will be used across all participants of the NHIN takes time and has multiple HIPAA implications. This agreement provides the privacy and security aspects of health information exchange as well as the governance necessary to join the NHIN. HHS has released a tool kit to help HIOs write DURSAs.[11] Another similar concern is whether private partners will have to adhere to regulations based on the Federal Information and Security Management Act (FISMA). This would add burdensome regulations on the private sector that could slow interoperability.[12]

The Social Security Administration (SSA) was the first federal agency in 2009 to use the NHIN to connect to MedVirginia HIO in order to request patient information for disability determinations. They were the first agency because they request between 15-20 million medical records each year as part of disability determinations. The time to retrieve the necessary information has been reduced from an average of 84 days to 46 days. It was announced in February 2010 that the SSA had released $17.4 million to 15 HIOs to expand their ability to receive disability-related patient information electronically. [13] Given the fact that the majority of veterans and active duty service members receive medical care outside their respective systems, VA and DOD initiatives with civilian entities have begun.[14-15] A HIO was created in Pensacola, Florida to connect a large civilian hospital with the local Navy hospital. Live testing of patient data was completed in 2009. [16] Also, in 2009 a pilot program was launched to share data between the electronic health records of Kaiser Permanente and the Veterans Health Administration in the San Diego, California. Specifically, allergies, medications and laboratory results were shared. [17]

The current NHIN is known as the NHIN Exchange. Participants have completed an application, signed a DURSA, completed the validation testing and have been accepted by a coordinating committee. Non-federal entities can participate only through a federally sponsored contract, grant or cooperative agreement. Details on the "NHIN Today" and the "NHIN Tomorrow" can be found at http://www.healthit.hhs.gov .

The Federal Health Architecture Group (FHA), part of the ONC as well as an eGovernment Initiative under the Office of Management and Budget (OMB), released the code for an open source NHIN gateway known as CONNECT into the public domain in March 2009. The intent of this release was to incentivize and promote adoption of the NHIN by releasing a basic "reference implementation" of NHIN standard services. With this tool, federal agencies can use the same gateway to access the NHIN as opposed to each agency developing their own. CONNECT utilizes service oriented architecture (SOA) on a Java-based platform (Fig 4.2). (SOA is discussed further in the chapter on architectures of information systems). CONNECT is free to download and can be used to: set up a health information exchange within an organization; tie a health information exchange into a regional network of health information exchanges or tie a health information exchange into the NHIN. Version 2.4 was released in April 2010 and is smaller, requiring less memory and faster. Later in 2010 they have plans to offer web services as part of CONNECT to support core services such as secure messaging and patient look-ups. This will allow developers to create new healthcare applications to augment HIE (analogous to iPhone apps). FHA CONNECT consists of three elements:

- **NHIN Gateway** implements the core services such as locating patients at other health organizations within the NHIN and requesting and receiving documents associated with the patient. It also includes authenticating network participants, formulating and evaluating authorizations for the release of medical information and honoring consumer preferences for sharing their information

- **Enterprise Service Component (ESC)** provides enterprise components including a Master Patient Index (MPI), Document Registry and Repository, Authorization Policy Engine, Consumer Preferences Manager, HIPAA-compliant Audit Log and others. This element also includes a software development kit (SDK) for developing adapters to plug in existing systems such as electronic health records to support exchange of health information across the NHIN
- **The Universal Client Framework** enables agencies to develop end-user applications using the enterprise service components in the ESC [18]

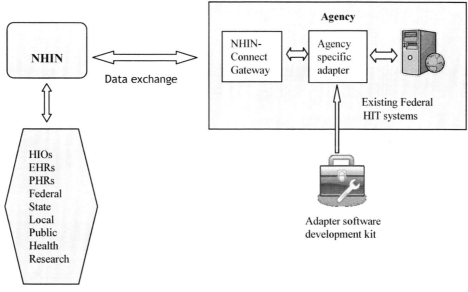

Figure 4.2 Federal Gateway Overview (Adapted from Federal Health Architecture) [19]

NHIN Direct

The original vision of the NHIN was a "network of networks" built upon existing HIT standards that would provide interoperability between HIOs and government agencies. By leveraging existing HIO's and the standards with which they were built, the NHIN could be built on a tested backbone of reliable core services. Even with a mature set of standards, implementation can run into issues ranging from technical (deciding on how much of the standard to support), procedural (agreeing upon vocabularies for proper semantic interoperability, and political (reconciling patient privacy and consent laws between locales).

The NHIN model was created to respond to a mobile patient base. With travel as prevalent as it is in present day, the need for timely responses to possible queries for patient information across multiple organizations and locations requires a robust network with a sturdy trust model. Towards that goal, we have seen the slow growth of HIOs across the country. With the advent of the HITECH Act there has developed a need for more interoperability for the average clinician, in order to meet Meaningful Use.

The NHIN concept has been adjusted or modified by the HIT policy committee's NHIN Working Group to function more like a simple HIE and has been renamed NHIN Direct (some refer to it as NHIN Lite and others the Health Internet). The newer model is a set of standards, policies and services that support the exchange of patient data. However, the goal now is a simpler, scalable, more direct exchange focused on achieving Stage 1 Meaningful Use. Towards that end, NHIN Direct focuses on the deployment of functionality using the lowest cost of entry from a technical and operational perspective. Web services can be deployed in RESTful and SOAP based manner;

both are discussed in a later chapter. The goal for NHIN Direct would be to help clinician A communicate with clinician B better and faster with patient summaries, reconciliation of medications and lab and xray results. It is likely that NHIN Direct will involve push and not pull technologies to provide information between clinicians.

Much work remains to be done before NHIN Direct is fully at an operational readiness state. The current timeline dictates stable pilot programs performing meaningful exchange in 2010, with a larger scale adoption pattern moving into 2011.[20-22]

Integrating the Healthcare Enterprise (IHE)

Integrating the Healthcare Enterprise (IHE) is a group of healthcare professionals whose efforts mean to solve healthcare data exchange problems. The workgroups within IHE discover interoperability gaps through healthcare workflow and the analysis of exchange use cases. These gaps are then covered via profiles that dictate the actors and transactions involved in sharing the data defined in the use case. IHE profiles leverage existing IT and HIT standards wherever possible, and extend or constrain them (while still inside of the spirit of the standard) to meet the scenario in question.

HITSP, when still in existence, would adopt IHE profiles for content and data exchange and constrain them to meet the goals set forth for that standards panel. While a case can be made towards standards being birthed as grandchildren of IHE, they are not created directly by that organization.

One of the four proposals for implementation out of the NHIN Direct workgroups involves the adoption of a set of profiles from IHE that are currently in use across the world for the secure exchange of patient information. The core exchange centers around a suite of profiles for the delivery of content exchange transactions implemented on a SOAP web services stack. These profiles stem from the Cross Enterprise Document Sharing (XDS) profile, and are typically referenced as XD*.

IHE is an open, vendor-neutral and transparent organization. Anyone is welcome to interact on the profile lifecycle, and both committees and workgroups are open to join.[23]

Health Information Organizations

In the early 1990s Community Health Information Networks (CHINs) began springing up across the US. Approximately 70 pilot projects were created but all eventually failed and were terminated.[24] In spite of this early failure, it became apparent that not only would electronic health records need to be adopted but there would be a need for new HIOs to exchange data and eventually connect to the NHIN. In 2006 the National Coordinator for Health Information Technology made the following suggestions as to how HIOs might proceed:

- Leverage the Internet as the foundation and think web-based
- Build upon existing successes; take advantage of any existing infrastructure
- Have a realistic implementation plan; build incrementally or by phases or modules
- Develop strong physician involvement; involve medical schools and medical societies
- Obtain hospital leadership commitment; much of the information to be shared comes from hospital IT systems
- Do not exclude any stakeholders; HIOs should consist of multiple types of healthcare organizations
- Seek inclusion of local public health officials; the goal is to also develop a public health information network or PHIN

- Obtain support from the business community; vendors who have networking experience will be valuable partners
- Establish a neutral managing partner; a commission or network authority [25]

By 2010 there were 73 HIOs actually exchanging clinical information (operational), out of an estimated 234 known entities that have organized to accomplish HIE.[26] There has been tremendous growth of HIOs in the United States, largely due to federal government support, but it is unknown how many HIOs have started and failed. The Santa Barbara County Care Data Exchange was a high visibility RHIO that folded in 2007 due to legal, technological and financial issues.[27] An excellent monograph describes the lessons learned from this project.[28] The Pennsylvania RHIO also closed in 2007 due to lack of short and long term financial support.[29]

Most HIOs begin with a collaborative planning process that involves multiple stake holders in the medical community. "Social capital" or an atmosphere of trust must occur before HIE can begin. The web based infrastructure can be built by local expertise or a HIE-specific vendor.

HIOs start with simple processes such as clinical messaging (test results retrieval) before tackling more complicated functionality. The planning phase generally takes several years and generally necessitates federal or state grant support.

Several models of health information exchange data storage have appeared:

- **Federated**: data will be stored locally on a server at each network node (hospital, pharmacy or lab). Data therefore has to be shared among the users of the HIO
- **Centralized**: the HIO operates a central data repository that all entities must access
- **Hybrid**: a combination of some aspects of federated and centralized model

The following table outlines some of the pros and cons of the federated and centralized models.

Table 4.2 Pros and Cons of RHIO models (Adapted from Scalese) [30]

	Centralized	Federated
Pros	<ul><li>Simplicity</li><li>Data appearance is uniform</li><li>Faster access to data</li><li>Easier to create because it is web-based</li></ul>	<ul><li>Greater privacy</li><li>Good examples exist</li><li>Buy-in may be easier if data is local</li></ul>
Cons	<ul><li>Higher hardware costs</li><li>Higher operating costs</li><li>More difficult with very large HIOs</li></ul>	<ul><li>Data display might not be uniform</li><li>Data retrieval delays from others</li></ul>

In order for a HIO to succeed, multiple participants need to be involved in the planning phase:
- Insurers (payers)
- Physicians
- Hospitals
- Medical societies
- Medical schools
- Medical Informatics programs
- State and local government
- Employers
- Consumers
- Pharmacies and pharmacy networks
- Business leaders and selected vendors
- Public Health departments

Multiple functions need to be addressed by a HIO according to the consulting group HealthAlliant such as:

- Financing: it will be necessary to obtain short term start up money and a long term business plan to maintain the program
- Regulations: what data, privacy and security standards are going to be used?
- Information technology: who will create and maintain the actual network? Who will do the training? Will the HIO use a centralized or de-centralized data repository?
- Clinical process improvements: what processes will be selected to improve? Claims submission? Who will monitor and report the progress?
- Incentives: what incentives exist to have disparate forces join?
- Public relations (PR): you need a PR division to get the word out to healthcare workers and the public regarding the potential benefits of creating a HIO
- Consumer participation: in addition to the obvious stakeholders you need input from consumers/patients [31]

The expectation is that HIOs will save money once they are established. It is presumed that the network will reduce office labor (faxing, etc) and duplication of orders. Many people feel that insurers are likely to benefit more from HIE than clinicians. It is clear that one of the potential benefits of health information exchange is more cost-effective electronic claims submission. As reported by the Utah Health Information Network, a paper claim costs $8, an electronic claim $1 and the charge by the HIO of 20 cents; therefore a savings of $6.80. [32]

Health Information Organizations may be operated by state governments, private entities or a private-public hybrid organization. They can be for-profit or not-for-profit. Most HIOs derive operating capital by charging hospitals, physician offices, labs and imaging centers. Some HIOs charge clinicians a subscription fee or transaction fee, while others charge nothing. HIOs can address the entire medical arena or simply a sector such as Medicaid patients. Moreover, they can be managed by a HIE vendor such as Axolotl or Medicity (covered in another section). HIOs can cover a city, region, an entire state, multiple states or an entire country.

HIOs are relatively new so many regions have little experience with the concept and further education is necessary for clinicians and administrators. Studies so far have shown that clinicians and patients are not very knowledgeable about HIOs but support the concept of sharing secure medical information. [33-34]

There are open source tools available for evolving HIOs. The California HealthCare Foundation donated server software for the master patient index and records locator services. These tools are available through Open Health Tools, an international consortium dedicated to open projects across the healthcare information technology domain. Alongside these open source offerings are a wide range of services and toolkits covering the gap from the core services to the edge system nodes in a HIO. This technology assists the EHR vendors attending the yearly Connectathon held at the annual HIMSS conference. [35-36] Furthermore, Misys Open Source Solution uses an open source platform for HIE, in spite of the fact that they are a commercial entity. [37]

According to eHealth Initiative's *2010 Seventh Annual Survey: Migrating Toward Meaningful Use: The State of Health Information Exchange*

- 73 HIOs reported being operational, up from 57 reported in 2009. 199 initiatives responded to the survey. Of the operational HIOs, 17 have reached the highest level (7) of functionality
- About half of operational HIOs charge physicians a subscription (not transaction) fee

- 18 initiatives were termed sustainable: operational, not dependent on federal funding in the past year and at least broke even through operational revenue alone
- Most HIOs are not ready for Meaningful Use
- Documented cost savings among the operational HIOs (number): reduced admin time (33), reduced staff time handling reports (30), decreased redundant tests (28), reduced medication errors (16), decreased chronic care costs (16), reduced staff time handling prescriptions (15)
- Opt-out policy (patient's data automatically included unless they opt out) most common
- Top clinical and non-clinical functions of sustainable HIOs are listed in table 4.3
- 44 initiatives allow patients to view data and 31 allow patients to contribute information
- Multiple challenges were listed: sustainable business plan, privacy issues, defining value, government mandates (Meaningful Use), technological issues, governance issues and engaging health plans
- Three most common sources of sharable information: hospitals, primary care physicians and community/public health clinics [26]

Table 4.3. Data exchange functionality (Courtesy eHealthinitiative)

Functionality	Functionality
Results delivery	Quality reporting
Connectivity with EHRs	Results distribution
Clinical documentation	Electronic health record (EHR) hosting
Alerts to clinicians	Assist data loads into EHRs
Electronic prescribing	EHR interfaces
Health summaries	Drug-drug alerts
Electronic referral processing	Drug-allergy alerts
Consultation/referrals	Drug-food allergy alerts

We mentioned in chapter 1 that the American Recovery and Reinvestment Act (ARRA) included funding for health information organizations. As of mid-2010, fifty-six states, US territories and state designated entities (e.g. Alabama Medicaid) received a total of $548 million as part of the State Health Information Exchange Cooperative Agreement Program. These agreements have a four year performance period and in the second year onward states must provide matching funds. ONC has provided a State HIE Toolkit on their web site (www.healthit.hhs.gov) that provides guidance in the following areas: general planning, governance, technical infrastructure, finance, NHIN and grants management.

As you will see in the following section, there are many different models of health information exchanges. Most rely on web-based service oriented architecture (SOA) and most link to actual lab and x-ray results. Other payer-based models use claims data and may refer to themselves as having a regional or state electronic health record. In reality, they are payer-based personal health records that may be linked to a HIO. At this time, they do not offer true EHR functions (patient notes, order entry, etc).

Examples of Health Information Organizations

Utah Health Information Network. Created in 1993, it has been one of the most financially successful non-profit statewide HIOs in existence. Figure 4.3 demonstrates the diverse entities associated with this HIO. They provide administrative (billing and eligibility), clinical (lab reports, medication histories, allergy histories, immunization records and discharge summaries) and credentialing information for physicians and dentists. Their web site is highly educational and includes their standards and specifications.[38]

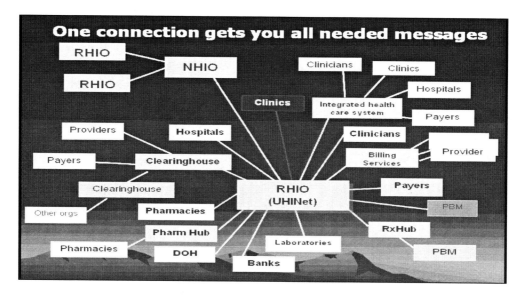

Figure 4.3 Utah Health Information Network (Published with permission)

Nebraska Health Information Initiative (NeHII)
- Statewide roll out began July 2009
- Not-for-profit organization managed by the HIE vendor Axolotl
- Offers a dynamic virtual health record (VHR) for users when they log on that resembles a CCD document
- Also offers an EMR-Lite for clinicians who desire an EMR as part of the HIE. Does not include practice management software
- Has a hybrid federated data storage architecture
- They have experienced a low opt-out rate of about 2%
- E-prescribing available as part of HIE
- $52 per month per clinician for all services [39]

Maine Statewide Health Information Exchange

- One of the largest state-based HIOs
- The network known as HealthInfoNet was launched August 2009
- An initial demonstration project will link 15 hospitals and 2000 physicians who will share Continuity of Care Records [40]
- Other states that plan a state wide HIO include Florida, Kentucky, New York, Connecticut, Maryland and Massachusetts

Indiana Health Information Exchange. Multiple partners helped create this RHIO in 1999, including the Regenstrief Institute that is part of the Indiana University School of Medicine. The RHIO includes twenty one hospitals from five hospital systems and sixty six percent of outpatient physician offices in the Indianapolis area. One of the busiest HIOs, they have approximately 2.5 million transactions per day. It is their goal to eventually cover the remainder of Indiana. They opted to use a centralized approach to storing data in one location. They also wanted to be an example for the rest of the country, employ more workers and create more data for better research. The network includes state and local public health departments and homeless shelters. They link to two other HIOs (HealthBridge and Michiana). Although they are considered one of the most robust of RHIOs with a successful business plan, they have had their share of financial issues. [41-44] Their HIO offers the following functions:

- Clinical abstracts
- Physician profiling data
- Results review: radiology results, discharge summaries, operative notes, pathology reports, medication records and EKG reports
- Clinical quality reports
- Research
- Electronic laboratory reports for public health: childhood immunization information and tumor registry
- Syndromic surveillance (looking for syndromes like flu like illnesses to track epidemics or bioterrorism)
- Adverse Drug Event (ADE) detection
- Disease management known as QualityHealth First™ supplies monthly reports, alerts and reminders to clinicians, for no charge
- They plan to launch medication reconciliation, diabetes and cholesterol management and breast cancer and colorectal cancer screening

North Carolina Health Information Exchange was developed by the North Carolina Healthcare Information and Communication Alliance, Inc (NCHICA). This state-wide innovative HIT organization enjoys the support of over 200 member stakeholders. NCHICA was awarded the NHIN contract for North Carolina. The architecture was created by IBM. Figure 4.4 of the North Carolina HIE Council provides a good look at the types of committees and sub-committees necessary to operate a HIO. [45]

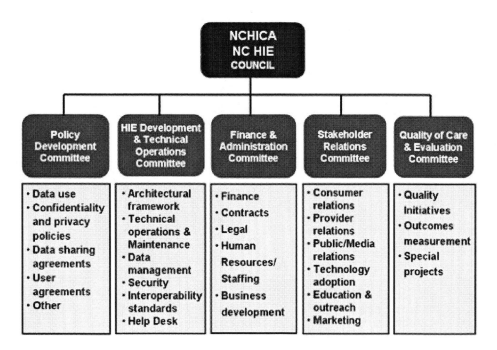

Figure 4.4 NC HIE Committees (Courtesy NCHICA)

HealthBridge is a not-for-profit HIO serving greater Cincinnati, Ohio as well as parts of Kentucky and Indiana that was founded in 1997. It has been quite successful financially with income not based on federal grants, but rather monthly subscription fees. HealthBridge provides information exchange for 29 hospitals and 5500+ physicians. They provide access to imaging, fetal heart monitoring and hospital-based EHRs.They were an early NHIN trial participant and in 2010 they were selected to be a regional HIT extension center. Their early technology partner was Axolotl who offered EMR Lite to integrate with their HIE. They have selected Mirth Meaningful Use Exchange (Mirth MUx) as their next interoperability platform. They have been able to demonstrate an annual return of 5-8% over the past 8 years. Fourty nine percent of connections to the HIE are with the EMR Lite option, 38% with other EHRs, 2% print content and 1% faxes. Physicians are not charged with this model for core services. Figure 4.5 demonstrates the architecture used to create the community infrastructure by HealthBridge. [46]

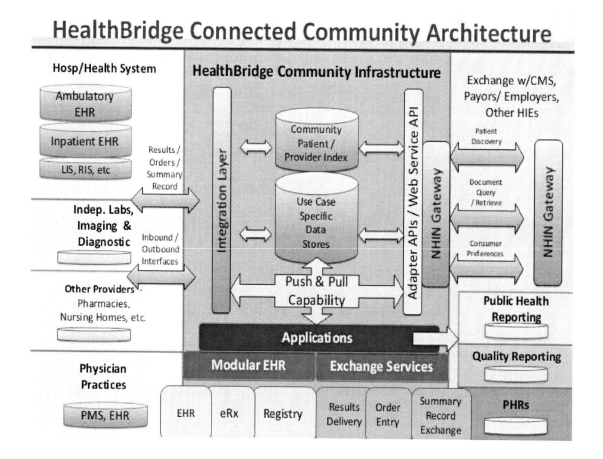

Figure 4.5 HIE Architecture used by HealthBridge (Courtesy HealthBridge)

Claims-Based HIOs

SharedHealth. This interesting entity is a large public-private HIO. They cover 2.6 million privately insured and Medicaid customers and their system is based on claims data. Claims data is limited by the fact that it will only tell you that a test was done and not the actual results and is not available real time. The service is free for participating clinicians. This is not a true EHR in the sense that it does not include either clinical notes or x-ray results. Some have termed this a payer-based health record (PBHR). They do offer integration with known HIOs as one of their features. In addition:

- Their Clinical Health Record® (CHR) provides medical histories from claims data: patient demographics, problem summary lists, lab results, medication histories, allergies, immunization histories, preventive medicine reminders and electronic prescribing
- In 2009 they added disease management tools
- Patients can also have access to their CHR
- CHR is being used by East Tennessee State University Family Medicine Department
- In April 2009 they announced that they would release HIO tools to hospital systems in the rest of the nation [47]

Availity Health Information Network. This is the first multi-payer based health information exchange. This network uses claims data for patients insured by Blue Cross/Blue Shield of Florida and Humana with customers in all 50 states. They claim to integrate with EHRs, practice management systems and hospital information systems and most services are available for free. Users can access this site for eligibility/benefit questions, claims status, treatment authorizations, referral status, payment collections and to review medications, diagnoses, treatments and lab orders. They claim 600 million transactions per year and offer the following features:

- Availity Care Profile® that includes:
  - A Continuity of Care Record (CCR) that shows services rendered, lab and x-rays ordered, diagnoses, procedures performed, hospitalizations and immunizations
  - CarePrescribe®, an e-prescribing service
  - An optional patient portal (RelayHealth)
- Availity CareCost®, a cost estimator for patients
- Availity CareCollect®, a payment processing service
- Availity CareRead® is a card with all of the member's ID information [48]

It is uncertain how rapidly this type of information exchange will catch on and whether new services will be added. Payers such as Blue Cross and Humana stand to gain a lot from electronic data collection and analysis. It should also be noted that the following limitations are associated with this model:

- Model only covers insured patients in the network
- If a patient does not file a claim for a service (pays out of pocket), there will be no record
- A patient can opt-out of sharing data on the HIO
- Patient's employer can opt-out from sharing claims
- Data older than 24 months cannot be retrieved
- There is a lag time between when the test was taken and when the results are posted

NaviNet. Originally, a large insurance clearing house (reported 800,000 clinicians in their network) they announced in 2009 that they would make their claims-based HIO service free to all "state designated entities". A transaction fee would be charged to users who exchange clinical or financial information. They offer access to multiple insurers on one web portal as well as patient access. A patient portal is an option offered by RelayHealth. In late 2010 they plan to offer an integrated EHR and practice management system as part of their platform. [49]

Medicaid-Based HIO

South Carolina Health Information Exchange. In 2008 South Carolina launched this claims based HIO for its 800,000 Medicaid recipients. Clinicians have free access to the HIO and at this time they can access medication, diagnoses and procedural histories. In the future they will add lab results, immunizations, discharge summaries, e-prescribing, operative and clinical notes. Data is encrypted and only stored on the network.[50]

Health Information Exchange Vendors

Axolotl. According to a KLAS report on HIOs, this vendor and Medicity have the largest market share of live, validated health information exchanges.[51] Axolotl provides a variety of interoperability services: Elysium ® EdgeServer, Elysium® Interoperability Hub, Elysium® EMR and Elysium® Virtual Health Record. According to KLAS, Axolotl has seven live HIEs in the acute-to-acute market that actively exchange patient data.[52]

Medicity. Based on the same KLAS report, Medicity has 22 valid HIOs in the acute-to-ambulatory market, making it the largest vendor in this space. [51] The vendor claims that they have 700 hospitals and 250,000 end-users as clients. Later in 2010 Medicity will offer a new HIE platform based on the iNexx® platform that is open and modular. They invite developers to create new applications or plug-ins for their platform by opening their application programming interface (API). In 2010, they announced that they would collaborate with Emdeon, a large national financial and administrative exchange provider so they can integrate financial with clinical information. [53]

Verizon Health Information Exchange. In 2010 Verizon's business arm announced that it would offer an open and standards based HIE solution. They will partner with MEDfx and Oracle to offer a complete package. Their first client will be MedVirginia, a Richmond based HIO. Verizon will have the following features: clinical messaging, referral management, virtual records, NHIN connectivity, a central datat repository, medication reconciliation, CCD/CCR, public health reporting, event alerts, an EMR lite and a clinical dashboard. [54]

Mirth Meaningful Use Exchange (Mirth MUx). Mirth is known as the "Swiss Army Knife" of interoperability. As an open source application it supports all major health data standards and incorporates NHIN Connect. Like many of the HIE vendors in this section, Mirth is appealing to medical practices, hospitals and healthcare organizations who need to comply with "Meaningful Use" criteria and are looking for a less expensive interoperability solution. Figure 4.6 demonstrates how this platform could connect a hospital or medical office to other HIEs, external laboratories and the NHIN, in order to achieve Meaningful Use. Figure 4.7 demonstrates how Mirth MUx can operate as a simple HIE to connect area hospitals and medical offices with access via the Internet and connectivity to the NHIN. This platform is less expensive than most HIE platforms because it is open source and there is no charge for the software license but installation, maintenance and support charges pertain. They also offer Mirth Connect (interoperability), Mirth Results (central data repository for lab) and pending Mirth Match (master patient index). [55]

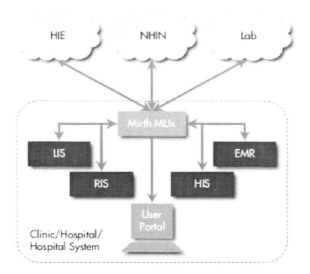

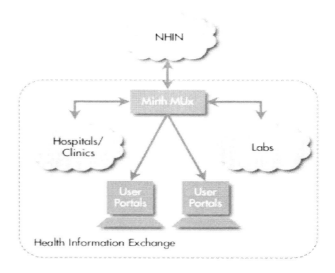

Figure 4.6 Mirth MUx for hospital or clinic

Figure 4.7 Mirth MUx standalone HIE (courtesy Mirth Corp)

Health Information Organization Concerns

There are multiple concerns surrounding the creation and sustainment of a health information organization. The following are just few of the reported concerns:

- Everyone has a different business model. Is this a public utility with no public funding?
- Who will fund HIOs long term? Insurers? Clinicians? Employers? Consumers?
- Is a subscription model better than a transaction model?
- Roughly $550 million from the ARRA will go towards establishing more HIOs. Have we learned enough at this point to decrease the failure rate?
- Will we have universal standards or different standards for different HIOs?
- What should be done with geographical gaps in HIOs and what regions should they cover? Should they be based on geography, insurance coverage or prior history?
- Are poorer cities, states and regions at a disadvantage?
- How can you create a NHIN if multiple HIOs fail and the adoption rate of EHRs stays low?
- What are the incentives for competing hospitals and their CIOs/CEOs in the average city or region to collaborate?
- Will HIOs have to comply with FISMA regulations?
- Will trying to comply with Meaningful Use stages 1-3 actually be an impediment to future HIOs?
- Will the newest HIPAA regulations or state specific personal health information-related laws become impediments to HIO implementation and operation? Clearly, many HIE privacy and security issues remain
- There are different patient consent models. Which one should be adopted? George Washington University reported on this topic for ONC in March 2010 regarding the current consent choices:
    - No consent: no opt out clause
    - Opt out: patient's data is included but they can opt out
    - Opt out with exceptions: only select patient data is included
    - Opt in: nothing is included without permission. All or none choice
    - Opt in with restriction: only a subset of patient information is included [56]

- Who will accredit HIOs: CCHIT or the Electronic Healthcare Network Accreditation Commission (EHNAC)?
- How will patient privacy and security rules under Meaningful Use come into play in the HIO domain?
- Will clinicians be comfortable making care decisions based on discrete data elements imported from an external record source?
- Will timely access to patient documentation be realized in the face of technical and procedural hurdles?

Future Trends

While the success of HIOs continues to be uncertain even with extensive ARRA funding several trends are appearing from the more mature and successful HIOs. First, many are attempting to achieve Meaningful Use by providing HIE to include quality reporting and other advanced functionalities. Second, clinical messaging is being combined with administrative and financial data to give users more of an Expedia.com experience where multiple disparate functions are available on one web page. It seems likely that we will eventually experience seamless integration of EHRs, practice management systems and claims management as part of a HIO. This would offer a single web based platform to conduct all clinical and financial business and the ability to generate a wide range of reports. Third, we can expect more efforts to use data secondarily for research and as a means of financially supporting HIOs.

Health Information Organization Resources

It can be argued that creating the technology architecture is the easy part in the life of a HIO. Far more time must be spent planning the governance and financing. It is therefore critical that localities do their homework to research the lessons learned from others who have successfully built a HIO. The following are valuable resources:

- *Privacy and Security Solutions for Interoperable Health Information Exchange. Report* for the AHRQ, December 2006 [57]
- *Guide to Establishing a Regional Health Information Organization.* Publication by the Healthcare Information and Management Systems Society. 144 page step-by-step resource. Cost $78 for non-members [58]
- *eHealth Initiative Connecting Communities Toolkit.* This non-profit site offers a wealth of web based information organized into these sections: value and financing module, practice transformation module, information sharing module, technology module and public policy and advocacy module. A resource library and glossary are also part of the toolkit [59]
- *Privacy, Security and the Regional Health Information Organization.* June 2007. California HealthCare Foundation [60]
- *Common Framework: Resources for Implementing Private and Secure Health Information Exchange* is published by Connecting for Health that is part of the Markle Foundation. The Framework consists of multiple documents that help organizations exchange information in a secure private manner, with shared policies and technical standards. Using their protocols a tri-state prototype HIO was created. The Common Framework with 9 policies guides and 7 technical guides is available free for download on their site [61]
- *Characteristics associated with Regional Health Information Organization viability.* Authors analyzed data from a large 2008 survey of HIOs. Two factors for success stood out: simplicity in terms of not trying to do too much and early financial commitment from a wide variety of participants [62]

- *Electronic Personal Health Information Exchange.* February 2010. Report to Congressional Committees. GAO report on healthcare entities' reported disclosure practices and effects on quality of care [63]
- *Statewide Health Information Exchange: Best Practice Insights From The Field.* Bates M, Kheterpal V. March 2010. White Paper. Provides 10 best practices and case studies for those who plan to build a statewide HIE/HIO [64]

Key Points

- Health information exchange is critical for achieving Meaningful Use of electronic health records

- In order to create a Nationwide Health Information Network (NHIN) multiple data standards will need to be reconciled and adopted

- Creating the architecture for a Health Information Organization (HIO) is not difficult; developing the long term business plan is

- Important interoperable demonstrations of the NHIN have taken place with multiple participating civilian and federal partners

- NHIN Direct is a very new "fast-track" approach to accomplishing Meaningful Use

Conclusion

Sharing of health-related data is a critical element of healthcare reform and Meaningful Use. Health information exchange among disparate partners is becoming more common in the United States due to evolving HIOs and the NHIN. Federal programs support the creation of exchanges as well as the services, standards and policies that make HIE possible. HIOs are proliferating, largely due to government support but they are often impeded by a lack of a sustainable business model, as well as privacy and security issues. The federal government is moving forward with the Nationwide Health Information Network in an effort to accelerate standards creation and adoption, in spite of the many problems with creating HIOs across the nation and the low adoption rate of EHRs. With the new monies from ARRA for EHRs and HIOs and the new direction of NHIN Direct, the immediate future should be very interesting. At the same time, insurance companies and claims clearinghouses are creating new models based on claims data. It is too early to know what a HIO of the future will look like but it seems clear that we can expect more features and better integration.

References

1. Commonwealth Fund. Perspectives on Health Reform. http://www.commonwealthfund.org/Content/Publications/Perspectives-on-Health-Reform-Briefs/2009/Jan/The-Federal-Role-in-Promoting-Health-Information-Technology.aspx (Accessed July 7 2009)

2. Defining Key Health Information Terms. National Alliance for Health Information Technology www.nahit.org (Accessed May 21 2008)
3. Executive Order: Incentives for the Use of Health Information Technology and Establishing the Position of the National Health Information Technology Coordinator http://www.whitehouse.gov/news/releases/2004/04/20040427-4.html (Accessed February 18 2006)
4. Summary of Nationwide Health Information Network (NHIN). Request for Information (RFI) Responses June 2005. www.hhs.gov/healthit/rfisummaryreport.pdf (Accessed January 30 2006)
5. Executive Order: Promoting Quality and Efficient Health Care in Federal Government Administered or Sponsored Health Care Programs http://www.whitehouse.gov/news/releases/2006/08/20060822-2.html (Accessed January 2 2008)
6. Summary of the NHIN Prototype Architecture Contracts. www.hhs.gov/healthit/healthnetwork/resources/ (Accessed November 15 2007)
7. Summary of the NHIN prototype architecture contracts. Gartner. http://www.hhs.gov/healthit/healthnetwork/resources/summary_report_on_nhin_Prototype_architectures.pdf (Accessed January 3 2008)
8. Monegain B. NHIN contracts awarded to nine exchanges. 10/5/07 Healthcare ITNews. www.healthcareitnews.com Accessed October 6 2007
9. Ferris, N. Federal Agencies begin to build a mini-NHIN. www.govhealthit.com February 2008 (Accessed March 1 2008)
10. Ferris, N. Six more organizations join NHIN demonstration project www.govhealthit.com May 1 2008 (Accessed May 8 2008)
11. Secretary Leavitt announces new principals; tools to protect privacy, encourage more effective use of patient information to improve care. December 15 2008 www.hhs.gov (Accessed March 3 2009)
12. FISMA-a roadblock for EHRs? http://ohmygov.com/blogs/general_news/archive/2009/06/30/fisma-a-roadblock-for-ehrs.aspx (Accessed July 9 200)
13. Mosquera M. SSA Awards 15 contracts to expand HIE nationwide. Government HIT. www.govhealthit.com (Accessed February 2 2010)
14. National Health Information Network Begins First Exchange March 2 2009 www.iheathbeat.org (Accessed March 2 2009)
15. Mosquera M. Network connects medical dots. Federal Computer Week February 23 2009. p32-33
16. Cogon Completes Successful Live User Testing of Pensacola Health Information Utility. January 27 2010. Pensacola Jazz. www.pensacolajazz.com (Accessed March 21 2010)
17. Monegain B, Merrill M. VA, Kaiser plan to Link Electronic Medical Records. November 25 2009. www.govhealthit.com (Accessed December 1 2009)
18. Connect Open Source www.connectopensource.org (Accessed March 28 2010)
19. Sankaran, V. The role of SOA on improving health quality http://www.omg.org/news/meetings/workshops/HC-2008/15-02_Sankaran.pdf (Accessed March 26 2010)
20. Halamka J. Introducing NHIN Direct March 10 2010 www.geekdoctor.blogspot.com (Accessed March 23 2010)
21. NHIN Direct. www.nhindirect.org (Accessed February 20 2010)

22. NHIN Direct. Health IT HHS.
http://healthit.hhs.gov/portal/server.pt?open=512&objID=1142&parentname=CommunityPa
ge&parentid=2&mode=2&in_hi_userid=10741&cached=true (Accessed March 26 2010)
23. Integrating the Healthcare Enterprise http://www.ihe.net/About/ (Accessed June 6 2010)
24. Waegemann P. Wrong National Strategy for EMRs? MRI July 2008 Newsletter. Vol 3,
issue 1b http://medrecinst.blogspot.com/2008_07_01_archive.html (Accessed July 7 2009)
25. Office of the National Coordinator for Health Information Technology
http://www.dhhs.gov/healthit/ (Accessed October 10 2005)
26. Seventh Ehealth Initiative Survey. 2010. Connecting the Nation to Achieve Meaningful Use
http://www.ehealthinitiative.org (Accessed July 22 2010)
27. What Killed the Santa Barbara County Care Data Exchange? March 12 2007.
www.ihealthbeat.org (Accessed March 13 2007)
28. The Santa Barbara County Care Data Exchange: Lessons Learned. August 2007.
www.cfcf.org. (Accessed September 1 2007)
29. Robinson B. Pennsylvania RHIO to close. June 12 2007 Government Health IT.
http://govhealthit.com. (Accessed June 19 2007)
30. Scalese D. Which way RHIO? Hospitals and Health Network
http://www.hhnmag.com/hhnmag_app/jsp/articledisplay.jsp?dcrpath=HHNMAG/PubsNewsArti
cle/data/2006June/0606HHN_InBox_Technology2&domain=HHNMAG (Accessed March 3
2007)
31. HealthAlliant http://www.healthalliant.org/ (Accessed February 18, 2008)
32. Sundwall D. RHIO in Utah, UHIN HIMSS Conference Presentation June 6 2005
33. Shapiro JS, Kannry J, KushniruK W, Kuperman G. Emergency Physicians Perceptions of
Health Information Exchange. JAIMA 2007;14:700-705
34. Wright A, Soran C, Jenter CA et al. Physician attitudes toward HIIE: Results of a statewide
survey. JAMIA 2010;17:66-70 doi 10.1197/JAMIA.M5241
35. Open Health Tools www.openhealthtools.org (Accessed July 7 2009)
36. Vendors Test Open Source HIE Apps. February 26 2009 www.healthdatamanagement.com
(Accessed March 2 2009)
37. Misys http://www.misys.com/corp/OpenSource (Accessed June 1 2010)
38. Utah Health Information Network http://www.uhin.com/ (Accessed February 18 2010)
39. Nebraska Health Information Initiative www.nehii.org (Accessed June 12 2010)
40. Maine Health Information Network. http://www.hinfonet.org/ (Accessed March 15 2010)
41. Indiana Health Information Exchange http://www.ihie.com/default.htm (Accessed
December 30 2009)
42. McDonald CJ et al. The Indiana Network for Patient Care: A Working Local Health
Information Infrastructure. Health Affairs 2005; 24:1214-1220
43. Hayes HB. A RHIO that works and pays. Government Health IT. April 2007; pp 38-9
44. QualityHealthFirst www.qualityhealthfirst.org. (Accessed December 30 2009)
45. North Carolina Healthcare Information and Communications Alliance www.nchica.org
(Accessed March 26 2010)
46. HealthBridge www.healthbridge.org (Accessed March 26 2010)
47. SharedHealth. www.sharedhealth.com (Accessed March 26 2010)
48. Availity. www.availity.com (Accessed November 20 2008)
49. NaviNet www.navinet.net (Accessed March 27 2010)
50. South Carolina Health Information Exchange www.schiex.org (Accessed March 27 2010)
51. Goedert J. KLAS Counts 89 Real HIEs. February 9 2010. www.healthdatamanagement.com
(Accessed February 10 2010)
52. Axolotl www.axolotl.com (Accessed March 26 2010)

53. Medicity www.medicity.com (Accessed March 26 2020)
54. Verizon Health Information Exchange. www.verizonbusiness.com/solutions/healthcare (Accessed July 20 2010)
55. Mirth Meaningful Use Exchange www.mirthcorp.com/MirthMUx (Accessed March 27 2010)
56. Goldstein MM, Rein AL. Consumer Consent Options For Electronic HIE: Policy Considerations and Analysis. March 23 2010. www.healthit.hhs.gov (Accessed March 24 2010)
57. *Privacy and Security Solutions for Interoperable Health Information Exchange.* Report for the AHRQ. December 2006 (http://healthit.ahrq.gov/portal/server.pt/gateway/PTARGS_0_1248_241358_0_0_18/IAVR_ExecSumm.pdf (Accessed December 10 2007)
58. *Guide to Establishing a Regional Health Information Organization.* HIMSS. http://marketplace.himss.org (Accessed December 7 2007)
59. eHealth Initiative Connecting Communities Toolkit www.ehealthinitiative.org (Accessed December 7 2007)
60. Privacy, Security and the Regional Health Information Organization. June 2007. www.chcf.org (Accessed January 4 2008)
61. Connecting for Health http://www.connectingforhealth.org/ (Accessed June 5 2008)
62. Adler-Milstein J, Landefield J, Jha AK. Characteristics associated with Regional Health Information Viability. J Am Med Inform Assoc 2010;17:61-65
63. Electronic Personal Health Information Exchange. Health Care Entities' Reported Disclosure Practices and Effects on Quality of Care. February 2010. www.gao.gov/new.items/d10361.pdf (Accessed March 10 2010)
64. *Statewide Health Information Exchange: Best Practice Insights From The Field.* Bates M, Kheterpal V. March 2010. White Paper. http://interest.healthcare.thomsonreuters.com/content/HIEWhitepaper (Accessed March 29 201

5

Architectures of Information Systems

ROBERT E HOYT
ROBERT W. CRUZ

Learning Objectives
- After reading this chapter the reader should be able to:
- Discuss why web services are used by HIOs
- List of the components of service oriented architecture
- Understand the importance of networks in the field of medicine
- Compare and contrast wired and wireless local area networks (LANs)
- Describe the newest wireless broadband networks and their significance

We believe that the average reader of this book should understand basic architectures and technologies that are commonly part of health information technology. This chapter will focus on two areas: web services and networks

Web Services

Prior to the advent of the Internet, disparate businesses and health care entities were not able to exchange data; instead data resided on a local PC or server. *Web services* are task specific applications which are deployed in a platform independent manner via a series of transactions to and from other web-aware applications/services over a network (such as the Internet). Web services can reduce the cost of converting data with external partners, by allowing for a modular component of a larger system to be invoked with little up front effort.

Web services can be broken down into two categories. Representational State Transfer (or RESTful) services are lightweight services which use existing Internet infrastructure and World Wide Web (WWW) concepts as their backbone. Simple Object Access Protocol (SOAP) web services utilize a potentially complex series of XML based ontologies to describe and invoke services over a network. There are obvious pros and cons to each concept, but most often the tradeoff between ease of implementation versus technical depth of field is the main point of comparison struck between the two.

RESTful Services

REST, as a concept, is an aggregate description of the functional model of how HTTP allows for the deployment of the WWW over the Internet. It can be utilized to provide non-WWW content delivery over any application protocol, not solely trapped in the realm of the HyperText Transfer Protocol (HTTP). It is important to realize that REST is an architecture, not a standard. As such, there are endless possibilities as to how REST can be applied to act as a service bus. Even though REST itself is not a standard, many standards are utilized when it is used for service interaction.

Communication with a RESTful service is a relatively quick to develop process and can utilize any existing content standard for packaging its messaging. Most commonly, a RESTful service will use XML or JavaScript Object Notation (JSON) for this content delivery. RESTful web services require three basic aspects:

- **URI** (Uniform Resource Identifier): a set of characters defining a specific object, resource, or location. One of the more common uses for a URI is in providing a Uniform Resource Locator (URL) for an object on the WWW. In a RESTful service, a URI can describe the service being invoked or a component within said service
- **Operation Type** (GET, DELETE, POST, PUT): these HTTP methods can be extended past their WWW function to provide 4 different points of access to a RESTful service. If a URI identifies an object, the HTTP operation type defines an accessor method to that object (e.g. GET a list, POST an update, PUT a new record, DELETE a purged record)
- **MIME Type** (Multipurpose Internet Mail Extensions): a means of communicating the content type used within a message transferred over the internet. Typically, in a RESTful service, this would be XML or JSON, but it could be any other type

Web Services using SOAP

SOAP is a protocol standard for interacting with web services. These services require a set of standards for content and a service oriented architecture (SOA) stack, a collection of services. The most common standards used in web services transactions are HTTP, as the Internet protocol, with XML as the delivery language (covered in the data standards chapter). SOAP Web services require three basic platform elements:

- **SOAP** (Simple Object Access Protocol): a communication protocol between applications. It is a XML based platform neutral format for the invocation and response of web services functions over a network. It re-uses the HTTP for transporting data as messages
- **WSDL** (Web Services Description Language): a XML document used to describe and locate web services. A WSDL can inform a calling application as to the functionality available from a given service, as well as the structure and types of function arguments and responses.
- **UDDI** (Universal Description, Discovery and Integration): a directory for storing information about web services, described by WSDL. UDDI utilizes the SOAP protocol for providing access to WSDL documents necessary for interacting with services indexed by its directory

So how does this work? SOAP acts as the means of communicating, UDDI provides the service registry (like the yellow pages) and WSDL describes the services and the requirements for their interaction. We can begin the process acting as a service requester seeking a web service to provide a specific function. Your application would search a service directory for a function that meets your needs using a structured language. There is a service requester seeking a web service. You search using a search engine that uses a structured language. Once the service provider is located, a SOAP message can be sent back and forth between the service requester and service provider. In reality, a service provider can also be a service consumer so it is helpful to view web services like the "*bus*" in a PC, where you plug in a variety of circuit boards.

HIOs often require a Master Patient Index (MPI) service to locate and confirm patients and a Record Locator Service (RLS) to identify documentation on those patients. For connecting multiple HIOs you may also require gateways (a network point that acts as an entrance to another network) and adapters (software that connects to applications). [1-3] A valuable recent article *Improving Performance of Healthcare Systems with Service Oriented Architecture* describes how SOA is the logical backbone for HIOs and electronic health records.[4] Another resource for understanding SOA and healthcare was published in March 2009 by the California HealthCare Foundation, *Lessons from Amazon.com for Health Care and Social Service Agencies.*[5]

Networks

A network is a group of computers that are linked together in order to share information. Although a majority of medical data resides in silos, there is a distinct need to share data between offices, hospitals, insurers, health information organizations, etc. A network can share patient information as well as provide Internet access for multiple users. Networks can be small, connecting just several computers in a clinician's office or very large, connecting computers in an entire organization in multiple locations.

There are several ways to access the Internet: dial-up modem, wireless fidelity (WiFi), a Digital Subscription Line (DSL), cable modem or T1 lines. The most common type of DSL is Asymmetric DSL (ADSL) which means that the upload speed is slower than the download speeds, because residential users utilize the download function more than the upload function. Symmetric DSL is also available and features similar upload and download speeds. Cable modem often begins with fiber optic transmission to the building, with coaxial cable run internally. The following are data transfer speeds based on the different technologies. Multiple factors influence these speeds, so that theoretical maximum as well as more typical speed ranges are listed.

Table 5.1. Data transfer rates

Transmission method	Theoretical max speed	Typical speed range
Dial-up modem	56 Kbps	56 Kbps
DSL	6 Mbps	1.5 Mbps download/128 Kbps upload
Cable modem	30 Mbps	1-6 Mbps download/128-768 Kbps upload
Wired Ethernet (Cat 5)	1000 Mbps	100 Mbps
Fiber optic cable	100 Gbps	2.5-40 Gbps
T-1 line	1.5 Mbps	1-1.5 Mbps
Wireless 802.11g Wireless 802.11n	54 Mbps 300 Mbps	1-20 Mbps 40-115 Mbps
WiMax	70 Mbps	54-70 Mbps
LTE	60 Mbps	8-12 Mbps
Bluetooth	24 Mbps	1-24 Mbps

Information Transmission via the Internet

Given the omnipresent nature of the Internet and faster broadband speeds, the Internet is the network of choice for transmission of voice, data and images. It is important to understand the basics of transmission using packets of information. The Internet Protocol (IP) is a standard that segments data, voice and video into packets with unique destination addresses. Routers read the address of the packet and forward it towards its destination. Transmission performance is affected by the following:

- **Bandwidth**: is the size of the pipe to transmit packets (a formatted data unit carried by a packet mode computer network). Networks should have bandwidth excess to operate optimally
- **Packet loss**: packets may rarely fail to reach their destination. The IP Transmission Control Protocol (TCP) makes sure a packet reaches its destination or re-sends it. The User Datagram Protocol (UDP) does not guarantee delivery and is used with, for example, live streaming video. In this case the user would not want the transmission held up for one packet
- **End-to-end delay**: is the latency or delay in receiving a packet. With fiber optics the latency is minimal because the transmission occurs at the speed of light
- **Jitter**: is the random variation in packet delay and reflects Internet spikes in activity

Packets travel through the very public Internet. An encryption technique such as the Federal Information Processing Standard (FIPS) encodes the content of each packet so that it can't be read while being transmitted on the Internet. Encryption, however, adds some delay and increase in bandwidth requirements. [6]

Network Types

Networks are named based on connection method, as well as configuration or size. As an example, a network can be connected by fiber optic cable, Ethernet or wireless. Networks can also be described by different configurations or topologies. They can be connected to a common backbone or bus, in a star configuration using a central hub or a ring configuration. [7-10] In this chapter we will describe networks by size or scale.

Personal Area Networks (PANs): A PAN is a close proximity network designed to link phones computers, PDAs, etc. The most common technology to create a PAN is Bluetooth. Bluetooth technology has been around since 1995 and is designed to wirelessly connect an assortment of devices at a maximum distance of about 30 feet. It does have the advantages of not requiring much power and connecting automatically, but has the disadvantage of being slower, with speeds of 1 Mbps. It operates in the 2.4 MHz frequency range so it can interfere with 802.11g wireless networks. Clearly, the most common application of Bluetooth today is as a wireless headset to connect to a mobile phone. Many new computers are Bluetooth enabled and if not, a Bluetooth USB adapter known as a dongle can be used or a Bluetooth wireless card. This technology can connect multiple devices simultaneously and does not require "line of sight" to connect. In an office Bluetooth can be used to wirelessly connect computers to keyboards, mice, printers, PDAs and smartphones. This will avoid the tangle of multiple wires. Bluetooth can connect in one direction (half duplex) or in two directions (full duplex). Security must be enabled due to the fact that even though the transmission range is short, hackers have taken advantage of this common frequency. In the near future it is anticipated that low energy Bluetooth devices such as heart monitors will be available with very long battery lives. In addition, faster Bluetooth 3.0 devices will be available in the 2010 timeframe with speeds in the 24 Mbps range that piggyback on the 802.11 standard. [7]

Local Area Networks (LANs): Generally refers to linked computers in an office, hospital, home or close proximity situation. A typical network consists of nodes (computers, printers, etc), a connecting technology (wired or wireless) and specialized equipment such as routers and switches. LANs can be wired or wireless:

- **Wired networks**: To connect to the Internet through your Internet Service Provider (ISP) you have several options.

  o **Phone lines** can connect a computer to the Internet by using a modem. The downside is that the connection is relatively slow. Digital subscription lines (DSL) also use standard phone lines that have additional capacity (bandwidth) and are much faster network connection than modems. DSL also has the advantage over modems of being able to access the Internet and use the telephone at the same time. Home or office networks can use phone lines to connect computers, etc. Newer technologies include frequency-division multiplexing (FDM) to separate data from voice signals. This type of network is inexpensive and easy to install. Speeds of 128 Mbps can be expected even when the phone is in use. Up to 50 computers can be connected in this manner and hubs and routers are not necessary. Each computer must have a home phone line network alliance (PNA) card and noise filters are occasionally necessary. The downside is largely the fact that not all home rooms or exam rooms have phone jacks

  o **Power lines** are another option using standard power outlets to create a network. A newer product (PowerPacket®) is inexpensive to install and claims data transfer speeds of 14 Mbps. All that is needed is a power outlet in each room

  o **Ethernet** is a network protocol and most networks are connected by fiber or twisted-pair/copper wire connections. Ethernet networks are faster, less expensive and more secure than wireless networks. The most common Ethernet cable is category 5 (Cat 5) unshielded twisted pair (UTP). A typical wired LAN is demonstrated in Figure 5.1.

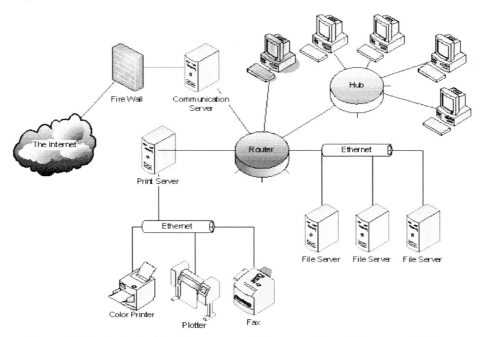

Figure 5.1 Typical wired local area network (Courtesy Dept of Transportation)

To connect several computers in a home or office scenario, a hub or a network switch is needed. Routers direct messages between networks and the Internet; whereas, switches connect computers to one another and prevent delay. Unlike Hubs that share bandwidth, switches operate at full bandwidth. Switches are like traffic cops that direct simultaneous messages in the right direction. They are generally not necessary unless you are running multiple computers on the same network. To handle larger enterprise demands Gigabit Ethernet LANs are available that are based on copper or fiber optics. Cat5e or Cat6 cables are necessary. Greater bandwidth is necessary for many hospital systems that now have multiple IT systems, an electronic medical record and picture archiving and communication systems (PACS). [7-10]

- **Wireless (WiFi) networks (WLANs):** Wireless networks are based on the Institute of Electrical and Electronics Engineers (IEEE) 802.11 standard and operate in the 900 MHz, 2.4 GHz and 5 GHz frequencies. These frequencies are "unlicensed" by the FCC and are therefore available to the public. Figure 5.2 shows the radio frequency portion of the electromagnetic spectrum where wireless networks function.

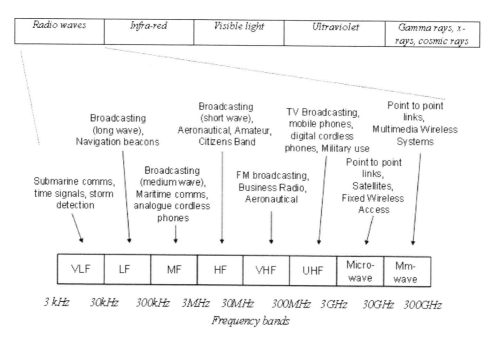

Figure 5.2 Radio frequency spectrums (Courtesy Commission for Communications Regulation)

Wireless networks have become much cheaper and easier to install so many offices and hospitals have opted to go wireless. This allows laptop/tablet PCs and smartphones in exam and patient rooms to be connected to the local network or Internet without the limitations of hardwiring but it does require a wireless router and access points. If an office already has a wired Ethernet network then a wireless access point needs to be added to the network router. The mode that utilizes a wireless router or access point is known as the infrastructure mode. The other mode is known as ad hoc or peer-to-peer mode and means that a computer connects wirelessly to another computer and not to a router.

In general, wireless is slower than cable and is more expensive, but does not require hubs or switches. The standards for wireless continue to evolve. Most people have used 802.11g networks that operate on the 2.4 GHz frequency at peak speeds of 54 Mbps with a range of about 100 meters. Keep in mind that this frequency is vulnerable to interference from microwaves, some cordless phones and Bluetooth. 802.11n is the newest standard that can operate at speeds up to 300 Mbps with a range of about 300 meters. This is accomplished with multiple input/multiple output (MIMO) or multiple antennas that send and receive data much faster and at greater distances. Actual data transfer speeds may be slower than the theoretical max speeds for several reasons. Most modern laptop computers have wireless technology factory installed so a wireless card is no longer necessary. In Figure 5.3 a simple WLAN is demonstrated. Access to the Internet is via a cable modem with the possibility of both Ethernet and wireless connectivity to different client computers demonstrated. A wireless router will connect the computers, server and printers and has a range of about 90-120 feet. For a larger office or hospital multiple access points will be necessary. The network router is usually connected to the Internet by an Ethernet cable to DSL or a cable modem. Security must be established using wired equivalent privacy (WEP) 128-bit encryption, a router firewall and a unique media access control (MAC) addressee. Each device has a unique address (MAC) and routers can have security lists that only allow known devices or MACSs into the network.[11-13]

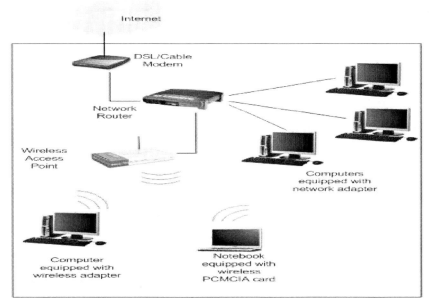

Figure 5.3 Wireless Local Area Network (WLAN) (Courtesy Home-Network-Help.com)

An emerging trend for hospitals is to use Voice over IP on a wireless network, referred to as **VoWLAN**. Hospitals can use existing wireless networks to contact nurses, physicians and employees with any wireless enabled device. Devices such as the Nortel **VoWLAN** phone or Vocera are frequently used. The chief advantage of this approach is saving local and long distance phone call charges. Using this technology, a patient could directly contact a nurse making rounds so a nurse is not forced to be located near a central nurse-call system. While in the hospital this system could replace landlines, pagers, cell phones and 2-way radio. The downside is that a strong signal is necessary for this system and is more important than that needed with just data.[7]

Another wireless option is wireless mesh networks that rely on a single transmitter to connect to the Internet. Additional transmitters transmit signals to each other over a wide area and only require a power source. Municipalities, airports, etc are using this type of technology to cover larger defined areas.[9]

Wide Area Networks (WANs): Cross city, state or national borders. The Internet could be considered a WAN and is often used to connect LANs together.[9]

Global Area Networks (GANs). GANs are networks that connect other networks and have an unlimited geographic area. The problem with broadband technology is that it is expensive and the problem with WiFi is that it may result in spotty coverage. These shortcomings have created a new initiative known as Worldwide Interoperability for Microwave Access (WiMax), using the IEEE 802.16 standard. This is a new 4G network and will not only be about 10 times (optimal) faster than 3G, it will have much greater capacity which is equally important. The network will be known as a global area network (GAN) with operating speeds in the 54-70 Mbps range. The goal is to be faster than standard WiFi and reach greater distances, such that it might replace broadband services and permit widespread wireless access to the Internet. A user would be able to access the Internet while traveling or from a fixed location. It would require WiMax towers, similar to cell towers, and a WiMax card in computers, similar to wireless cards. A tower could conceivably cover 3,000 square miles. This would permit a user to connect "non-line of sight" with a weaker lower frequency antenna or line of sight with a higher frequency and stronger dish antennae. WiMax technology also permits voice over IP (VoIP) or phone calls over the Internet. WiMax may replace several 3G networks and could be considered a 4G strategy or "broadband wireless". A WiMax network is illustrated in Figure 5.4. In 2008 several companies including Sprint, Intel, Comcast, Cisco and Google collaborated to develop a new company known as Clearwire to promote and develop WiMax technology. This company plans to cover 120 million Americans by the end of 2010 and is currently being offered in 27 markets.[7, 14] Sprint released a 4G phone (HTC Evo) in 2010 that is necessary to take advantage of the 4G network.

The second 4G wireless network being rolled out in several US cities is known as Long Term Evolution or LTE and offered by Verizon, AT&T and US Cellular. Verizon plans to start widespread implementation in 2010 with completion in about 5 years. Operating in the 700 MHz range it is touted to have maximum download rates of 100 Mbps and upload rates of 50 Mpbs. It is unknown how expensive WiMax or LTE will be.[15]

Both 4G wireless approaches will transport voice, video and data digitally via Internet Protocol (IP) rather than through switches which will reduce delay and latency. 3G phones will not work on 4G networks.

The Commerce Department will establish a lab to test 4G networks so that it can be used for a national public safety network. The lab will specifically test LTE networks because they are supported by a larger number of cellular vendors (80% of cellular market). They plan for the public safety network to be established in the 700 MHz band.[16]

Figure 5.4 WiMax networks (Courtesy How Stuff Works)

Virtual Private Networks (VPNs): If a clinician desires access from home to his/her electronic health record, one option is a VPN. In this case your home computer is the client and the computer at work you are trying to access is the VPN server. The Internet is the means of connection and VPN will work with wired or wireless LANs. Authentication and overall security are key elements of setting up remote access to someone else's computer network. (Figure 5.5) "Tunneling protocols" encrypt data by the sender and decrypt it at the receiver's end via a secure tunnel. In addition, the sender's and receiver's network addresses can be encrypted.[7, 10]

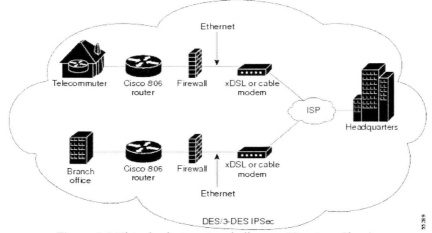

Figure 5.5 Virtual private network diagram (Courtesy Cisco)

Key Points

- Most HIOs use web services and service oriented architecture to exchange health related information

- Clinicians who use client-server based electronic health records need to understand wired or wireless office networks

- Wireless networks have become more attractive due to faster speeds and lower prices

- Wireless broadband is around the corner and will make Internet access faster and more widely available

Conclusion

Disparate services can be integrated by using web services as part of SOA. This platform provides the greatest degree of flexibility for many businesses, to include HIOs.

Hospitals' and clinicians' offices rely on a variety of networks to connect hardware, share data/images and access the Internet. In spite of initial cost, most elements of the various networks discussed continue to improve in terms of speed and cost. Many clinicians' offices will require a network expert to ensure proper installation and maintenance. Wireless technology (WiFi) has become commonplace in many medical offices and hospitals. When wireless broadband (WiMax-LTE) becomes cost effective and widely available it may become the network mode of choice. Network security will continue to be an important issue regardless of mode.

References

1. Sankaran V. The role of SOA on improving health quality http://www.omg.org/news/meetings/workshops/HC-2008/15-02_Sankaran.pdf (Accessed March 26 2010)
2. Web services. http://www.service-architecture.com/web-services/articles/index.htm (Accessed June 1 2008)
3. Ananthamurthy L. Introduction to web services. www.developer.com/services/article.php/1485821 (Accessed June 1 2008)
4. Juneja G, Dournaee B, Natoli J, Birkel S. InfoQ. March 7 2008 www.infoq.com/articles/soa-healthcare (Accessed April 21 2009)
5. Lessons from Amazon.com for Health Care and Social Service Agencies. March 2009. California Healthcare Foundation www.chcf.org (Accessed March 20 2009)
6. Gemmill J. Network basics for telemedicine J Telem and Telecare 2005;11:71-76
7. Wikipedia www.wikipedia.com (Accessed May 20 2009)
8. About.com www.about.com (Accessed May 18 2009)
9. How stuff works www.howstuffworks.com (Accessed May 21 2009)
10. Chambers ML. PCs for Dummies. Wiley Publishing. Indianapolis, Indiana. 2003
11. Smith C, Gerelis C. Wireless Network Performance Handbook. McGraw-Hill, Columbus, Ohio, 2003
12. Smith JE. A primer on wireless networking essentials. EDI. www.ediltd.com (Accessed May 24 2009)
13. Lewis M. A Primer on Wireless Networks. Fam Pract Management. Feb 2004. www.aafp.org (Accessed May 25 2009)
14. WiMax www.wimax.com (Accessed May 22 2009)
15. LTE. Gizmodo Feb 18 2009 and Mar 11 2009 www.gizmodo.com (Accessed May 25 2009)
16. Jackson W. New Lab to Pub 4G to the Test. Government Computer News. January 18 2010. p 7

6

Data Standards

ROBERT E. HOYT
ROBERT W. CRUZ

Learning Objectives

After reading this chapter the reader should be able to:

- Enumerate the reasons data standards are necessary for interoperability
- Understand the importance of the clinical summaries or Continuity of Care Documents (CCDs) or Continuity of Care Records (CCRs)
- Compare and contrast ICD-9 and SNOMED CT
- Discuss data standards used for billing

According to the Institute of Medicine's 2003 report *Patient Safety: Achieving a New Standard for Care* [1]

"One of the key components of a national health information infrastructure will be data standards to make that information understandable to all users"

In order for EHRs, HIOs and the NHIN to succeed there needs to be a standard language; otherwise you have a "*Tower of Babel*". We use standards every day but often take them for granted. All languages are based on a semantic language standard known as grammar. The plumbing and electrical industries depend on standards that are the same in every state. The railroad industry had to decide many years ago what gauge railroad track they would use to connect railroads throughout the United States.

In practice, standards can come in many flavors. Standards that focus on the communication of data between multiple systems can be referred to as transport standards. The rules that dictate the format of information as it packaged for transport are known as content standards. Individual segments within a content package are governed as a vocabulary. All of these standards are discovered over careful study of real world use cases.

The first National Coordinator for HIT believed that standards should come first to promote interoperability of electronic health records and not follow adoption of EHRs. Although we have come a long way towards universal standards, we are not there yet. The progress has been slow in part due to the fact that participation in standards determining organizations is voluntary. [2]

The next sections will discuss the major data standards and how the standards facilitate the transmission of data. Not all data standards have been included in the following sections and many standards are still a "*work in progress*".

Extensible Markup Language (XML)

- Although XML is not really a data standard it is a data packaging standard that has become a programming markup language standard for health information exchange. In order for disparate health entities to share messages and retrieve results, a common data packaging standard is necessary

- XML is a set of predefined rules to structure data so it can be universally interpreted and understood
- XML consists of elements and attributes
- Elements are tags that can envelop data and can be organized into a hierarchy
- Attributes help describe the element
- Below is a simple example where automobile is the root element and car is the child element. Ford and Chevy are attributes [3]

```
<automobile>
<car id= "Ford" model="2008">
    <phone id ="1"> All phone information
        <number>9216604</number >
    </phone>
</car>
<car id="Chevy" model="2008>
    <phone id = "2"> All phone information
        <number>9335676</number>
    </phone>
</car>
```

Health Level Seven (HL7)

- A not-for-profit standards development organization (SDO) with chapters in 30 countries
- Health Level Seven's domain is clinical and administrative data transmission and perhaps is the most important standard of all
- "Level Seven" refers to the seventh level of the International Organization for Standardization (ISO) model for Open System Interconnection. This serves to communicate that HL7 messaging lives in the application layer of the stack, with subordinate layers serving as items in the overall toolkit
- HL7 provides a set of standards for interactions between healthcare data services
- HL7 is a data standard for communication or messaging between:
    o Patient administrative systems (PAS)
    o Electronic practice management systems
    o Lab information systems (interfaces)
    o Dietary
    o Pharmacy (clinical decision support)
    o Billing
    o Electronic health records (EHRs)

```
MSH|^~\&|EPIC|EPICADT|SMS|SMSADT|199912271408|CHARRIS|ADT^A04|1817457|
EVN|A04|199912271408|||CHARRIS
PID||0493575^^^2^ID 1|454721||DOE^JOHN^^^^|DOE^JOHN^^^^|19480203|M|
NK1||CONROY^MARI^^^^|SPO||(216)731-4359||EC|||||||||||||||||||||||||||
PV1||O|168 ~219~C~PMA^^^^^^^^^||||277^ALLEN FADZL^BONNIE^^^^||||||||||
```

Figure 6.1 HL7 example (The vertical bars are called pipes and separate the bits of data)

- The most current version of the HL7 standard is 3.0 but version 2.x is still widely in use
- HL7 version 2.x separates messages into processable chunks known as segments.

- HL7 version 2.x segments are sewn together into messages of a given type (e.g. Admit Discharge and Transfer [ADT] or Pharmacy Administration [RAS])
- HL7 version 3.0 uses XML for packaging its content
- The Clinical Document Architecture (CDA)
  - A HL7 v3.0 content standard that makes documents human readable and machine processable through the use of XML. CDA is used in EHRs, personal health records, discharge summaries and progress notes. CDA delineates the structure and semantics of clinical documents, consisting of a header and body [4-5]
  - In 2007 HL7 recommended (and HITSP endorsed) the use of the Continuity of Care Document (CCD) standard. The CCD is the marriage of the Continuity of Care Record (CCR) (developed by ASTM International) and the CDA (developed by the HL7 organization)
- The CCD and CCR:
  - The electronic document exchange standard for the sharing of patient *summary* information between physicians and within personal health records
  - The CCD has the advantage over CCR of being able to accept free text and being capable of vocabulary specific semantic interoperability. It contains the most common information about patients in a summary XML format that can be shared by most computer applications and web browsers. It can printed (pdf) or shared as html
  - In 2008 CCHIT required EHRs to generate and format CCD documents using the C32 specification for patient registration, medication history and allergies.[6-8] The CCD and CCR are both currently listed as interchangeable content standards for achieving stage 1 Meaningful Use
  - The CCD has 17 data content/component modules as part of the C32 standards as noted in the table below. Each module will have additional data elements
  - Dr John Halamka has posted his CCD as an example for others to view [9]

Table 6.1 Data modules of the C32 standard for the CCD

Header	Payers	Vital signs	Social history
Purpose	Advance directives	Functional status	Medical equipment
Problems	Alerts	Results	
Procedures	Medications	Encounters	
Family history	Immunizations	Plan of care	

Digital Imaging and Communications in Medicine (DICOM)

- DICOM was formed by the National Electrical Manufacturers Association (NEMA) and the American College of Radiology. They first met in 1983 which suggests that early on they recognized the potential benefits of the storage, sharing, and transmission of digital x-rays
- As more radiological tests became available digitally, by different vendors, there was a need for a common data standard. Similarly, as more EHRs had PACS functionality, DICOM became the standard for images in EHRs
- DICOM supports a networked environment using TCP/IP protocol (basic Internet protocol)
- DICOM is also applicable to an offline environment [10]
- "I do Imaging" is a web site that promotes open source DICOM viewers, DICOM converters and PACS clients [11]

Logical Observations: Identifiers, Names and Codes (LOINC)

- This is a standard for the electronic exchange of lab results back to hospitals, clinics and payers. HL7 is the *content* standard, whereas LOINC is the *vocabulary* standard
- The LOINC database has more than 30,000 codes used for lab results. This is necessary as multiple labs have multiple unique codes that would otherwise not be interoperable
- LOINC is divided into lab, clinical and HIPAA portions
- The lab results portion of LOINC includes chemistry, hematology, serology, microbiology and toxicology
- The clinical portion of LOINC includes vital signs, EKGs, echocardiograms, gastrointestinal endoscopy, hemodynamic data and others
- The HIPAA portion is used for insurance claims
- As an example:
  - The LOINC code for serum sodium is 2951-2; there would be another code for urine sodium
  - The formal LOINC name for this test is: SODIUM:SCNC:PT:SER/PLAS:QN (component:property:timing:specimen:scale)
- LOINC is accepted widely in the US, to include federal agencies. Large commercial labs such as Quest and LabCorp have already mapped their internal codes to LOINC
- Other standards such as DICOM, SNOMED and MEDCIN have cross references to LOINC
- RELMA is a mapping assistant to assist mapping of local test codes to LOINC codes
- LOINC is maintained by the Regenstrief Institute at the Indiana School of Medicine [12] LOINC and RELMA are available free of charge to download from www.regenstrief.org/loinc
- For more detail on LOINC we refer you to an article by McDonald [13]

EHR-Lab Interoperability and Connectivity Standards (ELINCS)

- ELINCS was created in 2005 as a lab interface for ambulatory EHRs and a further "constraint" or refinement of HL7 standards
- Traditionally, lab results are mailed or faxed to a clinician's office and manually inputted into an EHR. ELINCS would permit standardized messaging between a laboratory and a clinician's ambulatory EHR
- Standard includes:
  - Standardized format and content for messages
  - Standardized model for transport of messages
  - Standardized vocabulary (LOINC)
- The Certification Commission for Healthcare Information Technology (CCHIT) has proposed that ELINCS be part of EHR certification
- HL7 plans to adopt and maintain the ELINCS standard
- California Healthcare Foundation sponsored this data standard [14]

RxNorm

- RxNorm is a standard for drug coding; developed by the National Library of Medicine
- The standard includes three drug elements: the active ingredient, the strength and the dose
- RxNorm is the standard for e-prescribing
- RxNorm encapsulates other drug coding systems, such as National Drug Code (NDC)
- The standard only covers US drugs at this point [15]

National Council for Prescription Drug Programs (NCPDP)

- A standard for exchange of prescription related information
- The standard facilitates pharmacy related processes
- Newest (2010) standard is NCPDP Script 10.6
- It is the standard for *billing* retail drug sales [16]

Accredited Standards Committee (ASC) X12

- A standard for electronic data exchange (EDI) or the computer-to-computer exchange of business data
- Standard is used in healthcare, transportation, insurance and finance industries [17]

Systematized Nomenclature of Medicine: Clinical Terminology (SNOMED-CT)

- SNOMED is the clinical terminology or medical vocabulary commonly used in software applications, including EHRs
- Provides more clinical detail than ICD-9 and felt to be more appropriate for EHRs
- SNOMED is also known as the International Health Terminology
- This standard was developed by the American College of Pathologists. In 2007 ownership was transferred to the International Health Terminology Standards Development Organization www.ihtsdo.org
- SNOMED will be used by the FDA and the Department of Health and Human Services
- This standard currently includes about 1,000,000 clinical descriptions
- Terms are divided into 19 hierarchical categories
- The standard provides more detail by being able to state condition A is due to condition B
- SNOMED concepts have descriptions and concept IDs (number codes). Example: open fracture of radius (concept ID 20354001 and description ID 34227016)
- SNOMED CT also defines two types of relationships
    - "Is-a" connects concepts within the same hierarchy. Example: asthma "is a" lung disease
    - "Attribute" connects concepts in different hierarchies. Example: asthma is associated with inflammation
- SNOMED links (maps) to LOINC and ICD-9/10
- SNOMED is currently used in over 40 countries
- EHR vendors like Cerner and Epic are incorporating this standard into their products
- There is some confusion concerning the standards SNOMED and ICD-9; the latter used primarily for billing and the former for communication of clinical conditions [18-20]
- A study at the Mayo Clinic showed that SNOMED-CT was able to accurately describe 92% of the most common patient problems [21]
- SNOMED-CT Example: Tuberculosis

    D E – 1 4 8 0 0

    . . . .

    . . . .

    . . . Tuberculosis

    . . Bacterial infections

    . E = Infectious or parasitic diseases

    D = disease or diagnosis

MEDCIN®

MEDCIN ® was developed by Medicomp in the 1980s as a proprietary medical vocabulary. In 1997 it was released as a national standard. MEDCIN® cross-references to many of the other standards already discussed. The nomenclature consists of about 270,000 clinical concepts organized into categories: symptoms, history, physical exam, tests, diagnosis and therapy. Each finding is associated with a numerical code, up to 7 digits, so the results are structured or codified. Unlike SNOMED, MEDCIN® findings can link to symptoms, exam, therapy and testing. The knowledge base also includes 600,000 synonyms, allowing look-ups under different terms. MEDCIN® is used by several EHR systems, to include the DOD's AHLTA.

The disadvantages of this system are the fact that there is a substantial learning curve to be able to search for all of the necessary MEDCIN® terms in order to create a completely structured note. Second, the note that is created tends to be poorly fluent and not like dictation (Figure 6.2). For that reason, Medicomp developed CliniTalk™ which is a voice to text option that means that a clinician can dictate and the end result is structured data. [22]

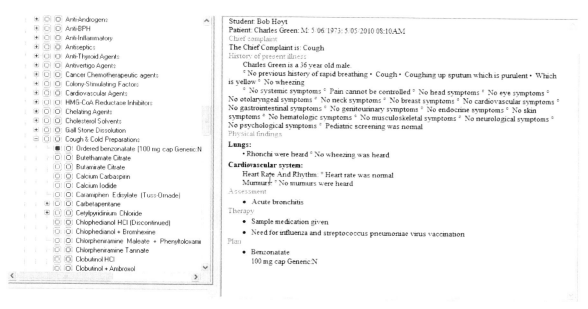

Figure 6.2 Simple note created with MEDCIN® (courtesy MEDCIN® Student Edition)

International Classification of Diseases 9th revision and ICD-10

ICD-9 is published by the World Health Organization to allow mortality and morbidity data from different countries to be compared. The basic ICD-9 code, for example diabetes, is 250.00 (3 digits). If it covers more detail such as diabetes with kidney complications, it becomes 250.40. (4 or 5 digits). An online web site lists common ICD-9 CM codes. [23] Although it is the standard used to provide a diagnosis for billing over the past 30 years, it is not ideal for distinct clinical diseases.

ICD-10 was endorsed in 1994 but not used in the US. The Federal government set October 2013 as the launch date for ICD-10.[24] ICD-10 will provide a more detailed description with 7 rather than 5 digit codes. ICD-10 CM (clinical modifications) was developed by the CDC for use in all healthcare settings. ICD-10 PCS (procedure coding system) was developed by CMS for inpatient settings only and will replace ICD-9 CM, volume 3. [25] ICD-10 would result in about 68,000 codes instead of the 24,000 codes currently used. A study by Blue Cross/Blue Shield estimated that adoption of ICD-10 would cost the US healthcare industry about $14 billion over the next 2-3 years. The more digits included generally results in higher reimbursement. [26-27] Figure 6.3 demonstrates the differences between ICD-9 and ICD-10 formats.

ICD-9-CM Code Format

XXX.XX

Category etiology
anatomic site,
manifestation

ICD-10-CM Code Format

XXX.XXX X

Category etiology,   extension
anatomic
site or
severity

Figure 6.3 ICD-9 and ICD-10 Code Formats

Current Procedural Terminology (CPT)

- CPT is used for billing the level and complexity of service rendered
- Standard was developed, owned and operated by the American Medical Association (AMA) for a fee
- A CPT code is a five digit numeric code that is used to describe medical, surgical, radiology, laboratory, anesthesiology, and evaluation/management services of physicians and hospitals. There are approximately 7,800 CPT codes ranging from 00100 through 99499. Two digit modifiers may be appended to clarify or modify the description of the procedure. The most recent CPT book is version 4
- CPT Codes are published in two versions: the first is for physicians and the second is for hospitals.
- Example CPT code: office visit, established patient, moderate complexity = 99214 [28]

Healthcare Common Procedure Coding System (HCPCS)

- Are codes used by Medicare and based on CPT codes
- They document medical, surgical and diagnostic services
- HCPCS Level I codes are identical to CPT codes
- HCPCS Level II codes are used by medical suppliers and not clinical services
- More information is available from CMS [29]

Evaluation & Management (E&M) Codes

- In order to bill for a patient visit, ICD-9 CM and CPT codes are selected to best represent the visit. It is up to the clinician to provide documentation to prove the level of the visit
- As an example, if a clinician chooses to select CPT code 99204 for a new outpatient patient visit, they must document that the problems are of moderate to high severity, the physicians spends about 45 minutes face-to-face and the E&M requires these key components: comprehensive history and physical exam and medical decision making of moderate complexity. This implies that an excellent history and physical exam are documented and the problems discussed were moderately complex
- Many EHRs have E&M calculators to help assist the clinician in determining the level of service. This is made easier if templates are used because clicking on history and physical exam elements can calculate an E&M code in the background [30]
- Figure 6.4 shows a typical E&M calculator that is part of an EHR. Note: this is an established patient, the E&M level is in the upper left, the diagnosis and ICD-9 code (462) are in the upper right. Multiple fields are available to input the complexity of the visit so the E&M code can be manually or automatically calculated

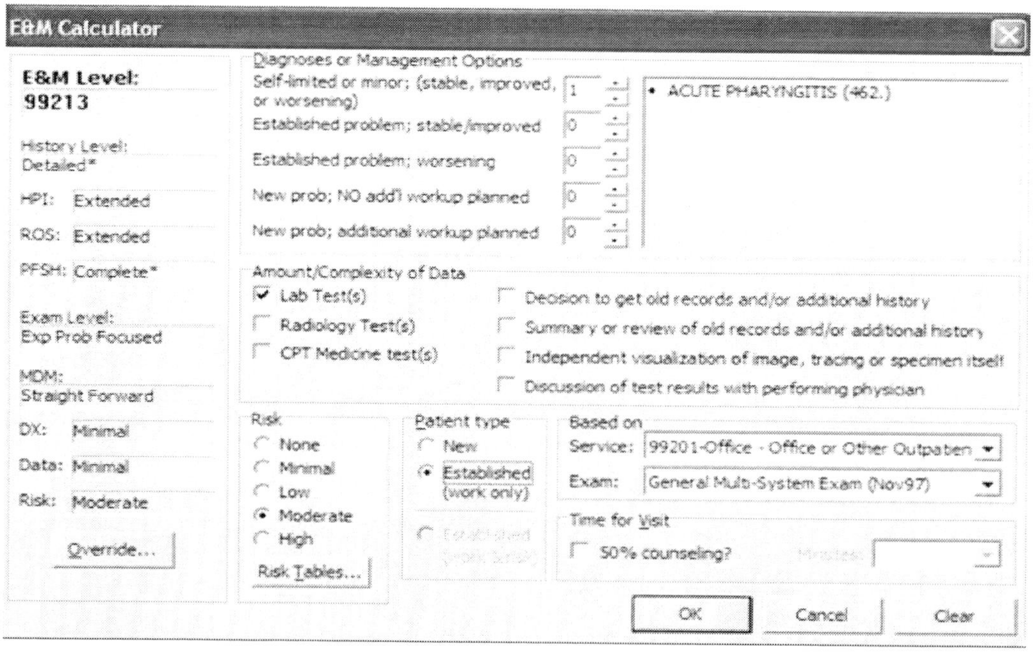

Figure 6.4 E&M calculator as part of Healthmatics EHR (courtesy www.network-systems.com)

Key Points

- Data standards play a major role in accomplishing interoperability

- In order for health information exchange to occur multiple data standards need to be reconciled and adopted

- We are slowly moving towards industry wide standards such as the Continuity of Care Document (CCD) and Continuity of Care Record (CCR)

Conclusion

Data standards are critical for interoperability between disparate technologies and organizations. Multiple standards developing organizations have proposed standards that are being tested, harmonized and updated for application in the field of medicine. Standards are important to exchange clinical data, as well as, administrative and financial data. Standards are essential for exchange of information between electronic health records, health information organizations and the Nationwide Health Information Network.

References

1. IOM. Patient Safety: Achieving a new standard of care. 2004. http://www.nap.edu/books/0309090776/html/ (Accessed February 22 2005)
2. Kim K. Clinical Data Standards in Health care: Five case studies. California Healthcare Foundation July 2005 www.chcf.org (Accessed January 1 2006)

3. Chitnis M, Tiwari P, Ananthamurthy L. Introduction to web services Part 3: Understanding XML www.developer.com/services/article.php/1557871 (Accessed June 20 2008)
4. Health Level Seven HL7 www.hl7.org (Accessed July 7 2009)
5. Dolin RH et al. HL7 Clinical Document Architecture, Release 2. JAIMA. 2006;13:30-39
6. HL7 Implementation Guide: CDA Release 2-Continuity of Care Document. November 1 2007 http://www.himssehrva.org/ASP/CCD_QSG_20071112.asp (Accessed May 1 2008)
7. Bazzoli F. Continuity of Care Document is approved by HL7 and endorsed by HITSP. February 14 2007 www.healthcareitnews.com (Accessed May 1 2008)
8. The Continuity of Care Document. Corepoint Health www.corepointhealth.com (Accessed February 29 2010)
9. PPHC Continuity of Care Document. November 3 2007. http://services.bidmc.org/geekdoctor/johnhalamkaccddocument.xml (Accessed March 24 2010)
10. Digital Imaging and Communication in Medicine. NEMA. 2009 ftp://medical.nema.org/medical/dicom/2009/09_01pu.pdf (Accessed March 23 2010)
11. I Do Imaging www.idoimaging.com (Accessed March 24 2010)
12. LOINC www.loinc.org (Accessed January 3 2010)
13. McDonald CJ, Huff SM, Suico JG et al. LOINC, a Universal Standard for Identifying Laboratory Observations: A 5 –Year Update Clin Chem 2003;49 (4):624-633
14. ELINCS. http://www.elincs.chcf.org (Accessed June 8 2009)
15. RxNorm http://www.nlm.nih.gov/research/umls/rxnorm/index.html (Accessed July 8 2009)
16. National Council for Presription Drug Programs (NCPDP) www.ncpdp.org (Accessed July 8 2009)
17. ASC X12 www.x12.org (Accessed March 14 2010)
18. SNOWMED-CT http://en.wikipedia.org/wiki/SNOMED_CT (Accessed July 8 2009)
19. Joch A. A blanket of SNOMED Federal Computer Week Nov 14 2005;s46-47
20. Lundberg C. SNOMED CT: An Introduction. CAP STS. Presentation. February 22 2010
21. Elkin PL et al. Evaluation of the content coverage of SNOMED-CT: ability of SNOMED Clinical terms to represent clinical problem lists Mayo Clin Proc 2006;81:741-748
22. Medicomp. www.medicomp.com (Accessed March 4 2010)
23. ICD-9 Search Engine. http://www.eicd.com/EICDMain.htm (Accessed March 4 2010)
24. ICD-9 CMS http://www.cms.hhs.gov/ICD10/01k_2010_ICD10PCS.asp#TopOfPage (Accessed March 20 2010)
25. ICD-10 www.who.int/whosis/icd10/othercla.htm (Accessed July 8 2009)
26. Featherly K. ICD-9-CM: An uphill struggle Healthcare Informatics Oct 2004: 14-16
27. Weier S. Letter Encourages Congress to promote ICD-10. www.ihealthbeat.org May 19 2006 (Accessed February 22 2005)
28. CPT www.ama-assn.org/ama/pub/category/3113.html (Accessed December 20 2008)
29. Healthcare Common Procedure Coding System. www.cms.hhs.gov/MedHCPCSGenInfo (Accessed March 24 2010)
30. Evaluation & Management Services Guide July 2009 CMS. www.cms.hhs.gov/MLNGenInfo (Accessed March 20 2010

Privacy and Security

BRENT HUTFLESS

Learning Objectives

After reading this chapter the reader should be able to:

- Describe privacy and security measures contained in HIPPA, the HITECH Act and Meaningful Use final rule
- Describe the importance of data security and privacy as related to public perception
- Identify the benefits and pitfalls of local vs. Software-as-a-service (SaaS) technical security solutions
- Recognize the importance of confidentiality, availability and integrity related to health IT security

The Health Insurance Portability & Accountability Act (HIPAA) passed in 1996 laid much of the groundwork for the privacy and security measures being adopted within HIT today. The original intent was to direct how patient data was used and made available when patients switched physicians or insurers, and included two major rules covering privacy and security of that data. The American Recovery and Reinvestment Act of 2009 (ARRA) and the HITECH Act which accompanied it, both brought about changes designed to improve privacy and security measures required by modern technologies and close loopholes within the original law. The final ruling on Meaningful Use in July, 2010 also impacts future health IT implementations. Before discussing the current state of privacy and security regulation and intent, a primer on HIPAA is needed to show what the original law provides.

HIPPA for the Consumer

Certain organizations, known as covered entities, are required to follow the HIPAA Privacy Rule: [1]

- Health Plans
  - Health insurers
  - HMOs
  - Company health plans
  - Government programs such as Medicare and Medicaid
- Health Care Providers that conduct business electronically
  - Most doctors
  - Clinics
  - Hospitals
  - Psychologists
  - Chiropractors
  - Nursing homes

- o Pharmacies
- o Dentists
- Health Care Clearinghouses

A number of organizations do not have to follow HIPAA law despite using personal health data: [1]

- Life insurers
- Employers
- Workers compensation carriers
- Many schools and school districts
- Many state agencies like child protective service agencies
- Many law enforcement agencies
- Many municipal offices

For those organizations that are required to abide by HIPAA, patient data and personal information must be protected according to the Security Rule. Protections apply to all personal health information (PHI), whether in hard copy records, electronic personal health information (ePHI) stored on computing systems, or even verbal discussions between medical professionals. Covered entities must put safeguards in place to ensure data is not compromised and that it is only used for the intended purpose. The HIPAA rules are not designed to and should not impede the treatment of patients.[2] Covered entities must comply with certain consumer rights; specifically a patient may: [1]

- Ask to see and get a copy of their health records
- Have corrections added to their health information
- Receive a notice that discusses how health information may be used and shared
- Provide permission on whether health information can be used or shared for certain purposes, such as for marketing
- Get reports on when and why health information was shared for certain purposes
- File a complaint with a provider, health insurer and/or the U.S. Government if patient rights are being denied or health information is not being protected

HIPAA Privacy for Covered Entities

Covered entities have a significant responsibility to protect the privacy and security of patient data and personal information. The U.S. Department of Health & Human Services (HHS) has an excellent website, www.hhs.gov/ocr/privacy/hipaa/understanding/coveredentities/index.html designed to serve this population and inform entities about subjects ranging from patient consent, incidental disclosures and contracts with business associates to the proper disposal of protected information. For detailed information regarding the HIPAA Privacy and Security rules, HHS, the Office of Civil Rights, and others provide formalized guidance. The following is a summary of highlights.

The Privacy Rule strictly limits how a covered entity and their business associates can use patient data, but there is a method that can be employed to use and release the data without restrictions. The Privacy Rule permits organizations to de-identify the data by removing 18 identifiers, which reasonably precludes the resulting information from being attributed to a patient. The 18 identifiers follow: [3]

- Names

- All geographic subdivisions smaller than a state, including street address, city, county, precinct, ZIP Code, and their equivalent geographical codes, except for the initial three digits of a ZIP Code if, according to the current publicly available data from the Bureau of the Census:
  - The geographic unit formed by combining all ZIP Codes with the same three initial digits contains more than 20,000 people
  - The initial three digits of a ZIP Code for all such geographic units containing 20,000 or fewer people are changed to 000
- All elements of dates (except year) for dates directly related to an individual, including birth date, admission date, discharge date, date of death; and all ages over 89 and all elements of dates (including year) indicative of such age, except that such ages and elements may be aggregated into a single category of age 90 or older
- Telephone numbers
- Facsimile numbers
- Electronic mail addresses
- Social security numbers
- Medical record numbers
- Health plan beneficiary numbers
- Account numbers
- Certificate/license numbers
- Vehicle identifiers and serial numbers, including license plate numbers
- Device identifiers and serial numbers
- Web universal resource locators (URLs)
- Internet protocol (IP) address numbers
- Biometric identifiers, including fingerprints and voiceprints
- Full-face photographic images and any comparable images
- Any other unique identifying number, characteristic, or code, unless otherwise permitted by the Privacy Rule for re-identification

Covered Entity Permitted Uses and Disclosures according to the Privacy Rule: [4]

- To the individual
- For treatment, payment or health care operations
- Uses and disclosures with opportunity to agree or object
  - Facility directories
  - For notification and other purposes
- Incidental use and disclosure
- Public interest and benefit activities
  - Required by law
  - Public health activities
  - Victims of abuse, neglect or domestic violence
  - Health oversight activities
  - Judicial and administrative proceedings

- o Law enforcement purposes
- o Decedents
- o Cadaveric organ, eye, or tissue donation
- o Research
- o Serious threat to health or safety
- o Essential government functions
- o Workers' compensation
- Limited data set

Administrative requirements were established by the Privacy Rule to ensure that all covered entities, regardless of size or organization, met a minimum standard for protecting patient privacy and permitting patients to exercise their rights. The scope of the solutions for each of the requirements is left up to the individual organization, but the required categories are as follows: [4]

- Develop and implement written privacy policies and procedures
- Designate a privacy official
- Workforce training and management
- Mitigation strategy for privacy breaches
- Data safeguards - administrative, technical, and physical
- Designate a complaint official and procedure to file complaints
- Establish retaliation and waiver policies and restrictions
- Documentation and record retention - six years
- Fully-insured group health plan exception

HIPAA Security for Covered Entities

Security techniques and safeguards are discussed elsewhere in the chapter due to recent changes in HIPAA through the ARRA and HITECH Act, in addition to Meaningful Use recommendations. There are three safeguards that are required by the HIPAA Security Rule that serve as the foundation for these changes, and they are as follows: [5]
Administrative Safeguards

- Security management processes to reduce risks and vulnerabilities
- Security personnel responsible for developing and implementing security policies
- Information access management - minimum access necessary to perform duty
- Workforce training and management
- Evaluation of security policies and procedures

Physical Safeguards

- Facility access and control limiting physical access to facilities
- Workstation and device security policies and procedures covering transfer, removal, disposal, and re-use of electronic media

Technical Safeguards

- Access control that restrict access to authorized personnel
- Audit controls for hardware, software, and transactions

- Integrity controls to ensure data is not altered or destroyed
- Transmission security to protect against unauthorized access to data transmitted on networks and via email

Privacy and Security Issues Today

As this chapter was being authored, a number of high-profile security breaches in the Healthcare community were coming to light. The FTC had just notified nearly 100 organizations of data compromise they incurred due to P2P file sharing; losses involving "health-related information, financial records, and drivers' license and social security numbers--the kind of information that could lead to identity theft," according to FTC Chairman Jon Leibowitz.[6] Meanwhile, Blue Cross and Blue Shield of Tennessee was still notifying nearly half a million members of identifiable data theft that occurred when 57 hard drives were stolen from one of its facilities. [7] This notice came just one day after Connecticut's Attorney General filed a lawsuit against Health Net of Connecticut, charging HIPAA privacy and security violations stemming from Health Net's loss of 1.5 million member records. [8] Additionally, network security at hospital and clinics remain a concern, as there are indications foreign hackers are increasingly targeting America's patient records.[9] These breaches illustrate the importance of securing these records systems and the data they contain. Data security is paramount to the success of an electronic health data management system that the public trusts.

Fourteen years after HIPAA was passed, emerging technologies continually force new concerns for security and privacy implications into the spotlight. New consumer products such as Google Health and Microsoft HealthVault offer no-cost personal health record systems to the masses, but both corporations challenge the assertion that either is bound by HIPAA privacy and security laws and regulations. Google still maintains a specific limited liability user agreement, a policy that could lead to security breaches with little recourse for users in the event of identity theft or worse. [9] The ARRA contains provisions to eliminate the loophole Google exploits in the HIPAA laws passed in 1996 regarding covered entities and data breach reporting. There are bound to be more challenges and legal rulings related to HIPAA and ARRA regulations as more vendors form partnerships with healthcare providers in order to meet emerging Meaningful Use requirements in the coming years.

Developing the Nationwide Health Information Network

The United States is undergoing a transformation to technology-driven healthcare through the use of electronic health records (EHRs), clinical decision support systems (CDSS) and other solutions broadly known as Health Information Technology (HIT), but it has been time consuming process and not without obstacles. This healthcare revolution is gaining traction through a convergence of driving forces; current and imminent government regulations, the need to cut rising insurance costs, calls for healthcare reform, improved technical capability, increases in available network bandwidth, advanced software solutions and a higher expectation from consumers to control and manage their own healthcare information. The ARRA provides financial incentives for EHR adoption and use, aiding hospitals, clinics and physicians in the push toward meeting the evolving requirements for electronic capture and tracking of patient data outlined in the Meaningful Use definitions.[11,12] A sample of the pending Meaningful Use guidance is shown in Table 7.1, which indicates the privacy and security goals.[13]

Table 7.1 Stage 1 criteria (Courtesy of Federal Register / Vol. 75, No. 8 / Wednesday, January 13, 2010)

Health outcomes policy priority	Care goals	Stage 1 objectives		Stage 1 measures
		Eligible professionals	Hospitals	
Ensure adequate privacy and security protections for personal health information.	Ensure privacy and security protections for confidential information through operating policies, procedures, and technologies and compliance with applicable law. Provide transparency of data sharing to patient.	Protect electronic health information created or maintained by the certified EHR technology through the implementation of appropriate technical capabilities.	Protect electronic health information created or maintained by the certified EHR technology through the implementation of appropriate technical capabilities.	Conduct or review a security risk analysis per 45 CFR 164.308(a)(1) and implement security updates as necessary.

TABLE 2—STAGE 1 CRITERIA FOR MEANINGFUL USE—Continued

The result of these congruent influences is an industry moving away from the paper-based record keeping. This data migration will help prevent portability and accessibility problems that are inherent with physical records, but simultaneously creates the potential for hundreds of millions of electronic patient records. This increases the size and complexity of healthcare network architectures and operating environments in thousands of organizations across the country. How are these medical records, personally identifiable data stores, and high assurance networks to be protected? This chapter will focus on the fundamental issues surrounding HIT security, discuss the principles of information technology security (with a spotlight on confidentiality), and show how organizations can apply this security philosophy to the medical community.

HIPAA, Meaningful Use and the ARRA of 2009 (HITECH Act)

Many of the fundamental concepts surrounding HIPAA were introduced at the beginning of the chapter, but HHS maintains an excellent 25 page summary of the HIPAA Privacy Rule in PDF format (www.hhs.gov/ocr/privacy/hipaa/understanding/summary/privacysummary.pdf) for quick reference. The Privacy and Security Rules established through HIPAA were designed for the healthcare system and processes that were in place in the mid-1990s. Today, technology has the opportunity to revolutionize health care, changing hospitals and clinical environments from isolated and unconnected islands of patient treatment records and knowledge to an interconnected system of healthcare. Hospitals and practitioners are taking note of studies advocating the use of EHRs and related systems that have the capability to quickly retrieve patient data and records, saving time, preventing duplication of treatment efforts, reducing drug interactions and contraindication situations; generally improving patient care and reducing administrative costs associated with paper records.[14] However, with new opportunities there often come new risks and in the case of medicine, an escalating chance to violate HIPAA privacy and security regulations. The number of companies and organizations that are offering data solutions for patients and providers is growing exponentially, increasing the challenge of finding a solution that meets the new reporting, use and billing requirements.

Certifying Compliance

The Certification Commission for Health Information Technology (CCHIT) was established to ensure that a product meets US Dept of Health and Human Services minimum specifications for compliance criterion. While this provides prospective customers with some assurance that approved products meet or exceed the emerging regulations, CCHIT has issued a notice on its website indicating that 2011 certified products may not meet the new Meaningful Use requirements, and that product retesting had begun February 12, 2010.[15] Table 2 includes a partial list of CCHIT 2011 certified software as retrieved from the website in June, 2010. Two important issues can be implied from the certification list. The first inference is that there are only 29 products on the list, several of which originate from the same vendor. The second is that *only 12 vendors* are currently capable of offering fully CCHIT 2011-certified products.[15] There are hundreds of commercial and open source EMR solutions available that are currently not certified.

Table 7.3 2011 Certified EHRs (Courtesy of CCHIT)

Company	Product	Date Certified	Domain
ABEL Medical Software Inc.	ABELMed EHR - EMR / PM 11 *Pre-Market*	11/23/2009	Ambulatory
AllscriptsMisys, LLC	Allscripts ED 6.3 Service Release 4 *Fully Certified*	05/10/2010	Emergency Dept.
AllscriptsMisys, LLC	Professional EHR 9.1 *Fully Certified*	05/26/2010	Ambulatory
Aprima Medical Software, Inc	Aprima 2011 *Pre-Market*	06/22/2010	Ambulatory
BizMatics Inc	PrognoCIS Version 2.0 *Pre-Market*	06/15/2010	Ambulatory
Compulink	Advantage/EHR 10 *Pre-Market*	12/16/2009	Ambulatory
CureMD Corporation	CureMD EHR Version 10 *Fully Certified*	04/13/2010	Ambulatory
Eclipsys Corporation	Sunrise Clinical Manager™ 5.5 2011 Suite 5.5 *Pre-Market*	06/11/2010	Inpatient
Eclipsys Corporation	Sunrise Emergency Care™ 5.5 2011 Suite 5.5 *Pre-Market*	06/08/2010	Emergency Dept.
EHS, Inc	EHS CareRevolution 5.3 *Fully Certified*	12/17/2009	Ambulatory

Data Storage and Defining Covered Entities

When the Health Insurance Portability & Accountability Act was passed in 1996, most healthcare organizations were still entrenched as paper-based systems. As technology evolved over the past decade, so too did the methods that healthcare entities used to share and store medical information. Electronic billing, patient records and personal data storage was becoming a more common practice, but high profile cases of data loss were increasingly in the news, as occurred when the VA lost a laptop containing more than 26 million veteran records, and later lost another with an additional 38,000 records.[16]

As a result of these and similar breaches, the HIPAA security standard enacted in 2003 needed amending to clarify the requirements for storing and sharing ePHI. In late 2006, the Department of Health and Human Services released its HIPAA Security Guidance, which identified various forms of remote use, data storage and the requirements for handling and reporting ePHI by covered entities.[17] While it was hoped that this guidance would lead to increased protections, loopholes would remain that would need to be addressed by the ARRA.

By 2009, HIPAA faced new challenges regarding the definition of covered entities. Major software companies had begun aligning new products to the burgeoning field of HIT with a variety of fee-based, open-source, and free solutions and services. Microsoft and Google both offer consumers no-cost, web-based electronic health records that allow users to share information with physicians, hospitals and pharmacies and store vast amounts of medical and personal data. Although this initially appeared to be a wonderful development, both companies asserted that HIPAA did not cover them. Through 2009, Google and Microsoft established data-sharing partnerships with hospitals and clinics, while maintaining that they were not bound by HIPAA regulations as covered entities or business associates of covered entities, arguing that they would bear no liability for patient data loss or any damage due to theft, breach or discovery of user personal and medical data.[18]

This inference was reinforced by each company's privacy notice which holds no party liable even in the event that the parties knew that there could be potential damage.[19,20] This attitude bodes poorly for patient rights and privacy, and could open users of either product to identity theft and leave organizations partnering with Google and Microsoft liable. This loophole is closed according to the American Recovery and Reinvestment Act of 2009, which states "each vendor that contracts with a covered entity to allow that covered entity to offer a personal health record to patients as part of its electronic health record" is bound by HIPAA through contract.[21] Much of the new guidance and reporting requirements fall under the HITECH Act, created through the ARRA.

As of this writing, there is still some dispute over the legal assertion that HITECH and ARRA close the HIPPA business association loophole. Google and Microsoft both maintain a policy that neither is a covered entity.[19,20] Text captured from each agreement is included in figures 7.1 and 7.2 – Google text in all caps is emphasis added by Google.

12. Limitation of Liability

NEITHER YOU NOR GOOGLE OR ANY OF ITS LICENSORS MAY BE HELD LIABLE UNDER THIS AGREEMENT FOR ANY DAMAGES OTHER THAN DIRECT DAMAGES, EVEN IF THE PARTY KNOWS OR SHOULD KNOW THAT OTHER DAMAGES ARE POSSIBLE OR THAT DIRECT DAMAGES ARE NOT A SATISFACTORY REMEDY. THE LIMITATIONS IN THIS SECTION APPLY TO YOU ONLY TO THE EXTENT THEY ARE LAWFUL IN YOUR JURISDICTION.

NEITHER YOU NOR GOOGLE OR ANY OF ITS LICENSORS MAY BE HELD LIABLE UNDER THIS AGREEMENT FOR MORE THAN $1,000.

The limitations of liability in this Section do not apply to breaches of intellectual property provisions or indemnification obligations.

Figure 7.1 Screen Capture of Google Health Terms of Service

10. LIABILITY LIMITATION.

You can recover from Microsoft only direct damages up to an amount you pay Microsoft for this Service. You cannot recover any other damages, including consequential, lost profits, special, indirect, incidental or punitive damages.

This limitation applies to anything related to:

- the Service,
- content (including code) on third party Internet sites, third party programs or third party conduct,
- viruses or other disabling features that affect your access to or use of the Service,
- incompatibility between the Service and other Services, software and hardware,
- delays or failures you may have in initiating, conducting or completing any transmissions or transactions in connection with the Service in an accurate or timely manner, and
- claims for breach of Service Agreement, breach of warranty, guarantee or condition, strict liability, negligence, or other tort.

It also applies even if:

- this remedy does not fully compensate you for any losses, or fails of its essential purpose; or
- Microsoft knew or should have known about the possibility of the damages.

Some states do not allow the exclusion or limitation of incidental or consequential damages, so the above limitations or exclusions may not apply to you.

Figure 7.2 Screen Capture of Microsoft HealthVault Account Service Agreement

Recent changes to HIT on a national scale will further the impact of privacy and security on the medical field. Medical organizations and physicians have been bound by HIPAA regulations for more than a decade, but compliance with HIPAA has the potential to impact the financial bottom line beyond fines and penalties now that the new standards have been adopted. The Health IT Policy Committee published their final recommendations in August of 2009 and proposed "that CMS withhold Meaningful Use payment for any entity until any confirmed HIPAA privacy or security violation has been resolved".[12] The committee has also made similar recommendations for state Medicaid administrators when HIPAA violations are suspected. These payment withholdings would be in addition to any potential fines and penalties attributed to the HIPAA privacy or security violations. These recommendations were reinforced throughout the Meaningful Use text published in the Federal Register.[13]

HIPPA, HITECH elements of the ARRA of 2009, and Meaning Use standards all serve to protect privacy and implement security consistency, but these tools alone are not enough to protect the systems, networks and data shares necessary for a national healthcare system. To be able to protect patient data and share medical records securely, other measures must be put in place. Fortunately, there is an entire field dedicated to protecting data and access to that data - information security.

Introduction to Information Security

The shift towards electronic health records creates a whole new problem of incredible proportions. How do you secure the most private of information, health data, for 300 million people? More difficult still, how does the industry instill confidence with a public, whose data is being collected and used? Without assurances by vendors, insurers and health care organizations, it may be difficult to win and keep the public trust. The solution will rely on a set of security principles that are the foundation for current solutions in other industries, such as banking or the airlines.

There are three pillars of information security; confidentiality, availability and integrity,[22] that are fundamental to protecting information technology solutions, including HIT. Security measures are instituted collectively to meet one or more of these primary goals, with the end result being one where confidentiality, availability and integrity are all covered. Confidentiality as it is termed here refers to the prevention of data loss, and is the category most easily identified with HIPAA privacy and security within healthcare environments. Usernames, passwords and encryption are common measures to ensure confidentiality. Availability translates to system and network accessibility and often focuses on power loss or network connectivity outages due to natural disasters or even man-made scenarios, such as a Denial of Service (DoS) attack or a malicious infection which brings down a network. Backup generators, continuity of operations planning, and peripheral network security equipment are used to maintain availability. Lastly, integrity covers the reliability and permanence of data, assurance that the lab results or personal medical history of a patient is not modifiable by unauthorized entities. Database security measures and data backups are implemented to prevent data manipulation, loss, or corruption - maintaining the integrity of patient data.

Access Security and Identifying Network Risks

The Architectures of Information Systems chapter briefly discusses and illustrates some of the measures put into place to maintain network security from point of entry to the end user. Although firewalls, intrusion detection and prevention systems and similar hardware devices currently work in conjunction with software solutions like antivirus, vulnerability scanners and host-based security systems, technologies like cloud computing and software as a service are changing the techniques and equipment used to secure networks. Most organizations will have staff dedicated to solving these issues, but a few security topics are more likely to be encountered and considered by end users.

Network and access security take on several connotations in a medical environment. One of the most prominent access controls are the physical security measures that are rigorously enforced in hospitals and clinics nationwide, where only authorized personnel have access to certain areas. While it is easy to identify a physical intrusion while a hulking figure strides unaccompanied down the wrong hallway, the same cannot necessarily be said of the attack occurring against the patient records database in the server room two floors down. In 2006, a 20 year old and several accomplices managed to use a bot-net (large number of hacker-controlled computers) to take over a portion of Seattle's Northwest Hospital network.[23] Although the attack was done to generate money through an online advertising scheme and did not compromise patient records, it did impact hospital operations.

Access controls pertain to a broad category of issues, but user-based controls are designed to prevent unauthorized access. Access security violations may bring to mind password cracking tools or something from the latest spy movie, but simple techniques such as shoulder surfing (watching as the password is entered) or social engineering (calling the help desk impersonating a distraught end-user) are successful techniques on gaining access.[22] Two access control issues that have been occurring at shared workstations and terminals for years are account sharing and workstations or terminals with usernames and password taped in prominent locations. These practices have to end to ensure that data and access to it remains guarded.

Additional methods of controlling access exist; there are role-based privileges (doctor, nurse, admin, etc) and rule-based privileges (read folder A, write folder B, read/write folder C) that limit what users can do following the technique known as least privilege, where users only have access to what they need to perform their job.[22] Another technique used is to constrain the interface so that it can execute only those functions that are needed to perform a task[22], but these may prove unpopular and to limiting. Beyond user name and password, there are a variety of newer techniques to identify users, from the smart cards like those employed by the Department of Defense, to biometric techniques employing voice recognition, fingerprint, palm, retinal, and even facial geometry scans.[22] If forced to use a password due to technical limitations, increasing the sophistication (through use of special characters, numbers and case) and length of the password in addition to occasionaly expiring them offers a reasonable level of protection.

Service Disruption and Data Breach

A recent paper by the Joint NEMA/COCIR/JIRA Security and Privacy Committee (SPC) identifies a number of known technical attack vectors for generic organizational networks and has applied medical scenarios to each, highlighting the systematic risk inherent within the healthcare community, particularly considering the types of information being collected and stored.[24] One of the most common assaults, the denial of service (DOS) attack, is designed to disrupt vital services and can be targeted or random in nature.[22] In an environment where so many interconnected devices are relied on for testing, monitoring and sustaining life-saving functions, the loss of these systems could have devastating consequences on a hospital.

Typical businesses experiencing downtime due to attacks lose revenue; medical facilities by comparison can lose lives. For example, during the attack on Seattle's Northwest Hospital network, the "doors to the operating room failed to open, pagers did not work, and computers in the intensive care unit were disrupted"[23]. The SPC paper offers recommendations such as "(1) Use additional staffing, (2) Use equipment available at other departments of the same healthcare facility, (3) Route patients to a nearby healthcare facility or (4) Deploy a secure wireless solution to communicate with the network backbone".[24] In the case of the Seattle hospital, they were able to employ backup systems to continue care. Had they not had another solution, the results could have been far worse.

Network risks beyond those designed to disrupt services are perhaps far more dangerous, as entities attempt to gain access to records and personal data held in the ever-enlarging medical databases around the country. HHS commissioned a study by Booz Allen Hamilton (BAH) that identified the incidence as well as prevention, detection and remediation of medical identity theft in the United States.[25] Although the study found that only 3% of identity theft victims had their information used for illegally acquiring medical care, that equates to 250,000 individuals, a situation that could increase in a poor economy or if higher healthcare costs make such options more desirable.[25] In an indication of how dire the situation has become, Federal Computer Week reported that organized groups of Chinese and Russian hackers have been increasingly targeting health records.[26] While the motivation is still open for debate and some hint at espionage and government backing, the use of stolen data to gain access to medical care or services is a possibility. Additionally, hackers from these countries are known to have run financial schemes where a computing system is held for ransom using software that locks data through encryption and prevents use until a fee is paid. The bounty could vary according to the size of the target and importance of the data. With large swaths of information at stake, this type of situation could bring healthcare to a standstill at hospitals and clinics, and destroy reputations and finances of both the organization and patient.

SAAS-Cloud vs. Client/Server Solutions

Recent changes in technology and product models have thrown an additional element into the mix for organizations to contend with - which type of solution to choose. The traditional practice management or electronic health record solution is based on software that runs on local network infrastructure and is delivered via a client terminal using terminal services or loaded on a workstation. Hospitals and practices maintain the system and equipment locally and worked with vendors for troubleshooting, software change requests, and upgrades. The latest contender is software as a service (SaaS), used to deliver the solution via a Web browser. Oftentimes SaaS solutions rely on another new technology, cloud computing, to store data and provide the back office computing power traditionally handled by servers and network storage devices. In this type of solution, the hospital contracts with a vendor to provide all of the services which are delivered to the end user.

Deciding which solution is appropriate for an organization is based on a number of factors that require careful deliberation and planning. Products like the Application Security Questionnaire (ASQ) from HIMMS can assist organizations performing their own research and planning for HIT solutions. The ASQ is a vendor-neutral, seven page capabilities checklist that hospitals, practices or medical organizations can request that software or services vendors complete for later comparative analysis of the various options being reviewed for selection.[27] Table 7.4 provides a comparison of the SaaS/Cloud and Client/Server models, indicating some of the advantages and disadvantages of each.[28, 29, 30, 31]

Table 7.4 Cloud versus client-server model

Feature or attribute	SaaS/Cloud	Client/Server
Integration with current systems	Web-based solution used with browser reduces client integration issues, but may have interoperability issues with other solutions in use.	Client software may have integration issues based on client configuration and may have interoperability issues with other solutions as well.
Software updates/upgrades	Software upgrades and updates are typically seamless, as they occur within the cloud before being delivered to the browser on the client end.	Software upgrades and updates require testing, may require downtime, and can be problematic if some systems are not available during the update window.
Costs	Infrastructure costs tend to be less than client/server, and SaaS solutions are less hardware dependent, but costs for bandwidth availability and service contracts can offset some of the savings.	Infrastructure costs associated with servers, storage, the solution product and support, in addition to life cycle costs of hardware and software that the solutions depend on for new features.
Reliability	Reliability is dependent of the product vendor and the quality and availability of the Internet connection to the provider.	Reliability is dependent of the product vendor and the capability of IT staff.
Availability	24/7 availability dependent upon Internet service	24/7 availability
Scalability	Easily scalable, but highly reliant on amount of bandwidth and signal latency, which serves as the performance bottleneck.	Scalability depends on capability of servers, storage and network infrastructure. Has less network latency to affect performance.
Security	Security in a cloud is still major sticking point, as data is on shared infrastructure and relies on virtual security methods and techniques.	Organization owns equipment and controls network security. Security dependent upon staff and defense measures.
Customization	Customization may be costly or limited due to support requirements in day-to-day operational environment.	Customization may be costly, but organization controls the implementation once complete.
Ownership	No ownership of solution, data is not located on site. Data may be difficult to obtain after contract ends, vendor is absorbed or goes bankrupt.	Organization owns data. Software is still usable in the event the vendor goes bankrupt.
Infrastructure	Requires no changes to infrastructure to support, unless additional bandwidth requirements dictate.	Requires more hardware; application servers and network storage. COOP solutions for redundancy require more equipment still.
Support	Support is almost entirely dependent upon vendor and service level agreement.	Support is dependent upon local IT staff and vendor when needed.

Recent Events - Meaningful Use and HIT Certification Rule

June and July have been met with anticipation by those awaiting the Meaningful Use and EHR Certification rules. The Federal Register published in June contained the temporary certification program for HIT and EHR applications in particular. The certification program addresses some of the comments related to the privacy and security certification criteria of both entire EHR systems and the individual modules submitted to certification. The Federal Register details the temporary certification program and specifics are presented in comment and answer format. These can be retrieved from the Federal Register at http://edocket.access.gpo.gov/2010/2010-14999.htm via GPO Access.

The Meaningful Use final rule does address security and privacy through use of certified EHR applications which comply with 45 CFR 170.302(d), which is in line with the previous Meaningful Use proposals. Liability issues surrounding security or privacy breach while using an EHR, including client or vendor responsibility, is still to be finalized in a later ruling. The proposed State Medicaid Health Information Technology Plan (SMHP) was made final, and includes a requirement for States to define a 5-year roadmap for HIT implementation. Part of the SMHP roadmap must include details of interstate data sharing, to include security.

The security rule was modified slightly for clarification as follows: "Conduct or review a security risk analysis per 45 CFR 164.308(a)(1) of the certified EHR technology, and implement security updates and correct identified security deficiencies as part of its risk management process".[32] Given the overall reduced compliance requirement for the initial phase, the security and privacy portion of Meaningful Use should be a smoother transition, as migrating to solutions incrementally provides less risk. Overall, the security objectives and measures proposed earlier in the year remained intact.

Key Points

- ARRA and the HITECH Act are designed to supplement the administrative, physical and technical safeguards implemented by HIPAA

- Despite hundreds of HIT products, fewer than 30 currently meet the 2011 certification requirements

- Data and records security will play a vital role in HIT success or failure

- Emerging technologies offer organizations a choice between traditional client-server or SaaS-Cloud product models

Conclusion

In conclusion, this chapter has sought to evaluate HIPAA and highlight its evolving impact on the future of HIT, the latest healthcare systems such as EHRs and continuing its intent in securing privacy data. The government continues to form and enforce healthcare privacy and security standards through the passage of the HITECH Act from the ARRA and the adoption of Meaningful Use standards. In addition to these administrative controls, information security concepts were introduced and both access control and network security concerns were identified. Although there will certainly be more questions and technical hurdles to face as EHRs gain prominence and medical data is collected on a national level, this section identified key topics of importance that should be considered for medical data privacy and security.

References

1. For Consumers. www.hhs.gov/ocr/privacy/hipaa/understanding/consumers/index.html (Accessed June 24 2010).
2. Touchet. http://ps.psychiatryonline.org/cgi/content/full/55/5/575 (Accessed June 24 2010).
3. HIPAA Privacy Rule and Its Impacts on Research. http://privacyruleandrearch.nih.gov/pr_08.asp (Accessed June 24 2010).
4. Summary of the HIPAA Privacy Rule. http://www.hhs.gov/ocr/privacy/hipaa/understanding/summary/index.html (Accessed June 24 2010).
5. Summary of the HIPAA Security Rule. http://www.hhs.gov/ocr/privacy/hipaa/understanding/srsummary.html (Accessed June 24 2010).
6. Widespread Data Breaches Uncovered by FTC Probe. http://www.ftc.gov/opa/2010/02/p2palert.shtm (Accessed June 24 2010).
7. Goedert J. http://www.healthdatamanagement.com/ Tenn. Blues Breach Affects 500,000. January 15, 2010 (Accessed January 20, 2010)
8. Goedert J. http://www.healthdatamanagement.com/ Health Net Sued for HIPAA Violations. January 14, 2010 (Accessed January 19 2010)
9. Ferris N. Foreign hackers seek to steal Americans' health records. http://fcw.com/articles/2008/01/17/foreign-hackers-seek-to-steal-americans-health-records.aspx January 17, 2008 (Accessed June 24 2010).
10. Protecting Your Information: Google Health and HIPAA. http://www.google.com/support/health/bin/answer.py?hl=en&answer=94610 (Accessed February 24 2010)
11. AMA - American Recovery and Reinvestment Act of 2009 (ARRA). http://www.ama-assn.org/ama/pub/advocacy/current-topics-advocacy/hr1-stimulus-summary.shtml (Accessed June 24 2010).
12. Meaningful Use Documents. http://healthit.hhs.gov/portal/server.pt?open=18&objID=888532&parentname=CommunityPage&parentid=5&mode=2&in_hi_userid=11113&cached=true (Accessed June 24 2010).
13. Frizzera C, Sebelius K. Proposed Rules. Federal Register 2010; 75.8: 1844-2011.
14. Health Information Technology: Can HIT Lower Costs and Improve Quality. http://www.rand.org/pubs/research_briefs/RB9136/index1.html August 14, 2009 (Accessed June 24, 2010).
15. Products with CCHIT Certified® Comprehensive 2011 certification http://www.cchit.org/products (Accessed June 24 2010)
16. Sullivan B. VA loses another computer with personal info. http://www.msnbc.msn.com/id/14232678/ August 8, 2006 (Accessed June 24 2010).
17. Centers for Medicare & Medicaid Services. http://www.cms.hhs.gov/SecurityStandard/Downloads/SecurityGuidanceforRemoteUseFinal122806rev.pdf (Accessed October 1 2009)
18. Google, Microsoft Say HIPAA Stimulus Rule Doesn't Apply to Them.

http://www.ihealthbeat.org/Articles/2009/4/8/Google-Microsoft-Say-HIPAA-Stimulus-Provision-Doesnt-Apply-to-Them.aspx (Accessed June 24 2010)

19. Google Health Terms of Service. http://www.google.com/intl/en-US/health/terms.html April 28, 2008 (Accessed February 24 2010)

20. Microsoft HealthVault Account Service Agreement. https://account.healthvault.com/help.aspx?topicid=ServiceAgreement (Accessed February 24 2010)

21. American Recovery and Reinvestment Act of 2009. http://frwebgate.access.gpo.gov/cgi-bin/getdoc.cgi?dbname=111_cong_bills&docid=f:h1enr.pdf (Accessed June 24 2010).

22. Tipton H, Henry K. Official (ISC)^2 Guide to the CISSP CBK. (ISC)^2 Press Series, 2007. Chicago: Auerbach.

23. SecurityFocus. http://www.securityfocus.com/brief/204 (Accessed June 24 2010).

24. Information Security Risk Management for Healthcare Systems. www.medicalimaging.org/policy/Information_Security_Risk_Management_Oct2007.pdf (Accessed September 29 2009).

25. Medical Identity Theft Environmental Scan. http://healthit.hhs.gov/portal/server.pt/gateway/PTARGS_0_10731_850701_0_0_18/HHS%20ONC%20MedID%20Theft_EnvScan_101008_Final%20COVER%20NOTE.pdf (Accessed June 24 2010).

26. Foreign hackers seek to steal Americans' health records. http://fcw.com/articles/2008/01/17/foreign-hackers-seek-to-steal-americans-health-records.aspx (Accessed June 24 2010).

27. HIMSS - Information Systems Security. http://www.himss.org/content/files/ApplicationSecurityv2.3.pdf (Accessed June 24 2010).

28. Private Clouds Take Shape. http://www.informationweek.com/news/services/business/showArticle.jhtml?articleID=209904474s/ApplicationSecurityv2.3.pdf (Accessed June 24 2010)

29. Software as a Service. http://www.himss.org/content/files/software_service_presentation091707.pdf (Accessed June 24 2010).

30. Federal CISOs worry they can't effectively secure cloud computing. http://searchsecurity.techtarget.com/news/article/0,289142,sid14_gci1511668_mem1,00.html (Accessed June 24 2010).

31. EMR and EHR – ASP Versus Client Server. http://www.emr-match.com/emr-complete-resource/emr-and-ehr-asp-versus-client-server/ (Accessed June 24 2010)

32. Meaningful Use Final Rule. www.ofr.gov/OFRUpload/OFRData/2010-17207_PI.pdf (Accessed July 16 201

8

Consumer Health Informatics

ROBERT E HOYT

Learning Objectives

After reading this chapter the reader should be able to:

- Identify the origin of consumer health informatics
- Discuss the role the Internet plays in patient education
- List the standard features of a patient web portal
- Compare and contrast the various types of personal health records and their projected benefits
- Describe the evolving role of secure e-mail and virtual visits as a new form of patient-physician communication and interaction

Consumer health informatics (also known as patient informatics) is a new aspect of medical informatics that largely reflects the empowered healthcare consumer. It was defined by the Agency for Healthcare Research and Quality as:

"any electronic tool, technology or electronic application that is designed to interact directly with consumers, with or without the presence of a healthcare professional that provides or uses individualized (personal) information and provides the consumer with individualized assistance to help the patient better manage their health or health care". [1]

Patients are aware that many non-healthcare businesses are automating and modernizing their business processes to attract a larger market share. ATM machines, as an example, can provide cash in a few minutes regardless where you are located worldwide. This innovation required re-engineering and the acceptance of universal standards, not unlike many aspects of health information technology. You do not have to look very far to find evidence of new patient-oriented healthcare:

- Walmart and other retailers are offering convenient in-store walk-in clinics with electronic care documentation and prescribing [2]
- Healthcare blogging through sites like Trusted.MD are becoming popular for patients. [3]
- An emergency room in Dallas, Texas has a self-service electronic check-in kiosk to expedite medical care [4]
- Dental practices can text message patients on their cell phones to remind them of upcoming appointments [5]

These consumer health informatics applications have occurred in just the past five to ten years. Unfortunately, little has been written in the medical literature about this topic so we must rely primarily on surveys and expert opinions. In table 8.1 results of a Harris Interactive survey of 2,624 adults show the informatics features desired by respondents and the percentages. Note that most of their desires are easily achievable technologically but unavailable in many healthcare systems. [6]

Table 8.1 Consumer desires and percentages

Desired informatics features by respondents	% Users
E-mail reminder of upcoming appointments	77%
Schedule medical appointments online	75%
Communicate with physician via e-mail	74%
Receive test results via e-mail	67%
Access to their electronic health record	64%
Home monitoring that would transmit information to physician's office	57%

In this chapter we will discuss four consumer health informatics sub-topics: the Internet for patient medical education, patient portals, personal health records and patient-physician e-mails and e-visits. It should become obvious after reading that these aspects are interrelated and not separate. In addition, many of these features are associated with electronic health records and health information organizations.

The Internet for Patient Medical Education

Multiple surveys have confirmed that the Internet is the premier medical resource for both patients and medical professionals. This has occurred in the incredibly short time span of just 10-15 years. According to a 2009 survey by the Pew Internet & American Life Project, entitled "*The Social Life of Health Information*" the following US trends were identified:

- 74% of adults go online
- 61% of adults look online for health information
- 57% of households have broadband connections
- 52% of online health inquires are for someone else
- Although social networking programs are popular they are rarely used for health-related issues
- 60% say online information affected their health decision making and 53% states it lead them to ask their physician new questions or seek a second opinion
- Highest percentage of "e-patients" fall into the 18-49 age range
- Percent of patients who look online is directly related to years of education and income
- The topics researched online are shown in Table 8.2 [7]

It should be pointed out that this is based on United States data as many countries are more "wired" than the US and many more are less connected.[8] Additionally, there is a significant digital divide between ethnic and racial groups in the United States. In a 2003 study by the National Center for Education Statistics, 54% of white students used the Internet at home as compared to about half that for Hispanic and African-American youth.[9]

In another survey by Opinion Research Corporation it was noted that the top three healthcare worries prompting Internet searches were: cost (41%), quality (25%) and medical errors (16).[10]

Recent events confirm that patients are becoming more discriminating in their choices of all aspects of healthcare. No longer do they automatically accept the opinion of their physicians. In a Harris poll it was shown that 57% of patients discussed their Internet search with their physician and 52% searched the Internet after talking to their physician.

Table 8.2 Health topics searched

Topics Searched	% Users who searched
Specific Disease	66
Medical Treatment	55
Diet and Nutrition	44
Exercise	52
Medication Issues	45
Alternative Medications	35
Insurance Companies	37
Depression	28
Doctor or Hospital	38

Eighty-nine percent felt their search was successful demonstrating confidence in the Internet as the new health library.[11] Excellent medical web sites exist but searches may yield low quality, non-evidence based answers, particularly when personal web sites are searched. As an example, in one study of Internet searching for the treatment of childhood diarrhea, 20% of searches failed to match the guidelines published by the American Academy of Pediatrics.[12]

Multiple reasons have been suggested for the increased use of the Internet in the healthcare arena:

- Healthcare is becoming more patient-centered in general. This has been promoted by the Institute of Medicine in all of their publications [13]
- Quality patient education web sites abound
- Most non-medical businesses are offering an online method to promote services
- The increasing use of medical blogs, podcasts and wikis as part of the Web 2.0 movement (the use of the Internet for free collaborative purposes) is very popular with patients [14]
- Patients are becoming impatient about finding the right answers, the best physicians, the best hospitals and the best medical care at the lowest cost
- Our population is "*graying*" with an estimated 78 million Baby Boomers (born 1946-1964) who are well educated and have more disposable income. This results in higher expectations
- Healthcare organizations are using information technology as a marketing tool to attract patients
- Computers are ubiquitous, as are broadband connections. Searching the Internet is now very fast and easy
- Patients receive less face time with their physicians than in years past, causing some patients to turn to the Internet for answers

Patient Education Web Sites

The following are only a sample of the many valuable patient education web sites available today.

WebMD. With more than 30 million people visiting this site monthly, it should be considered one of the true standard bearers. They have an extensive health library with top topics listed for men, women and children. Treatment and drug information is available, as is medical news. A symptom checker tool provides a patient with a simple differential diagnosis of what might be wrong with them based on their symptoms, age and gender. A daily e-mail newsletter is offered that can be customized to a patient's concerns. The only negative about this site is some commercial influence.[15]

Revolution Health. This free web site offers disease information, forums, CarePages, health calculators, a physician finder and symptom checker. Personal health records were discontinued in 2009. Members can rate physicians and hospitals, in addition to treatments and medications. They also offer an insurance marketplace to discuss and compare insurance options. Although these services are free, they also offer a fee-based membership that will allow a member to call and discuss a health concern 24 hours a day. There is limited commercial influence in the form of ads. [16]

MedlinePlus. This is the premier free patient education site developed by the National Library of Medicine and the National Institute of Health that links to the best and most respected web sites, such as the Mayo Clinic. MedlinePlus was ranked as the top information/news web site on the American Customer Satisfaction Index of federal government web sites.[17] In spite of its high marks, many patients and clinicians do not know about this site and many healthcare organizations pay for patient education content that could be obtained for free. Figure 8.1 shows the results of a search for obesity, showing the high quality references and the convenient folders on the left. Features of the web site include:

- 800+ health topics
- Drug information
- Health encyclopedia
- 165 tutorials
- Videos of surgical procedures
- Topics available in 40 languages
- Health check tools: quizzes, calculators and self-assessments
- Health dictionary
- Directories to locate physicians and hospitals
- Link to Clinical Trials.gov that can determine where experimental and new treatments are being studied in the USA
- Health news [18]

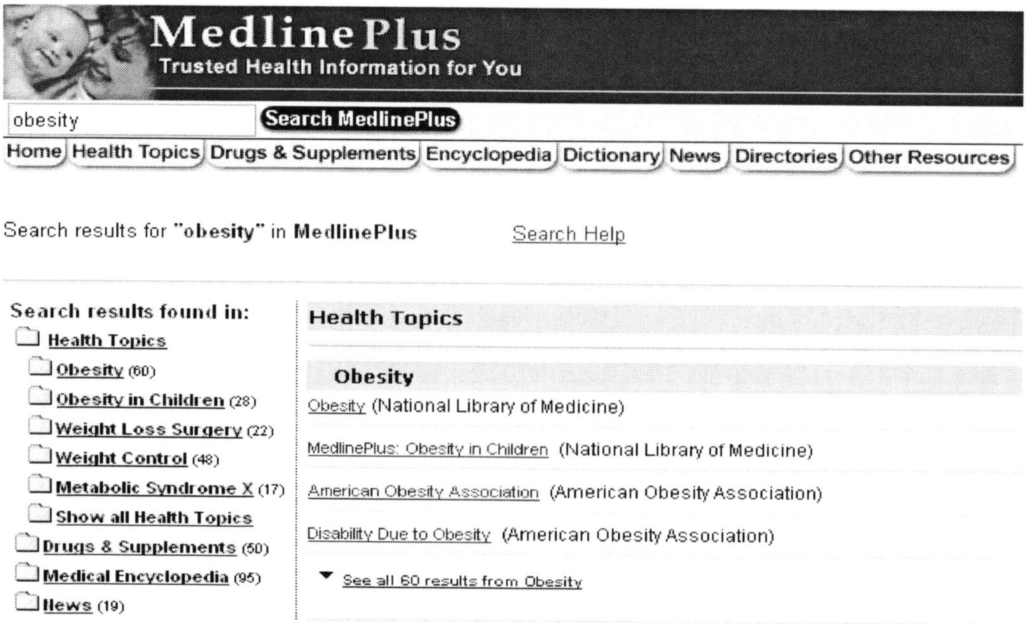

Figure 8.1 Search results for obesity (courtesy MedlinePlus)

Healthfinder. This government funded web site provides resources on a wide range of health topics selected from over 1600 government and non-profit organizations. It is coordinated by the Office of

Disease Prevention and Health Promotion and its health information referral service, the National Health Information Center. [19]

Everyday Health. This web site with more than 100 health centers provides information on the diagnosis, management and prevention of diseases and conditions, as well as on health lifestyles. It has an "ask the expert" question and answer section and multiple patient communities. It is part of Everyday Health, Inc. that has Everyday Health and multiple partnership web sites that also provide patient information and services. [20]

Healthwise. Multiple companies sell patient education for use on commercial medical web sites. Healthwise is a not-for-profit company that provides more than 6,000 medical topics in their knowledgebase. Other features include decision making tools, "take action tools" for chronic diseases and over 1,000 illustrations.[21]

UpToDate. This extremely popular physician education site also has a patient education site, aimed at college educated patients, unlike many sites that are aimed at high school educated patients. There is no charge for limited access to this site that covers more than 20 medical categories. For access to all articles posted, there is a charge. [22]

FamilyDoctor. The American Academy of Family Physicians sponsors this comprehensive free site. They cover all age groups as well as over the counter (OTC) drugs and a large library of health videos.[23]

Lab tests online. This free site allows for searching by test, disease condition or screening. The site is well organized with excellent resources for those seeking more information about clinical tests, why they are drawn, the results and what abnormal results mean. [24]

Patient Web Portals

Web portals are web-based programs that patients can access for health related services. A web portal can be a standalone program or it can be integrated with an electronic health record. Patient portals began as a web based entrance to a healthcare system for the purpose of learning about a hospital, healthcare system or physician's practice. They are clearly a marketing ploy to attract patients who are Internet savvy. Common features of patient portals are listed in Table 8.3.

Table 8.3 Web portal features and comments

Feature	Comments
Online registration	Allows patients to complete information before an office visit or hospitalization
Medication refills	Secure messages can be left for physicians to refill or renew medications, instead of telephone calls
Laboratory results	Patients can find results on recent tests as well as an explanation
Electronic visits	Portals exist that facilitate e-visits and the payment process
Patient education	Links to common educational sites
Personal health records (PHRs)	Allows patients and their families to create and update their PHR
Online appointments	Allows patients to see what appointments are available and when
Referrals	Patients can request referrals to specialists, e.g. OB-GYN
Secure messaging	More convenient than playing phone tag
Bill paying	Online payment using credit card is faster than "snail mail"
Document uploading	Several portals allow uploading of medical records to their site
Tracking function	Portal allows patients to upload diet, blood sugars, blood pressures, etc

Most patient portals offer multiple services, whereas others like TeleVox offer a specific service like lab results notification. This secure web based program, known as LabCalls™ enables patients to access a web site and obtain lab results. The nurse or doctor leaves the results along with a canned explanatory message. Patients can also receive a text message on their cell phone that lab results are ready. This program integrates with the practice management system or the EHR.[25]

A minority of web portals actually integrate with an EHR, which means that most patient data has to be manually inputted. In the future when EHRs become more widespread, selected patient lab results will automatically upload to the patient portal, thus saving time and money. Patients will also be able to access parts of their electronic records. A 2006 Harris Interactive study showed that 83% of patients wanted lab tests online and 69% wanted online charts to manage chronic conditions.[26]

Although several studies have shown patient interest in having access to lab results it remains to be seen if that would change consumer behavior or clinical outcomes. Are these patients primarily college educated and tech savvy? Do they desire results because physicians' offices are too slow to provide results? In a study at Beth Israel hospital, patients who accessed their portal PatientSite were younger and with fewer medical problems. They tended to access lab and x-ray results and use secure messaging more than a non-enrollee group.[27]

Figure 8.2 demonstrates the numerous methods patients now have to communicate with their physicians, besides face-to-face.

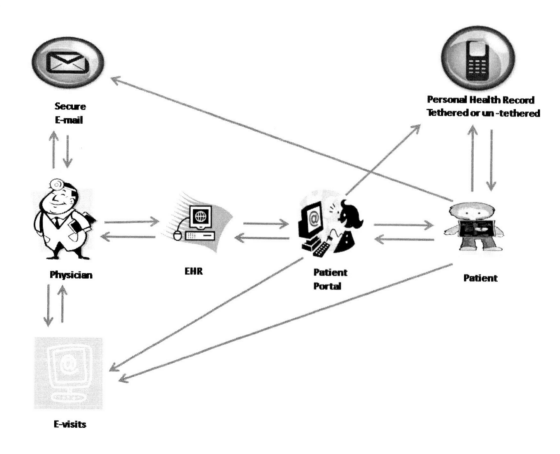

Figure 8.2 Patient, physician and chart interactions

In another survey by Connecting for Health, the following are the responses to "*I think that having my health information online would*" :

- Clarify doctor instructions (71%)
- Prevent medical mistakes (65%)
- Change the way patients managed their health (60%)
- Improve the quality of care (54%) [28]

Little is written about the benefits of patient web portals for the general consumer. McLeod Health System in Florence, South Carolina used online scheduling as part of the portal NexSched and was able to demonstrate fewer "*no shows*" and claims denials. They predicted a savings of about $1 million dollars yearly as a result of this program.[29]

Group Health Cooperative, a large mixed-model healthcare organization studied the effect of integrating its new comprehensive patient portal MyGroupHealth with its Epic electronic health record. As of December 2005, the highest monthly user-rates per 1,000 adult members were: test results, med refills, after-visit summaries and patient-provider messaging. A patient satisfaction survey revealed that the satisfaction rates were: 94% were satisfied overall, 96% for med refills, 93% for patient-provider messaging and 86% for test results. Although early use of the web portal was low there was a steady increase over time. Attrition rates were not reported.[30-31]

Robert Wood Johnson Foundation has awarded $2.45 million dollars to six organizations to study the effect of patient portals on chronic disease.[32]

Examples of Patient Web Portals

MySaintAls. This is the portal for Saint Alphonsus Medical Center located in Boise, Idaho. This comprehensive portal offers all of the standard features as well as the unique Patient Vault. They charge $10 monthly to upload (scan) and store patient records on their server. Lab results are accompanied by a separate program that explains the significance of the results and likely reduces the number of routine questions.[33]

Epic MyChart is a patient portal integrated with a well established electronic health record system. They also offer a standalone PHR known as "Lucy". MyChart functions: view test results, view and schedule appointments, pay bills, receive health maintenance reminders, view educational material, request refills, secure messaging, view child's record and manage care of elderly parents. An interactive demo is available on the web site.[34]

RelayHealth. This portal is owned by McKesson Corporation and is capable of being integrated with several EHRs. They have been very successful in getting insurance companies such as Aetna, Cigna and Blue Cross/BlueShield to utilize their platform as well as several well known HIOs. Their business model is to charge physicians a monthly fee to cover e-visits using their secure messaging site. They offer the following standard features:

- E-visits are part of the portal and the vendor handles eligibility, claims and collections
- They offer 100 interactive structured interviews to keep the e-visit focused and fact based
- Personal health records
- Access to lab results
- E-prescribing and refills
- Online appointments
- Ability to create a practice web site
- Colleague to colleague secure messaging

- Electronic data exchange to transmit lab, x-ray, discharge summaries, etc to physician offices or to the EHR
- The ability to store portal information on Microsoft's HealthVault
- Active notification of pending results is available [35]

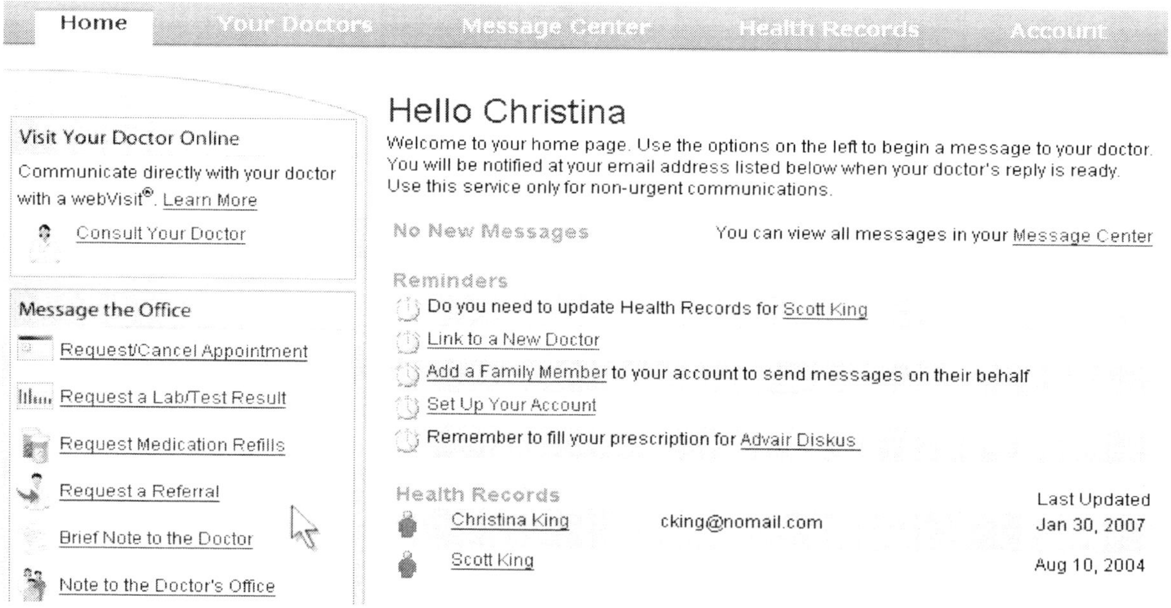

Figure 8.3 Welcome screen RelayHealth (courtesy RelayHealth)

MedFusion is a portal that offers many of the same features as RelayHealth with the additional unique feature of online bill payment. They were bought by Intuit in 2010. They claim to have a user-base of 4000 physician offices. They also have an extensive knowledge library of 6000 medical conditions to expedite an e-visit. A new area of involvement is the patient centered medical home model where they supply the technology to connect patient with physician. A free return on investment (ROI) calculator is available on their site. Additional features include:

- Front office solution to deal with patient registration, forms, appointments, check-in and patient messaging
- Back office solution for online bill payment, billing messaging and a virtual credit card payment system
- Clinical solution includes medication renewal and refills, secure messaging, personal health records, referral management, virtual visits, symptom assessment, laboratory results and reminders [36]

ReachMyDoctor. This site is aimed at improving communication with the doctor's office and offers two options:

- Free: schedule appointments, request medication refills, request a referral and address billing and insurance issues
- Subscription: for $8.95 monthly a patient can ask the physician non-urgent questions via secure e-mail. Physicians must be part of the network [37]

My HealtheVet. This portal integrates with the Veterans Health Administration's EHR (VistA) and offers lab results, wellness reminders, appointments, a personal health record (PHR), medication refills, patient education and online monitoring of: activity, food intake, oximetry, blood pressure, glucose and weight. In the 2010 time frame there are plans to include secure messaging, online appointments and lab results.[38]

Personal Health Records (PHRs)

PHR Definitions

According to the American Health Information Management Association (AHIMA) the personal health record (PHR) is:

> *"an electronic, universally available, lifelong resource of health information needed by individuals to make health decisions"* [39]

The National Alliance for Health Information Technology defines a PHR as follows:

> *"an individual's electronic record of health-related information that conforms to nationally recognized interoperability standards and that can be drawn from multiple sources, while being managed, shared controlled by the individual"* [40]

Introduction

The Institute of Medicine promotes PHRs by stating *"patients should have unfettered access to their own medical information"*. [41] The first principle endorsed by the Personal Health Technology Council is that *"individuals should be able to access their health and medical data conveniently and affordably"*. [42]

Interest in PHRs comes from multiple sources. In 2002 the Markle Foundation established *Connecting for Health*, a public-private collaboration to promote better information sharing between doctors and patients. In their July 2004 position paper they suggested: PHR development should be increased, PHRs will educate patients about their health and common data standards are a logical starting point. In one of their surveys 61% of respondents agreed that they should have access to their medical information *"anytime, any place"*. [43] A 2004 Harris Interactive survey of over 2,000 adults demonstrated that 42% kept personal or family health records but only 13% stored their records electronically. [44] Another survey conducted in 2004 by Bearing Point noted that over 50% of respondents would be interested in carrying their medical records in a portable device, accessible in an emergency. [45] The Centers for Medicare and Medicaid Services released a Request for Information (RFI) about PHRs in July 2005 to determine its future direction. They are well aware of the need for better patient information sharing and storage in older patients who are on multiple medications and have multiple physicians. [46] In 2005 attention was given to electronic health records and personal health records after hurricanes Katrina and Rita. As a result, Blue Cross/Blue Shield of Texas created personal health record-like summaries of care from insurance and pharmacy claims data, available to both physicians and patients. [47]

Ideal PHR Features

In spite of the fact that PHRs are new and are available in many formats, experts believe that PHRs should have the following features in order to be successfully adopted:
- Portable, i.e. information will transfer even when there is a job, insurer or clinician change
- Interoperable, i.e. standardized PHR format can be shared among disparate partners, such as the Continuity of Care Document (CCD)
- Autopopulated with clinical and test results that would be inputted automatically
- Controlled by the patient
- Longitudinal record and not just a snapshot
- Private and secure
- Integrated into the clinician's workflow and not be a separate process

The reality is that no organization has the ideal solution, with all of the above features.[48-50] In the following section we will discuss the current choices, organized by format.

PHR Formats

Tethered. The word tethered implies that the PHR is connected to one platform and not interoperable. The earliest and most common examples of this would be claims-based PHRs from insurers and healthcare organizations. Other examples would be PHRs tethered to an EHR or standalone patient portal. Payer-based PHRs have the advantages of being free to patients and easy to populate with claims data. They have the disadvantages of not being portable or interoperable and not controlled by patients. Moreover, claims data is usually several weeks old and indicates that a test was ordered but does not provide the actual results. Furthermore, the payer-based PHR is not longitudinal because it likely only covers patient encounters insured by their company.

In late 2006 America's Health Insurance Plans (AHIPs) and Blue Cross/Blue Shield Association announced a comprehensive plan to supply PHRs to their members. Importantly, the established core data standards determined that much of the information would come from claims and administrative data. With established data standards PHRs can be shared between different insurance companies, should a patient move or change coverage.[51-52] Aetna has offered its PHRs to 6 million of its 37 million members. They also offer their PHR to Medicare plan enrollees. Of interest, they will use *CareEngine,* a software rules engine that reviews the claims-based PHRs and gives personalized alerts (called *care considerations*) to patients and physicians about how to improve medical care.[53]

Untethered PHRs. Untethered PHRs imply they are not connected to one platform and there have more interoperability potential. PHR programs are available in multiple mobile and static platforms. There are more than fifty untethered personal health record products on the market, giving consumers many choices, but obvious limitations.[54]

- Web based. Most are commercial sites that are secure and can be accessed from a distant site. A minority of PHRs reside in patient portals that connect to an electronic health record system
- Mobile technology. Patient health information can be downloaded to:
  - Secure digital cards and USB drives: Most USB programs synchronize to a web based portal where patient information is also stored. Mobile technology offers several unique advantages. It is not dependent on the Internet for operation and is truly portable but not interoperable. In 2006 a new USB drive in the form and size of a credit card with memory up to 8 GB appeared and is now available as a PHR[55]
  - Smart Phones: Given the soaring popularity and expanding features of smartphones they may become the mobile storage of choice.[56] Blue Cross of Northeastern Pennsylvania will provide members secure electronic access to their medical records via a web enabled smartphone. The information will be derived from claims data and patient input. The program will use *MobiSecure Wallet and Vault* to access the Internet and authenticate the user.[57] Another example is MyRapidMD that downloads a client's emergency medical information to any Java enabled cell phone. Membership includes a wallet card, phone sticker, windshield sticker and key chain cards [58]
  - Smart Cards: The United States has been slow to use smart cards in the field of medicine, unlike countries like France and Germany. Most smart cards are used for patient authentication in the healthcare environment. Most smart cards can hold 64 KB-144KB of information (60 plus pages of single spaced text). These cards have a

small processor that can be programmed to do several tasks such as encryption and the cards are re-writable. Cards can be read by contact or be contactless, using radio frequency identification (RFID). They have the potential to speed up electronic claims submission by decreasing clerical errors. In 2007 Mt. Sinai Hospital began a pilot project and issued about 14,000 cards to patients in the New York City area. In 2008 they decided the cards should have open standards, such as the Continuity of Care Record (CCR), to record patient information on the card. This would make it interoperable with other entities. The card is updated every time it is placed in a card reader and autopopulates with demographics, lab results, etc.[59] Another issue with current healthcare cards is that they are not standardized and as such are not readable by all readers. Medical Group Management Association (MGMA) is promoting an industry-wide effort known as ProjectSwipeIT to standardize these cards, even though they are not truly smart cards.[60] For more details about smart cards in healthcare we refer you to this 2009 monograph [61]

- Personal Health Record Systems is an arbitrary term to indicate an untethered PHR that is interoperable. The following are the major examples of these systems:

  o **Google Health**: Google entered the PHR market in 2008 with a pilot project with the Cleveland Clinic and subsequently made their product available to the public in May 2008. Their product is known as Google Health and offers a very simple patient interface that includes a medical topics search engine, discussion boards, drug-drug interaction engine and a do-it-yourself PHR. You can opt to share your record with others (including physicians) and get an activity report. You can also print a wallet-sized PHR. More details about how developers and vendors can connect to Google Health using their application programming interface (API) can be found under Google Code. At this point they support the Continuity of Care Record (CCR). In 2009 Google announced that they will partner with IBM to design software so patients can upload device data (like blood sugar results) to their PHR.[62] Google maintains that there will be no advertising or selling of patient data. Once you create your personal profile you can link the profile to external services. As of June 2009 the following links were offered:

    - *Personal health services* offers connections to nine services. Each service requires you to register with a username and password and permission to link to your information. Each takes advantage of your profile. As an example, a link to TrialX.org provides you information about what investigational trials exist to match your medical history
    - *Import Medical Records* links to fourteen services; nine are pharmacy related, one is lab related and the remainders are healthcare organizations. These services allow you to import your records to Google Health. All services to convert paper records to Google Health are fee-based. One interesting service is Epocrates Patient Snapshot that allows an Epocrates participating physician (with the patient's permission) to access a patient's profile on Google Health. Unfortunately, there is no easy way for the primary care physician to upload data into patient's PHRs
    - *Find a Doctor* permits searching by name, specialty or location [63]

  o **Microsoft HealthVault**: Microsoft announced in 2007 that they would offer a free service known as HealthVault. It is not a PHR, but is instead a means to upload and store health information. You must register for other free or fee-based programs to upload information such as blood pressure or glucometer results. Microsoft has announced that it will release the source code of the HealthVault.NET Software Development Kit and the XML interfaces under the Open Specification Promise (OSP). This will enable third party developers to develop HealthVault compatible applications. The goal would be to interface with all electronic health records in the future. In mid-2008 Kaiser-Permanente began a PHR pilot project with its 156,000 employees. They will link their information with Microsoft's HealthVault using the Continuity of Care Document (CCD) standard (discussed in chapter on data standards). If this pilot project is successful, they will decide whether to offer it to their 8.7 million members.[64] The Cleveland Clinic will test HealthVault as well as Google Health. They plan to enroll 400 patients in a pilot study to test home devices to better manage chronic diseases. The PHR would then transfer information to the Cleveland Clinic's EHR Epic® system [65-66]

o Dossia: The system was founded by Applied Materials, BP America, Intel, Pitney Bowes and Walmart with data derived from insurers, pharmacies and physicians. The system known as Indivo is hosted by Childrens Hospital in Boston and consists of a free open source, open standards platform. Application programming interfaces (APIs) are available to developers for customization. Indivo handles both CCR and CCD documents. They plan to interface with EHRs, patients, physicians, researchers, health information organizations and public health services.[67] More details about this application were reported by Mandl et al. in the medical literature [68]

How are PHRs interoperable?

The following scenario is taken from the *HITSP Consumer Empowerment and Access to Clinical Information via Networks Interoperability Specification* monograph. A patient signs on to e.g. Google Health to establish his or her PHR. He/she also adds information on the spouse and states that the spouse and primary care physician (PCM) have permission to access his/her medical information. With the patient's permission Google Health establishes relationships with the patient's health information exchange (HIO), the primary care physician's EHR, the local drug store, the pharmacy benefits manager and any insurance company. When he arrives at his PCM's office, instead of filling out the standard paper registration forms, the office staff retrieves updates to their EHR by accessing his PHR via the HIO. Medications, allergies, etc are available to retrieve as well.[74] This implies data standards as well as document summaries such as the CCD are in place. This scenario also depends on ubiquitous EHRs and HIOs as well as mature information networks. It is likely that this conceptual model will take time and substantial resources to become a reality. Figure 8.4 demonstrates the potential interoperability of PHRs with the rest of the healthcare system. [69]

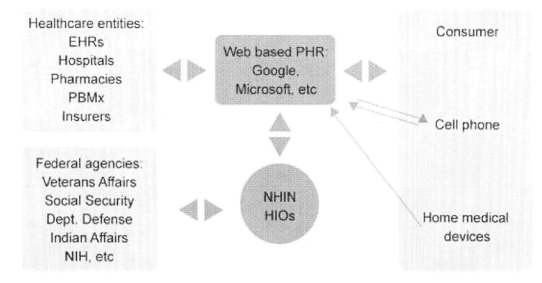

Figure 8.4 PHR Interoperability of PHRs with multiple federal and non-federal partners (PBM = pharmacy benefits manager)

PHR Research Questions

Kaelber et al published a very helpful article in December 2008 titled "A Research Agenda for Personal Health Records". They correctly point out that there is very little published about the impact of PHRs on patient behavior and outcomes. They believe research is needed to answer the following questions:

1. What PHR functionality is needed in the areas of data collection, sharing, exchange and self-management?
2. What is needed to improve adoption of PHRs by patients and clinicians? Research should focus on specific populations like the elderly, patients with chronic diseases, etc.
3. What is needed to ensure privacy and security?
4. What PHR architecture or model is likely to be most effective? Tethered? Untethered?
5. What is the business case for PHRs and are the incentives aligned for patients and physicians? [70]

Thus far, personal health records have been voluntary, placing the burden of downloading and maintaining health information on the patient. A busy physician's office is not likely to want this additional responsibility without reimbursement. The vision is to have records stored in repositories like Google Health in a format (XML) that is compatible with EHRs, HIOs, etc, but this will likely take years to accomplish, for those patients who are interested. As PHRs develop more user friendly features, perhaps the appeal to the average consumer will increase. Some PHRs, for example, will provide alerts such as "medications about to expire" or "upcoming medical appointments". An ideal business model for personal health records does not exist. Some studies suggest patients are willing to purchase their own PHRs if the price is low and others suggest insurance companies are the entity most likely to play a major role. Theft of personal health information (PHI) is a definite concern to the average patient and personal health records is just one more platform of concern.

At this time it appears that patients are lukewarm about personal health records, regardless of format or who picks up the bill. A 2006 survey of more than 11,000 American and Canadian consumers demonstrated that 22% of patients have not used the PHR feature offered by their insurance company and 53% had never accessed their health plan's web site. Thirty four percent interviewed did not trust the site's security and 29% did not see benefit from the concept of the PHR.[71] In a similar 2007 survey by Aetna, only 36% of those surveyed were familiar with PHRs and of those only 11% used one to track their medical history.[72] It will take time to determine the impact of PHRs and what incentives actually are important to patients and other members of the healthcare team. There are healthcare consultants who believe that PHR systems have a brighter future than HIOs for future data interoperability, but both suffer from inadequate business models and consumer enthusiasm.[52] A 2010 survey of 1849 adults showed that only 7% of those surveyed have used a personal health record, an increase from 2.7% in 2008. [73]

Other countries such as England are dabbling with the issue of PHRs for its citizens. They created HealthSpace in 2003 to store health notes, book appointments, store physiological parameters such as blood pressure and access National Health Service (NHS) contacts and health links. By late 2008 they allowed access to the NHS Summary Care Record that included allergies and medication histories.[74]

Several PHR research projects are worth mentioning:

- Medicare plans to determine whether PHRs change patient outcomes or save costs. The vendors selected are Google Health, Health Trio, NoMoreClipBoard and Passport MD. The one year program known as Medicare PHR Choice began in 2009 for patients located in Arizona and Utah. CMS will transfer up to 2 years of Medicare claims data upon request. Medicare will partner with HIP USA, Humana, Kaiser Permanente and the University of Pittsburgh Medical Center. Each plan will have a unique PHR that allows patients to access their own information. It is anticipated that the information will derive from hospital and physician claims data. Medicare will also pilot a PHR program in South Carolina in this same time frame using claims data [75-76]
- Robert Wood Johnson Foundation created Project HealthDesign to study PHRs that deal with chronic diseases such as breast cancer, diabetes, chronic pain, etc using nine research teams. The final report for round one is posted on their web site. Current projects include the monitoring of elderly activities with a PHR and a PHR for adults with asthma and depression and or anxiety [77]

Patient-Physician E-mail and E-visits

"Digital Rx: Take two aspirin and e-mail me in the morning"
New York Times 3/2/2005

Secure patient-physician e-mail communication

The vast majority of Americans today use e-mail, but few use it to communicate routinely with their physicians. [78] Manhattan Research reported in 2010 that a survey of 1900 US physicians showed that 39% of physicians used electronic messaging to communicate with their patients. According to this survey dermatologists and oncologists were the two specialties most likely to use this technology. [79] On the other hand, the 2009 National Health Interview Survey of over 7000 adult patients noted about 5% of adults had communicated with a clinician in the past 12 months. [80] Importantly, 41% of respondents (2001 Harris Interactive study) found it very frustrating to see a physician in person when they thought a telephone call or e-mail would suffice.[81] Physicians cite the following reasons for not using e-mail: liability (56%), poor compensation (45%), privacy issues (43%), staff not trained (30%) and the feeling that face-to-face visits are better (27%).[82]

Multiple benefits of e-mail communication have been pointed out. The communication is asynchronous so physicians can answer at their convenience and avoid *"phone tag"*. Overhead is lower for electronic messages compared to phone messages and they are self-documenting with a good audit trail that includes time stamps. Patients tend to lose less time from work for minor issues. It is preferable to communicate with patients using secure messaging through a web-based third party where authentication is validated; not true for routine e-mail. An October 2007 article in Pediatrics reported that physician-parent e-mail communication was 57% faster than telephonic communication and resulted in high consumer satisfaction. Forty percent of e-mails occurred afterhours but only amounted to 1-2 e-mails per day. It should be noted that the only physician involved in this study was a pediatric rheumatologist and not a generalist. Also, only one third of families offered e-mail communication took advantage of it. It is therefore difficult to generalize these results to the average physician's office.[83]

Potential disadvantages of e-mail messaging might include: indigent patients less likely to use the service; inability to examine the patient; potential for communication errors; possible slow responses; security issues and the potential to be overwhelmed.[84-85] Clinicians are likely to use secure messaging in the future if EHRs become widespread, there is clear evidence on the return on investment or there is a reimbursement strategy. There will need to be new guidelines to cover secure messaging. Cliff Rapp, a medical risk manager, recommends the following: develop clear cut policies, provide a disclosure statement, publish disclaimers about emergencies and privacy, use encryption, comply with HIPAA, designate a privacy officer and obtain confirmation of message delivery. [86]

Electronic visits

Electronic visits (e-visits or virtual visits) are an example of telehealth or telemedicine where medical care is delivered remotely (telemedicine is covered in much more detail in another chapter). Virtual visits are available as a continuum of care. (figure 8.5).

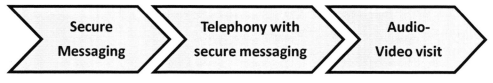

Figure 8.5 Remote patient communication continuum

The web-based choices for patient-physician communication include:
Secure messaging and templates to input a health concern and wait for a response by a clinician
Telephonic communication (audio) to communicate with clinicians along with secure messaging in a web application
Audio and video using standard web cam and secure messaging on a web application
Secure messaging: E-visits require secure messaging and not routine e-mail. In this section we are addressing a virtual visit and not just a simple e-mail communication. Virtual visits have the advantages of much better security and privacy and the ability to have a third party involved in the billing process. Patients and physicians must utilize a username and password to log onto a secure web site in order to conduct an e-visit. Numerous vendors such as RelayHealth and MedFusion provide the platform for e-visits in addition to their patient portal features. Some authorities feel the e-visits have a bright future. A Price Waterhouse study estimated that 20% of outpatient visits could be eliminated by using e-visits.[87]A new CPT code 0074T was developed specifically for e-visits.[88] Guidelines need to be established to define what constitutes an e-visit in order for insurance companies to reimburse for the electronic visit.

Several reports address how e-mail and e-visits might impact a physician's productivity.[89-91] A 2007 report from the Kaiser Permanente system suggested that e-mail communication decreased office visits by about 10%, compared to a control group. This would be good news for a health maintenance organization but bad news for fee-for-service practices, unless e-visits are reimburseable.[92] The consensus is that minor complaints can be dealt with more efficiently electronically, thereby allowing sicker patients to be seen in person. Furthermore, patients miss less time from work for minor issues. It has also been pointed out that if the patient provides a history during the e-visit and still has to be seen face to face, the physician has the advantage of knowing why the patient is there, therefore saving time.

In spite of the enthusiasm for e-mailing physicians, most patients are not willing to pay more than minimal co-pays for an e-visit.[93] In a study by RelayHealth they were able to demonstrate that, compared to controls, the patients who had e-visits had lower insurance claims. The profit more than paid for the $25 physician charge and the $0-10 patient co-pay. Importantly, 50% were less likely to miss work and 77% said it only took 10 minutes for the e-visit. Patient and physician satisfaction levels were good.[94] Pilot projects and studies are underway to evaluate e-visits. BlueCross/Blue Shield of Tennessee and other regions are reimbursing physicians for electronic visits.[95-98] The University of California at Davis Health System has been performing e-visits since 2001 and states that 80% of insurance companies in their region support the concept. Participating physicians seem to be more cost effective and physicians are receiving bonuses.[99] Cigna's HealthCare for Seniors program now covers electronic visits for older patients using the RelayHealth portal. Non-seniors must pay a fee for e-visits.[100] In January 2008, Cigna and Aetna announced that they would expand their virtual visit pilot programs to the rest of the nation. Cigna will pay $25-35 for an online consultation and the patient would pay their co-pay with a credit card via RelayHealth. Aetna will have a similar program that will include 30 specialties. These payers are also looking at discounts to physicians using RelayHealth as an incentive to offer virtual visits.[101]

A new free secure messaging service is available and known as HouseDoc. It would permit a virtual asynchronous visit, a request for medication refills, a request for appointments and test results. If the clinician charges the patient, the web service charges $2 and services are paid for by credit card.[102]

An excellent review of patient-provider communication can be found in a 2003 monograph by the First Consulting Group.[103]

Telephonic visits: The concept of virtual visits has spawned innovation in the delivery of healthcare. As an example, *TelaDoc* is a telephone-based consult service that is intended to supplement the care delivered by the primary care physician. This web-based application guarantees a clinician will return a phone call in 3 hours and the average charge is $35. They claim to have 1 million members and offer services 24/7. The clinician will prescribe and handle refills but not prescribe narcotics or order labs. Interestingly, they save the patient encounter as a Continuity of Care Record (CCR) that can be shared with others and accessed at the next visit.[104]

Audio-Video Televisits: Another innovative virtual visit service worth mentioning is *American Well*. Patients can interact with clinicians using web-based videoconferencing, as well as secure chat and telephonic communication. They are promoting 24/7 access for patients from home and aim to coordinate care with the primary care clinician (PCM) and insurance company. The service locates an appropriate clinician (including specialists), initiates a live audio-video conversation with a clinician and forwards the results to the PCM. For the clinician there is automatic claims submission and a per-consultation malpractice insurance coverage is offered. In addition, clinical practice guidelines are promoted for standardized care, known as *online care insight*. This vendor is promoting this application for the patient-centered medical home model. In April 2010, they introduced "Team Edition" with the goal of supplying on-demand specialty care as part of the team. Delta Airlines will make American Well services available to all employees and Rite-Aid will use the platform for in-store consultations with its pharmacist network. The approximate cost for an e-visit is $45.[105]

MDLiveCare is another telehealth initiative that provides real time virtual visits by secure messaging, audio or audio-video visits. Features include:

- Visits available 24/7 including for mental health
- Lab testing through Lab Corp
- E-prescribing with option to have drugs mailed to home
- Visit information can be shared with Google Health or Microsoft HealthVault
- Member card provides access to emergency medical information
- Physicians are encouraged to join to supplement practice incomes
- With membership costs of $9.95/month, a visit costs $39.95 and patients have unlimited free e-mail advice. A virtual visit by a non-member is $59.95 [106]

Little has been reported about the medical value of e-visits. A 2010 article did confirm that e-visits seem to be a successful alternative to standard care for the follow-up treatment of acne by dermatologists. The intervention group used RelayHealth, aided by digital photographs sent to the physician every six weeks. Patient and physician satisfaction was high. The intervention saved time for patients and was time neutral for Dermatologists. [107]

Key Points

- Healthcare consumers are becoming more sophisticated and more demanding

- For some healthcare processes patients would like to have the same convenience of an ATM machine

- Patients are using the Internet as the medical library of choice

- Patient web portals are now available that are standalone or integrated with electronic health records that offer a multitude of patient oriented services

- Everyone is talking about personal health records but it is unknown who will pay

- Secure patient-physician e-mail and e-visits have great potential to expedite acute care visits, once reimbursement becomes the standard

Conclusion

Many patients have cast their vote in favor of a more user-friendly healthcare system. They desire rapid access to medical answers via the Internet and rapid access to their healthcare system through their web portal. They would like to have storage of their personal health information somewhere but are reluctant to pay for it. Information technology giants like Microsoft and Google have entered the healthcare field with an uncertain long term effect. Lastly, patients want to communicate with their clinicians via secure messaging and if necessary, initiate an e-visit. It will take time to see if patients, clinicians and payers align to make this a reality. If patient satisfaction becomes associated with reimbursement or incentives then we can expect multiple consumer centric healthcare innovations to appear.

References

1. Gibbons MC, Wilson RF, Lehmann SL et al. Impact of Consumer Health Informatics Application. Executive Summary. Evidence report. No. 188. October 2009. www.ahrq.gov/clinic/tp/chiapptp.htm *(Accessed March 23 2010)*

2. Schmit J. Could walk-in retail clinics help slow rising health costs? USA Today August 28 2006. http://www.usatoday.com/money/industries/health/2006-08-24-walk-in-clinic-usat_x.htm. (Accessed December 29 2007)

3. Trusted.MD www.trusted.md. (Accessed December 29 2007)

4. Chest pain? Click here. Dallas ER goes self-service. MSNBC. http://www.msnbc.msn.com/id/20761116/ (Accessed December 29 2007)

5. Smile Reminder www.smilereminder.com (Accessed December 1 2007)

6. Harris Interactive. http://www.harrisinteractive.com/NEWS/allnewsbydate.asp?NewsID=1174PR Newswire. http://sev.prnewswire.com/health-care-hospitals/20060922/NYF08222092006-1.html (Accessed July 27 2010)

7. The Patient and the Internet. Pew Internet & American Life Project. May 2005 www.pewinternet.org   (Accessed January 5 2006)

8. Matthew Zook. Zooknic http://www.zooknic.com/ (Accessed September 23 2006)

9. Hayes, D. Midday Business Report. The Kansas City Star September 5 2006 (Accessed 20 September 2006)

10. Opinion Research Corporation—Siemens. Survey 2003. www.Informationtherapy.org/conf_mat05/hallppt.pdf (Accessed January 5 2006)

11. Harris Poll http://harrisinteractive.com/harris_poll/index.asp?PID=584 (Accessed January 10 2006)

12. McClung H. The Internet as a source for current patient information. Pediatrics 1998;101: p. e2

13. Institute of Medicine www.iom.edu (Accessed December 9 2007)

14. How Web 2.0 is changing medicine www.bmj.com/cgi/content/full/333/7582/1283 (Accessed March 15 2007)

15. WebMD www.webmd.com (Accessed March 15 2010)

16. Revolution Health www.revolutionhealth.com (Accessed February 15 2010)

17. American customer satisfaction index of fed govt web sites. December 2004. www.theacsi.org/index.php?option=com_content&task=view&id=27&Itemed=62 (Accessed January 5 2005)

18. MedlinePlus www.medlineplus.com (Accessed March 22 2010)

19. Healthfinder www.healthfinder.gov (Accessed March 22 2010)

20. Everday Health www.everydayhealth.com (Accessed August 8 2010)

21. Healthwise www.healthwise.org (Accessed March 22 2010)

22. UpToDate. www.uptodate.com/patients/index.html (Accessed March 22 2010)

23. FamilyDoctor.org www.familydoctor.org (Accessed March 22 2010)

24. Labtestsonline. www.labtestsonline.org (Accessed March 22 2010)

25. Televox. www.televox.com (Accessed March 22 2010)

26. First Health, HarrisInteractive. Consumer Benefits Health Survey. Executive Summary http://www.urac.org/savedfiles/URACConsumerIssueBrief.pdf (Accessed January 12 2006)

27. Weingart SN et al. Who Uses the Patient Internet Portal? The PatientSite Experience JAMIA 2006;13:91-95

28. Connecting for Health. Markle Foundation June 2003. www.connectingforhealth.org. (Accessed February 10 2006)

29. Egan C. Online Patient Scheduling Improves Time, Cost Efficiency www.ihealthbeat.org October 28 2004 (Accessed December 1 2004)

30. Group Health Patient Portal www.ghc.org/mygrouphealthpromos/onlinesvcs.jhtml (Accessed September 2 2008)
31. Ralston JD, Carrell D, Reid R et al. Patient Web Services Integrated with a Shared Medical Record: Patient Use and Satisfaction. JAMIA 2007;14:798-806
32. Broder C. Foundation Grants Fund Patient Web Portal, Disease Management Initiatives www.ihealthbeat.org October 4 2004 (Accessed December 1 2004)
33. Alphonsus Medical Center. www.MySaintAls.com (Accessed March 22 2010)
34. Epic. www.epic.com (Accessed March 22 2010)
35. RelayHealth www.relayhealth.com (Accessed March 22 2010)
36. MedFusion www.medfusion.net (Accessed March 22 2010)
37. ReachMyDoctor www.reachmydoctor.com (Accessed March 22 2010)
38. My HealtheVet http://www.myhealth.va.gov/ (Accessed March 22 2010)
39. American Health Information Management Association www.ahima.org (Accessed March 22 2010)
40. National Alliance for Health Information Technology. Defining Key Health Information Technology Terms April 28 2008 www.nahit.org (Accessed May 1 2009)
41. Crossing the Quality Chasm: A New Health System for the 21st Century Institute of Medicine 2001 The National Academies Press p. 8
42. Personal Health Technology Council www.markle.org (Accessed September 1 2006)
43. Connecting for Health. Working group on policies for sharing information between doctors and patients. July 2004 http://www.connectingforhealth.org/resources/wg_eis_final_report_0704.pdf (Accessed October 1 2005)
44. HarrisInteractive market research http://www.harrisinteractive.com/news/newsletters/healthnews/HI_HealthCareNews2004Vol4_Iss13.pdf (Accessed October 1 2005)
45. Press Release SanDisk www.sandisk.com/pressrelease/20050214b.htm (Accessed October 5 2005)
46. Centers for Medicare and Medicaid Services www.cms.hhs.gov and http://www.gcn.com/vol1_no1/health_IT/36422-1.html (Accessed November 1 2005)
47. Insurer makes electronic patient summaries available for hurricane evacuees http://www.healthcareitnews.com/NewsArticleView.aspx?ContentID=3710 (Accessed December 1 2005)
48. Terry K. Will PHRs rule the waves or roll out with the tide? Hospitals & Health Networks. www.hhnmag.com (Accessed September 2 2009)
49. Grossman JM, Zayas-Caban T, Kemper N. Information Gap: Can Health Insurer Personal Health Records Meet Patients' and Physicians" Needs? Health Affairs 2009;28(2):377-389
50. Kuraitis V. Birth Announcement: The Personal Health Information Network (PHIN). E-CareManagement Blog March 8 2008. http://e-caremanagement.com (Accessed July 30 2008)
51. Insurers to Provide Portable, Interoperable PHRs. www.ihealthbeat.org December 14 2006. (Accessed December 14 2006)
52. Industry Leaders Announce Personal Health Record Model; Collaborate with Consumers to Speed Adoption. http://bcbshealthissues.com December 13 2006 (Accessed December 14 2006)
53. Aetna Broadens PHR Availability. August 28 2008. www.healthdatamanagement.com (Accessed August 0 2008)

54. Personal Health Records in the Marketplace http://library.ahima.org/xpedio/groups/public/documents/ahima/pub_bok1_027459.html (October 20 2005)

55. Walletex www.walletex.com (Accessed March 22 2010)

56. Worldwide smartphone market soars in Q3. http://www.geekzone.co.nz (Accessed January 2 2006)

57. Diversinet's new mobisecure wallet and vault enable anytime, anywhere delivery of mobile services. January 2007. http://www.diversinet.com/doc/DVNT-Vault%20and%20Wallet-Jan29.pdf (Accessed January 10 2007)

58. MyRapidMD www.myrapidmd.com (Accessed July 5 2009)

59. Messmer E. Mt. Sinai Medical Center looks to open standards for patient smartcards. August 28 2008 (Accessed August 29 2008)

60. Project SwipeIT. www.swipeit.org (Accessed March 5 2009)

61. A Healthcare CFO's Guide to Smart Card Technology and Applications Smart Card Alliance February 2009. www.smartcardalliance.org (Accessed February 28 2009)

62. Google Partners with IBM on Online Personal Health Record Service. February 5 2009. (Accessed February 5 2009)

63. Google Health www.google.com/health (Accessed March 30 2010)

64. Lohr S. Kaiser Backs Microsoft Patient Data Plan. June 10 2008. The New York Times

65. Microsoft HealthVault. www.healthvault.com (Accessed March 30 2010)

66. Cleveland Clinic Tests Microsoft's HealthVault PHR System November 10 2008 www.ihealthbeat.org (Accessed November 10 2008)

67. Indivo. http://indivohealth.org (Accessed March 26 2010)

68. Mandl KD, Simons WW, Crawford WCR et al. Indivo: a personally controlled health record for health information exchange and communication. www.biomedcentral.com BMC Medical Informatics and Decision Making 2007;7:25

69. HITSP Consumer Empowerment and Access to Clinical Information via Networks Interoperability Specification www.hitsp.org (Accessed June 10 2008)

70. Kaelber DC, Jha AK, Johnston D et al. A Research Agenda for Personal Health Records JAMIA 2008;15:729-736

71. Survey: Consumers Have Concerns About Insurer-Provided PHRs. January 31 2007. www.ihealthbeat.org. (Accessed January 31 2007)

72. Aetna. www.aetna.com July 17 2007. (Accessed December 1 2007)

73. Urdem T. Consumers and Health Information Technology: A National Survey. April 2010. www.chcf.org (Accessed April 10 2010)

74. Pagliari C, Detmer D, Singleton P. Potential of electronic personal health records. BMJ 2007;335: 330-333

75. CMS and the Defense Department Pilot Projects Could Jump Start PHR Use. March 2009 www.hhnmag.com p.9

76. Centers for Medicare/Medicaid http://www.cms.hhs.gov/perhealthrecords/ (Accessed June 17 2009)

77. Project HealthDesign www.projecthealthdesign.org (Accessed March 30 2010)

78. Slack WV. A 67 Year Old Man Who e-mails his Physician JAMA 2004;292:2255-2261

79. Physicians in 2012: The Outlook on Health Information Technology. Manhattan Research. 2010. www.manhattanresearch.com (Accessed March 23 2010)

80. National Health Interview Survey. January-June 2009. Reported February 2010. NCHS. CDC. www.cdc.gov (Accessed March 23 2010)

81. Study Reveals Big Potential for the Internet to Improve Doctor-Patient Relations. Harris Interactive 2001 www.harrisinteractive.com (Accessed September 24 2006)

82. WSJ examines physician's reluctance to e-mail patients. www.ihealthbeat.org June 3 2003 (Accessed October 2004)

83. Rosen, P, Kwoh, CK. Patient-Physician E-mail: An Opportunity to Transform Pediatric Health Delivery. Pediatrics 2007; 120(4): 701-706

84. Car J, Sheikh A. E-mail consultations in health care: scope and effectiveness. BMJ 2004;329:435-438

85. The Changing Face of Ambulatory Medicine—reimbursing physicians for computer based care. American College of Physicians Medical Service Committee Policy Paper March 2003 www.acponline.org/ppvl/policies/e000920.pdf (Accessed March 15 2003)

86. Rapp C. Liability Issues Associated with Electronic Physician-Patient Communication. Internat Ped March 2007;22(1)

87. Healthcast 2010: Smaller world, bigger expectations. Price Waterhouse Cooper. November 1999 www.pwc.com (Accessed February 3 2006)

88. Broder C. What's in a code? www.ihealthbeat.org January 14 2004 (Accessed January 14 2004)

89. Liederman EM .Web Messaging: A new tool for patient-physician communication JAMIA 2003;10:260-270

90. Chen-Tan Lin .An Internet Based patient-provider communication system: randomized controlled trial JMIR 2005;7 (4): e47

91. Leong SL .Enhancing doctor-patient communication using e-mail: a pilot study J Am Board of Fam Med 2005;18:180-8

92. Zhou YY et al. Patient Access to an Electronic Health Record with Secure Messaging: Impact on Primary Care Utilization. Am J Manag Care 2007;13:418-424.

93. Juniper Research October 2003 www.juniperresearch.com (Accessed December 10 2005)

94. The RelayHealth web visit study: Final Report www.relayhealth.com (Accessed December 2 2007)

95. Tennessee Hospital Pilots two e-mail programs www.ihealthbeat.org March 14 2005 (Accessed March 14 2005)

96. Blue Shield of California web communications pilot to enroll 1,000 physicians www.ihealthbeat.org April 20 2004 (Accessed April 20 2004)

97. Health Plan to pay Doctors for web visits www.ihealthbeat.org May 24 2004 (Accessed June 30 2004)

98. Microsoft Pilot Project to Test Online Physician Visits www.ihealthbeat.org January 10 2006 (Accessed February 2 2006)

99. UC Davis Virtual Care Study. Eric Liederman. Presented at AMDIS/HIMSS 2004

100. Cigna Offers Seniors Free Online Health Services. April 11 2007. www.ihealthbeat.org. (Accessed April 12 2007)

101. Lowes R. Aetna and Cigna want to pay you for online visits. Medical Economics. January 25 2008. www.memag.com (Accessed January 29 2008)

102. HouseDoc. http://housedoc.us (Accessed March 30 2010)

103. Online Patient-Provider Communication Tools: An Overview. First Consulting Group. November 2003. California HealthCare Foundation www.chcf.org (Accessed September 20 2006)

104. TelaDoc www.teladoc.com (Accessed March 30 2010)

105. American Well www.americanwell.com (Accessed June 15 2010)

106. MDLiveCare http://mdlivecare.com (Accessed March 30 2010)

107. Watson AJ, Bergman H, Williams CM et al. A Randomized Trial to Evaluate the Efficacy of Online Follow-up Visits in the Management of Acne. Arch Derm 2010;146(4):406-411

Online Medical Resources

JANE A. PELLIGRINO
ROBERT E. HOYT

Learning Objectives

After reading this chapter the reader should be able to:

- State the challenges of staying current for the average clinician
- Describe the characteristics of an ideal educational resource
- Describe the evolution from the classic textbook based library to the online digital library
- Compare and contrast the different formats of digital libraries
- Describe the future of digital resources integrated with electronic health records
- Describe emerging Web 2.0 technologies in medicine
- Identify the most commonly used free and commercial online libraries

"The complexity of modern American medicine is
exceeding the capacity of the unaided human mind"

David Eddy

Trying to keep up with the latest developments in medicine is very difficult, primarily due to the accelerated publication of medical information and the significant time constraints placed on busy clinicians. It is likely that clinicians are in fact so busy that they have no idea what new educational resources are available to them. They would like to move from the "information jungle" to the "information highway" but who will show them the way? This chapter is devoted to those clinicians who are seeking rapid retrieval of high quality medical information.

What are the challenges clinicians face today?

- **Educational**. More than 460,000 articles are added to Medline yearly.[1] The 2006 Physicians Desk Reference (PDR) is over thirty four hundred pages long making it exceedingly cumbersome to search for drug information.[2] It seems obvious that this is a disincentive to search for drug information and therefore is a patient safety issue. Standard medical textbooks are expensive and out of date shortly after publication. In addition, some argue that the descriptions of diseases are not always updated or evidence based.[3] Moreover, Shaneyfelt estimated that a general internist would need to read 20 articles every day just to maintain present knowledge.[4] There is a transition that is taking place in medical education where emphasis is no longer on developing physicians that know everything, but rather, practitioners who can find and use information when it is needed. Physicians find it difficult not to think of themselves as experts and it often shakes their confidence, but a new confidence can be found in knowing how to locate needed information

- **Diffusion of information.** Recommendations from specialty organizations take time to trickle down to the generalists. There is no standard way to disseminate information that is either reliable or particularly effective. National guidelines, usually written by specialists, face the same challenges. Once there is a new standard of care for a disease such as diabetes, how do you get the word out, particularly to small or remote medical practices?
- **Translational.** Studies have shown that it may take up to ten or more years for research to be "translated" to the exam room (e.g., thrombolytics). [5] In a study by Antman, experts were also slow to make recommendations in textbooks even though high quality evidence was published many years prior.[6] On the other hand, many physicians are skeptical and wait for confirmatory studies. If they have been in practice for many years, they may have witnessed the pendulum sweep back and forth regarding, for example, the use of post menopausal estrogens. Recent studies often contradict older studies due in part to better study design and larger subject populations [7]
- **Evolutionary.** We can no longer teach "classic medicine," because diseases and their presentations change over time as demonstrated by new presentations for infectious diseases. Rocky Mountain Spotted Fever began to disappear as Lyme disease began to appear. Additionally, diseases were detected at more advanced stages in the older literature, because lab tests were lacking, making clinical presentations more dramatic. Currently we tend to diagnose diseases earlier, before the patient has advanced signs and symptoms due to better and earlier tests. Medical resources therefore must reflect new evidence
- **Retention.** According to many studies there is an inverse relationship between current knowledge and the year of graduation from medical school. Ramsey compared board scores of Internists and the number of years elapsed since certification and demonstrated this inverse relationship [8]

How often do we actually have patient related questions and how often do we find answers?

- Covell reported that on average internal medicine physicians had two questions for every three patients seen and found the answers for only 30% [9]
- A study by Ely showed that Family Medicine physicians had 3.2 questions per 10 patients seen. The answer was pursued in only 36% of cases [10]
- In a primary care survey Gorman noted 56% of physicians pursued answers where they thought an answer existed and 50% of answers dealt with an urgent issue. Most physicians turned to other physicians for answers and not the traditional medical library. Lack of time was the universal reason not to pursue answers in most studies [11]
- In another study by Ely, the most common questions dealt with drugs, Ob-Gyn and adult infectious disease issues. Answers to 64% of questions were not pursued and physicians averaged less than two minutes per search. The most common resources used were books and colleagues and only two physicians performed literature searches.[12] It is important to point out that all of the above studies evaluated primary care physicians, so the needs of other physicians such as surgeons are less clear. Also, after these studies were published software programs such as Epocrates appeared and significantly changed how we seek drug information

What is the state of medical libraries today?

In the article, "Quiet in the Library," Lee speaks of the quiet created as a consequence of physicians no longer needing to go to the library to do research. Although today's medical libraries provide timely, pertinent and authoritative knowledge-based information in support of patient care, education and research, a significant percentage of the journal and textbook literature has migrated from print to online during the last decade. Physicians can research their clinical questions from their desktops without going to the library. [13-14] Burrows in a paper reviewing electronic journal use at the Louis Calder Library at the University of Miami School of Medicine reports an 88% decrease in the use of print journals in the period from 1995-2004. [15] Libraries have moved their collections online and re-designed stack space into study areas, computer workstations and collaborative areas.

How have we evolved from the traditional library to online resources?

Within a very short time the Internet has become the educational resource of choice due to the speed of retrieval and depth of information. A 2001 study by the American Medical Association showed that 75% of physician practices had Internet access and 79% used it to research answers. Three out of ten medical practices had their own website. [16] These statistics continue to rise as does the availability of broadband access. Patients using the Internet as an online library has closely mirrored the pattern of physicians. The Internet now hosts more than 3 billion web sites. As an indication of growth, a Google search for the words "medical education" in 1995 by one of the authors yielded 760 results, [17] whereas a search in 2010 yielded about 126 million citations. Although the 21st century searcher is at the center of a virtual library, he or she must cope with the quantity of easily retrieved information and be capable of evaluating that information for reliability, currency and authority.

Before the advent of the Internet physicians used print resources for verifying factual information related to patient care, sought insights from their colleagues on difficult cases and performed library research on exceptional cases themselves or with the aid of medical librarians. The nature of physician information-seeking has not changed substantially in the last decade, but mode of access to resources has changed dramatically. To provide the best care for their patients, physicians still need to check drug information, differential diagnosis tools, current textbooks, or the journal literature, but instead of heading to the library they turn to the Internet for answers. The resources available online are far more extensive than the personal libraries or hospital libraries that physicians used in a print world.

Reference materials in the twenty-first century have been migrated to online formats. Drug information compendia, laboratory references and textbooks have been converted to electronic formats, although only the major references and texts are available online. Two types of online journals have emerged: the electronic version of the print journal and the born-electronic journal. Either type can be open access (free to all users) or available by subscription only. Recently publishers are experimenting with hybrid journals that offer their most important content online, while still publishing print issues. Although many predict the demise of the print journal, the transition phase may last another decade. Today medical library collections are a mosaic of print and online content, but the mission of the medical library remains the same. Medical libraries today provide more extensive offerings than the print collections found in most hospital libraries a decade ago, but these resources are expensive and strictly controlled by site licenses. To have access to the most authoritative information a well-informed physician still needs to have an affiliation with a medical library or be willing to read premium medical content on a subscription or pay per view basis.

Journal indexes were the pioneers of research online. The Medline database that today has over 16 million citations from over 5,000 journals searchable online began in 1966 as an electronic archive of citations to the medical literature that was only searchable by highly trained librarians.

When MEDLINE introduced end-user searching in the pre-Internet era citations previously accessed manually through the *Index Medicus* were available directly to clinicians at their desks. Although end-user searching revolutionized access to the journal literature, it was limited to titles and abstracts. With the advent of the Internet came online journals and the ability to link the full-text journal articles to Medline citations. Despite these improvements Medline searching still is not an easy path to high quality, quick answers to clinical questions. Several studies have shown that finding an answer is difficult and takes too much time for a busy clinician.[18-19] A pertinent abstract might be located, but it requires additional time to obtain the full-text whether it be from a free or fee-based online source or through the medical library. A Medline search should be reserved for rare medical problems, research, writing a paper or creating a clinical practice guideline (CPG). Medline will be discussed in more detail in the chapter on search engines.

In 1994 Shaughnessy stated that the usefulness of medical information is equal to the relevance times validity divided by the amount of work to access it.[20] A 2004 study in the journal *Pediatrics* comparing retrieval of information from online versus paper resources showed it took eight minutes for an answer via an online resource as compared to twenty minutes using traditional paper based resources.[21] There is little doubt about the tremendous potential of online resources for speed of access, but the quest to find the precise, authoritative answer to a clinical question within the limitations of a patient visit remains elusive. Turning to resources of known quality appeared to be an efficient choice, so converting traditional resources to online formats was the logical first step.

Harrison's Online (the online version of *Harrison's Principles of Internal Medicine*, 17th edition) and Scientific American Medicine (now known as ACP Medicine) were among the first online full-text resources. The online versions of these popular textbooks are continually updated and are accessible from anywhere. Many libraries offer online access to these textbooks and individuals may purchase subscriptions to the online versions at about the same cost as copies of the print textbooks. Recent online versions offer a variety of subscription options and offer their content through several portals. The print edition of *Harrison's Principles of Internal Medicine*, 17th edition, published in 2008, offers supplementary material on DVD and is using RSS feeds and podcasts to disseminate its updates. Although these textbooks make valuable expert knowledge easily accessible, their main drawback is that they tend to cover only the basics about any subject and therefore lack depth. In spite of the fact that they have a search engine, like a standard textbook a reader may have to review multiple book chapters to find the answer.

More comprehensive aggregated resources followed the advent of online textbooks. MDConsult, Medscape, StatRef, and OVID were created to offer multiple resources such as books and journal articles, patient education materials, medical calculators and medical news in one product. Searches of these, otherwise excellent, resources yield multiple references to the full-text of various documents that must be analyzed to find the answer to the clinical question. You might have to read twenty book or journal pages to finally find the answer. This is not optimal if you are seeking an answer while the patient is still in the exam room. Ideal medical resources are those that are:

- Evidence based with references and level of evidence (explained in the chapter on evidence based medicine)
- Updated frequently
- Simple to access with a single sign-on
- Available at the point of care
- Capable of being embedded into an electronic health record
- Likely to produce an answer with only a few clicks
- Useful for primary care physicians <u>and</u> specialists
- Written and organized with the end user in mind

According to Richard Smith the "best information sources provide relevant, valid material that can be accessed quickly and with minimal effort".[22] The need for a synthesized resource that can easily provide evidence-based answers to questions during the patient visit has given rise to several excellent, focused resources, often referred to as point of care resources or bedside information products. UpToDate, eMedicine, DynaMed, ACP-PIER and FirstConsult present their content, so that clinicians can answer clinical questions with current, comprehensive and rapid retrieval. These products focus on patient-oriented information and differ in the number of topics covered, the way the evidence is documented and the organization of the material. UpToDate, Essential Evidence Plus (formerly InfoRetriever), ACP Pier, Diseasedex, DynaMed and First Consult have been very well received, and clinicians develop their preferences among these offerings based on user interfaces and the ability of the database to answer questions. In an evaluation of five bedside information products, Campbell and Ash took a user-centered, task-oriented approach to testing the ability of these products to answer clinical questions. The study rated UpToDate the highest in ease of interaction, screen layout and overall satisfaction and found that users were able to answer significantly more questions quickly with UpToDate [23]; however, other researchers have found that users have preferred resources such as ACP Pier and Essential Evidence Plus, because of the way evidence levels are documented. [24] A 2004 study showed that 85% of medical students easily transitioned from traditional resources to primarily online medical resources (UpToDate and MDConsult). [25] In a report published in 2005, internal medicine residents were able to find answers 89% of the time and the information changed patient management 78% of the time. The most common resources accessed were UpToDate and PubMed. [26]

Use of the point-of-care tools discussed above begins with a diagnosis. In a controversial article in BMJ Tan and Ng[27] reported that Google could function as a useful diagnostic aid. Others would argue that specially designed tools such as the new generation of clinical decision support systems are more appropriately designed to improve medical diagnosis and reduce diagnosis errors by directing physicians to the correct diagnosis. Although clinical decision support systems have been around for years the new generation of tools as exemplified by *Isabel* (Isabel Healthcare) has been shown to suggest the correct diagnosis in approximately 96% of adult patients when tested with 50 consecutive internal medicine case records published in the New England Journal of Medicine.[28] Tools such as *Isabel* by assisting the physician in making the diagnosis provides an entry point into the literature and can link clinicians to resources such as UpToDate, PubMed and more in order to obtain information in depth on the case at hand.[29]

Several medical resource vendors are in the process of making the leap towards having the resource embedded into electronic health records. Examples would include iConsult, Dynamed, UpToDate and ACP-PIER to mention a few.

Figure 9.1 demonstrates the evolution from the traditional library to the online library and integrated libraries into electronic health records.

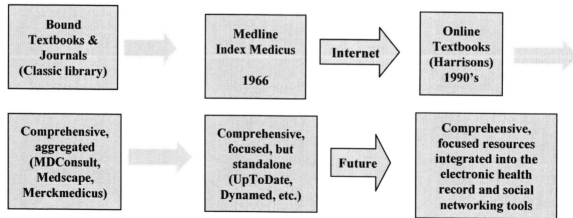

Figure 9.1. The evolution from traditional to online medical libraries

What new tools are available to help you stay abreast in the ever-expanding online library?

Many of the medical information resources described above mirror print resources and are written and designed to be read as questions arise. Lee mentions in his article "Quiet in the Library"[13] that "the flood of new information and the demands of simply getting through the day have become so overwhelming that many physicians no longer find the time for 'lifelong learning' through such activities as reading journals or attending grand rounds." To keep their medical practice current physicians need new tools. Interactive web technologies, known as Web 2.0, have emerged that allow knowledge sharing among users and allow customized content to be distributed to interested users. These tools can be harnessed to help physicians learn about the new developments that will improve their practice. [30-31]

Weblogs or blogs, websites that build content through dated entries, have generated large repositories of focused medical content. Web 2.0 technologies hold the promise of an enriched learning environment enabled by collaboration on a large scale. Blogs such as *Kevin MD, Clinical Cases and Images* have developed large readership. Through the dialog between bloggers and readers physicians are discovering new ways to learn. British Medical Journal has launched a collaborative site called doc2doc to foster peer-to-peer communication. [32]

Wikis (the name comes from the Hawaiian word meaning quick) allows authors to collaborate to create peer-reviewed content that is written and edited by participating users. The most famous general example is Wikipedia. Several medical references have been developed using wiki architecture. WikiDoc[33] and Ganfyd[34] were created by physicians and are being developed by participating health professionals. WikiDoc, a collaborative online textbook, boasts over 70,000 textbook chapters and continuously updated medical news. Ganfyd, "a free medical knowledge base that anyone can read and any registered medical practitioner may edit," only allows credentialed individuals to provide content.

Keeping up with changes to important websites can be challenging. Really Simple Syndication (RSS) can make that process manageable. RSS is a format that delivers changes from multiple websites to one place. RSS allows physicians to request content from various websites and read that content in single place, known as an aggregator. Subscribe to an aggregator such as MedWorm or Bloglines, the look for the RSS icon (🔲) at a website. Subscribe to RSS feeds of interest and read them with the aggregator. Physicians are using RSS to receive textbook and website updates, Pub Med search results, journal tables of contents and medical news.

Audiocasts (podcasts) and videocasts offer educational programs and updates in multi-media formats. New England Journal of Medicine now offers weekly article summaries via podcasts and Johns Hopkins University offers weekly health news podcasts. To subscribe to multiple podcasts you will need to subscribe to an audio aggregator, known as a podcatcher. Choosing a podcatcher can be confusing. Podcatcher Matrix will assist you in choosing a podcatcher that is compatible with your operating system and mobile device. [35]

Sponsored Medical Web Sites

Multiple excellent web sites are available that are either sponsored or fee-based. Most of the sites discussed in this section have multiple features that continue to improve. Medical education traditionally has been based on reading journals or textbooks, but can also involve the presentation of interesting and unique cases. A thorough discussion of this alternative approach appeared in the February 2007 Mayo Clinic Proceedings.[36]

Medscape

- An all purpose medical web site

- Covers 30+ medical specialties as well as sections for nurses, medical students and pharmacists
- Over 150 Resource Centers
- Provides updates, continuing medical education (CME), conference schedules, Medline, drug searches and multiple specialty articles and an eclectic selection of journal abstracts
- Weekly newsletters and updates (MedPulse) and Best Evidence; both are features unique to Medscape
- Drug and Device Digest providing the latest in alerts and approvals; helpful for patient safety concerns
- A free personal web site option
- Dermatology atlas
- Clinical practice guidelines and Cochrane connection
- Sponsored by advertising [37]

MerckMedicus

- Multipurpose site sponsored by Merck and Company, customizable for 20 specialties
- 60+ specialty textbooks
- 150+ full text journals
- Cochrane Reviews links
- Clinical podcasts
- Includes customized versions of MDConsult and OVID, DxPlain (differential diagnosis engine from Harvard), medical news and national meeting reports
- PDR Electronic Library
- Patient handouts
- Unique 3-D Atlas of the human body
- Professional development using CME, board reviews, medical meetings, medical school links and Braunwald's Atlas of Internal Medicine (1500 slides you can copy). Also, a slide image bank of other slides that can be copied
- PDA portal, formatted for use with Palm or Pocket PC, includes news, the Merck Manual, Pocket Guide to Diagnostic Tests, and TheraDoc antibiotic assistant for PDA
- Journal abstracts and the ability to do a Medline search (If your PDA is not connected to wireless Internet, searches will be done the next time you synchronize with your PC.)
- Sponsored by Merck and Company and state health professional license required for full access [38]

Amedeo

- This service will search major medical journals for a topic you select and then e-mail the results to you every week. It offers weekly webpage alerts displaying abstracts of selected journal tables of contents linked to PubMed
- Covers about 100 topics falling into 25 specialties
- Valuable if you are a subject expert and don't have the time to do a frequent journal search on your own
- Related websites include Free Books 4 Doctors www.freebooks4doctors.com and Free Medical Journals www.freejournals4doctors.com
- Similar tracking of articles also available through Google Alerts and NCBI (Pubmed)
- Self-supported non-profit site [39]

Resources Available in Sponsored and Non-Sponsored Versions

E-medicine

- 6,000 articles by 10,000 authors covering primary care and multiple sub-specialties
- Owned by WebMD and incorporated into Medscape
- Continually updated and peer reviewed Clinical Knowledge Base
- Articles are referenced and selectively cross-referenced
- References are presented at the end of each article with links to PubMed, but no footnotes in the text body
- Levels of evidence not given
- CME available
- Sponsored and institutional version available
- Institutional version, available at (http://www.imedicine.com) offers PDA downloads, an extensive image library and monographs in pdf
- Sponsored version and subscription versions available [40]

Online Epocrates

- Online Epocrates was the obvious next step after the successful PDA software program (see chapter on mobile technology)
- Both a free and fee based online version (up to $199 yearly) is offered
- Program covers 3300 drugs and 400 alternative medications
- Fee-based program includes local formulary information, pill identifier, MEDCALC 3000, alternative medications as well and an extensive drug library
- Free online program includes pill pictures and patient education
- Epocrates Linx is an online version that can integrate with an EHR
- The features online Epocrates offers over the PDA version: Ability to print or e-mail results, Medline search capability, pill pictures, MedCalc 3000 calculations and patient education sheets in English and Spanish
- Sponsored version and subscription versions available [41]

Government Medical Web Sites

National Library of Medicine

- **PubMed** (discussed in search engine chapter)
  - Provides free access to MEDLINE, NLM's database of citations and abstracts in the fields of medicine, nursing, dentistry, veterinary medicine, health care systems, and preclinical sciences
  - Links to many sites providing full text articles and other related resources
  - Provides a Clinical Queries search filters feature, as well as a Special Queries feature, which have recently combined in one interface
  - Links to related articles for a selected citation [42]

- **NLM Gateway**
  - Provide "one-stop shopping" for many of NLM's information resources
  - Offer citations, full text, video, audio, and images
  - Link within and across NLM databases [43]

- **Toxnet**
  - Cluster of databases covering toxicology, hazardous chemicals, environmental
  - Health and related areas [44]

National Guidelines Clearinghouse

- Comprehensive searchable database of evidence-based clinical practice guidelines and related documents
- Structured abstracts (summaries) about the guideline and its development
- Links to full-text guidelines, where available, and/or ordering information for print copies
- Palm-based PDA downloads of the complete NGC Summary for all guidelines
- Guideline comparison utility for a side to side comparison of multiple guidelines [45]

MedlinePlus

- Premier online patient education site
- Important to have in exam room
- Service of the National Library of Medicine and the National Institute of Health (NIH)
- Covers over 750 Health Topics in English and Spanish
- Drugs, Supplements, and Herbal Information
- Medical dictionary, encyclopedia and news
- 165 interactive video tutorials and surgical procedure videos
- Links to major patient education sites offered by health clinics, government and advocacy organizations such as, Mayo Clinic, National Institute of Health (NIH), American Heart Association, etc
- Links to Clinical Trials.gov to search research centers for specific diseases [46]

Other Excellent Free Patient Education Sites

- http://www.familydoctor.org/
- http://www.mayoclinic.com
- http://www.webmd.com
- http://kidshealth.org/

Free Medical Web Sites

HighWire Press

- Free site created by Stanford University to produce online peer-reviewed journals and scholarly content as open access or pay per view depending on the title
- Hosts 1270 journals with over four million full text and 2 million free full text articles
- Capability to search HighWire and Medline at same time with access to both free and pay-for-view articles
- E-mail alerts, PDA channels and RSS feeds available
- Site hosts 37 free trials of journals, 43 free journals, 249 journals that offer back issues free and approximately 1000 pay-for-view journals [47]

Medical Algorithms

- Developed by the Institute for Algorithmic Medicine, a non-profit organization that develops online medical algorithms
- Currently includes 13,500 scales, tools and assessments
- Algorithms are evidenced-based with multiple references
- Many algorithms are presented as an Excel spreadsheet so you can plug in actual patient numbers and get immediate results
- Covers many unusual calculations not found in MedCalc and other similar programs [48]

Medical Podcasts

- Several medical organizations offer podcasts for medical education; mostly in audio format with some in video
- The American College of Cardiology posts "Heart Sounds" in a MP3 format as a download
- The Arizona Heart Institute and Hospital provides podcasts as part of the Cardiovascular Multimedia Information Network
- The Journal of the American Medical Association, New England Journal of Medicine and other journals now offer audio article summaries as podcasts
- Medical school library websites offer links to podcasts from a variety of journals
- CME providers are expanding their use of podcasts [49]

Subscription (fee based) Resources

MicroMedex

- Micromedex offers multiple drug databases of unbiased drug information searchable with a single query
- New interface organizes the database into a point-of care tool
- Databases include Poisondex (toxicology), Diseasedex (Disease database), Lab advisor (laboratory information), DrugDex (drug interactions) , ReproRisk (human reproductive toxicology), CareNotes (patient education handouts in English and Spanish)
- Fully referenced drug database
- Handheld is available on all major platforms
- Unlike Epocrates it has:
  - Both renal and liver failure dosing
  - Drug-food interactions
  - Off label uses
  - Comparative efficacy
  - IV Compatability
  - Toxicology
  - Extensive references [50]

Lexi-Comp

- Comprehensive database of unbiased drug information
- Core pharmaceutical information includes population specific dosing, indication-specific dosing, IV Compatibility, drug identification, drug interactions, toxicology and more
- Diseases and disorders via Harrison's Practice
- Laboratory and diagnostic medicine

- Formulary information
- Specific modules available for medicine, dentistry and oral surgery
- Patient handouts available in 18 languages.
- Handheld version includes the five most requested databases and Harrison's disease database
- Handheld is available on all major platforms including the Ipad [50]

OVID

- Several hundred textbooks in most specialties including drug references
- Approximately two thousand full text medical journals
- Access is to journal articles is available by institutional subscription or pay per view
- Search interface supports natural language and Boolean searching
- Medline search capability linked to online full-text of journal articles
- Cochrane Library is available under the title Evidence-based Medicine Reviews
- Supports searching multiple databases simultaneously, i.e, Cochrane and Medline [52]

UpToDate

- Comprehensive resource containing over 97,000 pages of original, peer-reviewed text embedded with graphics and links to Medline abstracts
- Available online, on CD-ROMs or downloadable to handhelds
- Individual, educational and institutional subscriptions available
- Personal subscribers receive CME researching clinical questions
- Institutional subscribers may purchase online or single workstation CD-ROMs
- Emphasizes Internal Medicine, Women's Health and Pediatrics, but also covers Emergency Medicine, Neurology and Allergy/Immunology
- Logically organized for fast answers
- 4,400 authors review 440 journals
- Began grading recommendations for treatment and screening in 2006 and continues to expand that effort
- Continuously updated with about 40% of the content being edited each quarter
- Drug database includes drug-drug interactions
- Patient information topics in English
- Available for download to smartphones, Pocket PC or Palm
- Integrated into GE Centricity EHR [53]

MDConsult

- 60+ textbooks
- Over 50 full text journals
- 35 Clinics of North America
- Comprehensive drug database
- 1000 clinical practice guidelines
- 2500 Patient education handouts
- 50,000 medical images
- Online CME and medical news
- Medline search capability
- MDC Mobile is the PDA portal

- Excellent Search engine for entire site
- Individual and institutional subscriptions available [54]

StatRef

- Offers about 200 textbooks and Medline online in a cross-searchable reference tool that includes textbooks and evidence-based resources.
- ACP PIER, Journal Club & AHFS DI® Essentials™
- MedCalc3000
- PDA portal will allow downloads of ACP PIER content only
- Institutional subscriptions available [55]

Essential Evidence Plus (formerly InfoRetriever/Infopoems) is a program that was created by physicians for physicians. POEMS are "patient oriented evidence that matters". Specifically, this means the authors look for articles that are highly pertinent to patient care and patient outcomes.

- Consists of two products: DailyPOEMS and InfoRetriever
- DailyPOEMs are e-mailed to the subscriber M-F and are distilled from 100+ journals with only 1/40 accepted
- Site has 2000 POEMS
- POEM of the Week podcasts (RSS feeds available)
- Essential Evidence Plus (formerly InfoRetriever) available in online or handheld versions searches multiple resources simultaneously
- Essential Evidence Plus tools: EBM guidelines (1,000 primary care practice guidelines, 3000 evidence summaries and 1,000 photographs and images), Daily POEMS, Cochrane abstracts (2,193), Selected practice guidelines (751), Clinical decision rules (231)
- Number Needed to Treat (NNT) tool
- Derm Expert (photographic atlas)
- Diagnosis calculators (1180)
- History and physical exam calculators (1282)
- 5 Minute Clinical Consultant
- ICD-9 and E&M lookup tool
- Drug of Choice tool
- Searching results in a summary of resources on that topic categorized into typical quick reference categories like diagnosis, treatment, prognosis, etc. 5 Minute Clinical Consult monographs are listed first
- Available for desktop, Palm or Pocket PC
- Individual and institutional subscriptions available [56]

ACP Medicine

- Publication of the American College of Physicians and Web MD
- Previously known as *Scientific American Medicine*
- Evidence-based and peer-reviewed
- Covers most subspecialties plus Psychiatry, Women's Health, Dermatology and Interdisciplinary medicine
- Available in binder, CD-ROMs and Online
- Up to 120 hours CME available
- Binder version is 2800 pages

- Articles are dated and references are footnoted with PubMed links to the abstract
- Monthly updates (free) to be added to chapters
- Handheld point-of-care tool, *Best DX/Best Rx*
- Individual and institutional subscriptions available [57]

ACP PIER

- Organized into five topic types: diseases, screening and prevention, complementary and alternative medicine, ethical and legal issues and procedures
- Each of the 430 disease modules presents guidance statements and practice recommendations, supported by evidence of evidence
- PDA version available
- Drug resource, accessible from every module page
- Provides the medical resource content for Allscript's EHR
- What they cover they do well.
- Like an online textbook; updated frequently
- Disease modules continue to be added
- Available directly from the ACP and through Stat!Ref by individual or institutional subscription [58]

FirstConsult

- Synthesizes evidence from journals and other sources into one database
- Offers concise, readable summaries of evidence that relate to patient care
- Organized into medical topics, differential diagnoses and procedures
- Updated weekly; major releases quarterly
- 475 topics at this point
- 300 Patient education files in English and Spanish
- Procedure files and videos
- EHR ready
- New handheld version coming in 2011
- Lack of a drug database and limited topics are negatives
- Individual and institutional subscriptions available [59]

DynaMed

- Disease and condition reference
- Almost 3000 clinical topics commonly seen in primary care
- Peer-reviewed and continually updated.
- Information presented based on validity, relevance and convenience
- All topics are organized in the same categories such as, general information, causes and risk factors, complications and associated conditions, history, physical, diagnosis, prognosis and treatment
- Bottom line recommendations are presented first, along with level of evidence. Links to articles will take you to the full text article if available and free online. Other links take you to PubMed where some are linked through medical libraries to full text articles
- Weekly e-mail of important articles; also available as podcast
- Handheld version available on popular platforms and is free with subscription
- Can be linked to an EHR with the EBSCOhost Integration Toolkit
- Individual and institutional subscriptions available [60]

Table 9.1 is a matrix that compares many of the features of the online resources just covered. The speed of retrieval is an approximate estimate of how much time it takes to find an answer to a common medical question.

Table 9.1 Online resource comparison matrix. Speed (Slow =1, Fast = 4)

Source	Medline	CME	Books	Journals	Drugs	News Letters Updates (e-mail)	Expert Opinion	Patient Info	Speed (1-4)
Medscape	X	X	X	X	X	X	X	X	3
MerckMedicus	X	X	X	X	X		X	X	3
OVID	X		X	X					2
UpToDate		X			X		X	X	4
MDConsult	X	X	X	X	X		X	X	2
FirstConsult							X	X	4
StatRef	X		X						2
ACP Medicine		X				Updates to site only	X		3
eMedicine	X	X	X	X	X		X	X	3
DynaMed					X	X	X	X	4

Key Points

- Clinicians are overwhelmed by the amount of new information and the lack of time

- We have shifted from traditional print textbooks in our medical libraries to online libraries

- Multiple resources exist that are both free and fee-based to serve as rapid high-quality references

- Ideal medical resources should be easy to access and fast to retrieve the most current information

Conclusion

Online resources are becoming the medical library of choice for healthcare workers due to depth of content and speed of retrieval. Furthermore, subject matter can be updated more rapidly compared to standard textbooks. Many excellent resources are free and the subscription resources are competitive with traditional textbooks. Resources vary from a low of about 400 topics to a high of 8000 topics. Prices tend to correlate with the scope of the content offered. There are many free resources that should be considered by all clinicians such as Epocrates Online, MedlinePlus and Medscape. The authors want to stress that very extensive resources such as UpToDate, eMedicine and DynaMed offer the greatest possibility of finding an answer in a few clicks. Other resources may point you to multiple book chapters and journal articles where you must sift through the data to find the answer. Clinicians are strongly encouraged to "test drive" these resources, adopt the ones that make the most sense and add them as desktop icons in each exam room.

References

1. Medline Fact Sheet http://www.nlm.nih.gov/pubs/factsheets/medline.html (Accessed 2 July 2010)
2. Physician Desk Reference http://www.pdrbookstore.com/H (Accessed 3 July 2010)
3. Richardson WS, Wilson MC Textbook descriptions of disease—where's the beef? ACP Journal Club July/August 2002: A-11-12
4. Shaneyfelt T. Building bridges to quality. JAMA 2001;286: 2600-01
5. Contopoulos-Ioannidis DG, Ntzani E, Ioannidis JP. Translation of highly promising basic science research into clinical applications. Am J Med. 2003 Apr 15; 114(6):477-84.
6. Antman EM et al. A comparison of results of meta analyses of randomized control trials and recommendations of clinical experts JAMA 1992; 268: 240-248
7. Ioannidis JP.A Contradicted and Initially Stronger Effects in Highly Cited Clinical Research JAMA 2005;294: 219-228
8. Ramsey PG et al. Changes over time in the knowledge base of practicing internists JAMA 1991; 266: 1103-1107
9. Covell, DG, Umann GC, Manning PR. Information needs in office practice: are they being met? Ann Int Med 1985; 103: 596-599
10. Ely J, Osheroff J, Ebel M et al. Analysis of questions asked by family doctors regarding patient care BMJ 1999; 319: 358-361
11. Gorman PN, Helfand M. How physicians Choose Which Clinical Questions to Pursue and Which to Leave Unanswered. Med Decision Making 1995; 15: 113-119
12. Ely JW et al. Analysis of questions asked by Family doctors regarding patient care BMJ 1999;319:358-361
13. Lee T. Quiet in the Library NEJM 2005;352: 1068-70
14. Lindberg DAB, Humphreys BL. 2015-The Future of Medical Libraries NEJM 2005;352:1067-1070
15. Burrows S. A review of electronic journal acquisition, management and use in health sciences libraries JM:A 2006; 94(1):67-74
16. Technology usage in physician practice management AMA survey Dec 2001
17. Miccioli G. Researching Medical literature on the Internet---2008 Update www.llrx.com/features/medical2008.htm (Accessed 3 July 2010)
18. Gorman PN. Can primary care physicians' questions be answered using the medical journal

literature? Bull Med Libr Assoc 1994; 82:140-146

19. Chambliss ML. Answering clinical questions J Fam Pract 1996; 43:140-144.
20. Shaughessy A, Slawson D, Bennett J. Becoming an Information Master: A guidebook to the Medical Information Jungle J Fam Pract 1994; 39: 489-499
21. D'Alessandro DM, Kreiter CD and Petersen MW. An Evaluation of Information Seeking Behaviors of General Pediatricians Pediatrics 2004; 113: 64-69
22. Smith R. What Clinical Information do Doctors Need? BMJ 1996; 313: 1062-1068
23. Campbell R, Ash J. An evaluation of five bedside information products using a user-centered, task-oriented approach JMLA 2006; 94(4): 435-441
24. Trumble JM et al. A Systematic Evaluation of Evidence-Based Medicine Tools for Point-of-Care, presented at South Central Chapter, Medical Library Association Meeting, October, 2006, http://ils.mdacc.tmc.edu/papers.html (Accessed 3 July 2010)
25. Peterson MW et al. Medical student's use of information resources: is the digital age dawning? Acad Med 2004;79:89-95
26. Schilling LM et al. Residents' patient specific clinical questions: opportunities for evidence based learning Acad Med 2005; 80: 51-56
27. Tan H, Ng JHK. Googling for a diagnosis – use of Google as a d diagnostic aid: internet based study, BMJ 2006; 333: 1143-5
28. Graber ML. Performance of a Web-Based Clinical Diagnosis Support System for Internists, 2007; 23(Suppl 1) 37-40
29. Johnson C. What physicians don't know, Medicine on the Net 2008; 14(1): 1-5
30. Giustini D. How Web 2.0 is changing medicine, BMJ 2006; 333: 1283-4
31. Liesegang TJ. Web 2.0,Library 2.0, Physician 2.0, Am J Ophthalmology 2007; 114 (10):1801-3
32. http://doc2doc.bmj.org (Accessed 25 July 2010)
33. http://www.wikidoc.org (Accessed 25 July 2010)
34. http://ganfyd.org (Accessed 25 July 2010)
35. Podcatcher Matrix http://www.podcachermatrix.org/ (Accessed 3 July 2010)
36. Pappas G, Falagas ME Free Internal Medicine Case-based education through the World Wide Web: How, Where and With What? Mayo Clin Proc 2007;82(2):203-207
37. Medscape http://www.medscape.comH (Accessed 3 July 2010)
38. Merck Medicus http://www.merckmedicus.comH (Accessed 3 July 2010)
39. Amedeo http://www.amedeo.com (Accessed 4 July 2010)
40. E-Medicine http://www.emedicine.comH (Accessed 3 July 2010)
41. Epocrates http://www.epocrates.comH (Accessed 11 July 2009)
42. PubMed Fact Sheet http://www.nlm.nih.gov/pubs/factsheets/pubmed.html (Accessed 4 July 2010)
43. NLM Gateway Fact Sheet http://www.nlm.nih.gov/pubs/factsheets/gateway.html (Accessed 4 July 2010)
44. Toxnet http://toxnet.nlm.nih.gov (Accessed 4 July 2010)
45. National Guidelines Clearinghouse http://guidelines.gov (Accessed 4 July 2010)
46. MedlinePlus http://medlineplus.govH (Accessed 4 July 2010)
47. High Wire Press http://highwire.stanford.edu/ (Accessed 4 July 2010)
48. Medical Algorithms. www.medal.org (Accessed 4 July 2010)
49. Collaborate CME Using Web 2.0 Technologies. Almanac. Alliance for CME, August 2008. http://www.med.upenn.edu/cme/Almanc%20ACME_08%20(2).pdf (Accessed 25 July 2010)
50. Micromedex http://www.thomsonhc.com/(Accessed 4 July 2010)
51. Lexi-Comp http://www.lexi.com (Accessed 25 July 2010)

52. OVID http://gateway.ovid.com (Accessed 2 July 2010)
53. UpToDate http://www.uptodate.comH (Accessed 25 July 2010)
54. MDConsult http://www.mdconsult.comH (Accessed 4 July 2010)
55. StatRef http://www.statref.comH (Accessed 2 July 2010)
56. Essential Evidence Plus (formerly Inforetriever) http://www.essentialevidenceplus.com/ (Accessed 2 July 2010)
57. ACP Medicine http://www.acpmedicine.comH (Accessed 4 July 2010)
58. ACP Pier http://pier.acponline.org (Accessed 3 July 2010)
59. First Consult http://www.firstconsult.comH (Accessed 4 July 2010)
60. DynaMed http://www.ebscohost.com/dynamedH (Accessed 3 July 2010)

10

Search Engines

JANE A. PELLIGRINO
ROBERT E. HOYT

Learning Objectives

After reading this chapter the reader should be able to:

- State the significance of rapid high quality medical searches
- Define the role of Google in healthcare and its many search features
- Describe the meta-search engines and the features that are distinct from Google
- Describe the role of PubMed and Medline searches
- Identify the variety of search filters essential to an excellent PubMed search
- Enhance PubMed searching with third party PubMed tools
- Use NLM Mobile

"Getting information off the Internet is like taking a drink from a fire hydrant"

Mitchell Kapor

Introduction

The most rapid and comprehensive way to access information today from anywhere in the world is a search of the World Wide Web via the Internet. If we assume that the Internet is the new global library with more than 3 billion web sites, then it should come as no surprise that search engines are the gateway. Popular search engines such as Google provide successful searches for medical and non-medical issues. Although PubMed is the search engine of choice for formal searches of the medical literature, most inquires are informal so searches need to yield primarily rapid and relevant results. Given the prevalence of web surfing for answers, multiple articles have been written about "search wars". [1-2] it is unclear to what extent the use of search engines has changed human behavior and medical knowledge in the fifteen years. Previously, questions such as what is the difference between HDL and LDL cholesterol meant a trip to the library, the purchase of a book or a visit with a doctor. Now anyone can execute a search and have a reasonable likelihood that the search will be successful.

Just as important as selecting the search engine with which you are comfortable is learning to use all of the advanced search options. It is imperative to use filters to refine a search or you will become frustrated by the avalanche of information returned. In this chapter we will begin with a discussion of Google, followed by other less well-known search engines and finally a primer on PubMed searching.

Google

Google is by far the most widely used search engine in the world.[3] Its name is derived from the word googol which is the mathematical term for the number 1 followed by 100 zeroes.[4] Google's success is based largely on its intuitiveness, retrieval speed and productive results. Google is listed as one of the ten forces to flatten the world in Thomas Friedman's book "The World is Flat".[5]

Google has proven to be a fascinating company with a myriad of innovations on a regular basis. Google was developed by Larry Page and Sergey Brin in 1996 when they were graduate students at Stanford. They created the "backrub" strategy which meant that a search would prioritize the results by ranking the page that is linked the most first (page ranking).[6] Some could argue that it used a popularity contest as a strategy. A shortcoming of this approach would be that new web sites might take time to be linked. As the worlds largest and fastest search engine it performs one billion searches daily by utilizing thousands of servers (server farm) running the Linux operating system.[7] "Googling" has deeply influenced the users' expectations about the answers to their questions and the way of searching the web." [8] Google can be criticized for being a shotgun and not a rifle in terms of returning too many results but this has not diminished its popularity. Because Google yields so many results in an average search, it is very important to learn about how to narrow or filter a search.

Google can be an acceptable medical search engine for common as well as rare conditions that are not likely to be found in journals or textbooks. Google provides a very global review, returning articles from the lay press, medical journals, magazines, etc. In a 2006 article, Dr. Robert Steinbrook notes that Google (56%) was the most common search engine used to refer someone to find a medical article at High Wire Press; compared to PubMed (8.7%).[9]

Google will cite Medline abstracts and occasionally full text articles, so for an informal search it is not unreasonable to start with Google to see if you find an answer in the first few citations listed. It is likely you will find an acceptable answer in less time than it takes to use PubMed, particularly if you narrow the search with additional descriptors and use an advanced search strategy. Meats et al. showed that clinicians searching for medical information prefer to use a simple strategy of the disease term and the population in question.[10] Google makes that type of searching possible, albeit inefficient, if advanced search techniques are not employed. If you search with the terms *type 2 diabetes foot checks frequency* you will likely retrieve clinical articles that describe how often foot checks should be performed in diabetics. In a Google search the most important term should be listed first.

Several recent articles in the medical literature have confirmed that Google has become a common medical search engine; even at academic centers.[11-13] Google continued its foray into the medical world in 2008 by launching Google Health. Google is very aware that many patients use Google to search for medical answers to common and complex health questions. [14] Recognizing that 25% of the Internet searches are health related and that there is growing desire for universally available online medical records Google developed Google Health, a site that allows users to upload, store and manage personal health information in one place, at no cost. Although privacy advocates have voiced security concerns, Google states that it will never sell the information and that the information is secure and private.

Successful searching depends on maximizing Google search options.
- Begin by setting preferences
    - Language you prefer e.g. English
    - Number of search results per page e.g. 10 or 20
    - Whether you want your search to launch in new window (Recommended: When you exit the current page, you lose your search.)

- Select the *Advanced search* option on the main page.
  - o Under occurrences (Find webpages that have..) you can search for a term in the title only or in the body or both
  - o Search for synonyms using OR as the operator
  - o Select search by domain such as .org, .edu or .gov
  - o Select search by format: Word, Excel, PDF, PowerPoint, etc
  - o Put quotation marks around the words to search for an exact phrase, e.g. "University of West Florida" so you don't retrieve every citation with the words Florida, West or University
- Take a look at *Advanced operators* to refine the search
  - o Type *define*:before a word or phrase to have Google serve as a dictionary
  - o Enter an arithmetic string and Google will function as a calculator
- Essentials of a Google Search (www.google.com/help/basics.html) Provides helpful tips to improve the search process:
  - o Searches are not case sensitive
  - o The word "and" is not necessary, because Google, unlike PubMed, "and" is implied. Use Advanced search to use "or"
  - o Most popular web sites are listed first
  - o For a search of a common subject, select "I'm feeling lucky" on the main Google search page and you will be taken to the most popular web sites on the subject [15]

The following are Google features that, in our judgment are relevant to Medical Informatics:

- Google includes an *Image* search of over 880 million images (some copyrighted). Advanced image search filters are available
- *Google Talk* is a free proprietary voice over Internet protocol (IP) and instant message service. You can talk to another person via your computer and the Internet
- *Gmail* is free web mail. Unlike other web mail services it offers over 2 gigabytes of memory. Google maintains that you don't need to organize your e-mail into folders because of its excellent search engine
- *Google Docs & Spreadsheets* provides an alternative to Microsoft Word and Excel. You can collaborate with others and publish your work to a web site. Only drawback is a limit of 500 K for a document upload
- *Google Groups* creates a collaborative web site where you can post discussions and web content. Members can be by invitation only
- *Google Directory* organizes the Web into categories so the search may be more focused. If you search, for example, under a Health directory > Medicine > Informatics > Telemedicine, the search will yield page ranked web sites so you do not see citations from the lay press, PubMed, etc. The web site choices are actually selected by experts so this is why the number of sites returned is far smaller than a true web search
- *Google Health* is a personal health record portal for consumers. This is discussed in detail in the chapter on Consumer Health Informatics [16]
- *Google Health Topics* provides extensive references on health topics including overview, treatment, diagnosis, clinical guidelines, symptoms, complications, news, etc [17]
- *Google Custom Search Engines* under "co-op" you will find the ability to customize searches by limiting them to certain web sites. You can also integrate this search engine into your personal web site
- *Google Code* searches for open source codes and APIs (application programming interfaces)

- *Google Page Creator* creates simple web pages with easy to use tools so anyone can create their own web site
- *Google Scholar* (feature is described below) uses the Google search engine to search journal articles at publishers' websites
- The *Desktop search* program rapidly searches your personal computer files
- *Book Search (beta)* provides access to books online in the public domain. Potentially Google Books will search 50 million textbooks. If a book is currently copyrighted, there will be a link to purchase the book or borrow it from a library [18]
- Google changes so rapidly it is a good idea to look at *Google labs* often. It provides new web page alerts and news alerts, toolbar shortcuts, a glossary and discussion groups and the ability to create a web page. Alerts can track any topic and e-mail you any new information as it becomes available

Google Scholar

Google Scholar is an offspring of Google that searches the full text of peer-reviewed scholarly journal articles at publishers' websites and the citations and abstracts provided online by the National Library of Medicine through PubMed.[19-20] Google Scholar uses the same search technology as its parent. Because the search technology relies on an algorithm that weighs articles by their links to other relevant content, it is difficult to retrieve recent articles. Google Scholar delivers the quantity of retrieval, but not the quality, necessary to allow it to standalone. Because Google Scholar searches the full text of articles in contrast to PubMed that searches the title and abstract, Google Scholar enables the searcher to retrieve articles that contain words and phrases not found in the title or abstract. Google Scholar makes searching easier, but it should be used in conjunction with PubMed, not as a replacement for it. Google Scholar offers "cited by referencing" which is not available elsewhere for free. [21] Google Scholar is a good tool for accessing the open source literature, it provides access to unique content not in other search tools and it is free and openly available. [22]

Other Search Engines

Meta-search engines search more than one database or utilize more than one search engine.[23] It remains to be seen if this is necessarily an advantage or not. We could find no publications or reviews regarding their use in medical searches.

Bing

- Search engine introduced by Microsoft in May, 2009 as a challenger to Google
- Intuitive search box with enhanced search capability
- Bing targets four most important categories of search: shopping, travel, health and local with specialized results displays for each category
- Displays results in search categories on a left hand navigation bar [24]
- Displays search history

Clusty powered by Vivisimo

- Metasearch engine that searches the web, images, Wikipedia, blogs, government, etc.
- Available for mobile devices and as a Firefox Mozilla add-on for desktop searches
- "Clusters" the search into sub-topics in convenient folders

- Vivisimo now offers a *Velocity* option that is more specific for the life sciences and bioinformatics
- Powers USAgov.gov
- Adopted by the National Library of Medicine and MedlinePlus to use on their web sites [25]

Dogpile
- Metasearch engine that searches Google, MSN, Ask Jeeves and Yahoo
- Advanced search uses the filters of qualified words or phrases, language, dates and domain; similar to Google
- Searches sponsored and non-sponsored web sites
- Search the web, images, audio, video, news, white and yellow pages
- Note: a 2008 search for avian influenza returned 69 high quality citations, whereas a Google search returned 2.3 million! [26]

Omni Medical Search
- Although touted as a medical meta-search engine, it does not utilize different search engines and has very limited filters
- Tabs conduct searches of the web, news, images, forums and MedPro
- MedPro search is intended for serious medical inquiries, yet yielded very limited results and is associated with commercial influence. They now use Google as their search engine as well as ads by Google
- Available as a desktop toolbar search engine
- A reference desk includes searches for medical: acronyms, dictionaries, images, associations, databases, journals, conditions and diseases and forums [27]

PubMed Search Engine

PubMed is a web-based retrieval system developed by the National Center for Biotechnology Information (NCBI) at the National Library of Medicine (NLM).[28] PubMed is one of twenty-three databases in NCBI's retrieval system, known as *Entrez*. *Entrez* includes biological databases that index information in toxicology, bioinformatics and genomics and even includes textbooks.

MEDLINE is the primary *Entrez* database, containing 19 million citations from the world's medical literature from the 1940s to the present, covering the fields of medicine, nursing, dentistry, veterinary medicine, health care administration, the pre-clinical sciences and some other areas of the life sciences.[29] NLM licenses its data to vendors to be used through proprietary interfaces, but PubMed search interface for MEDLINE is only available directly from the National Library of Medicine.

For simple answers to common problems PubMed may not be the place to begin a search, but it is the primary search engine for physicians seeking information on unusual cases and research topics. Although some would argue that the PubMed search process is too labor intensive, all physicians seeking to retrieve evidence-based medical answers should learn to use PubMed. It is especially important in an academic or research environment. Without proper training PubMed searching can be challenging and frustrating. This section emphasizes the important features and shortcuts to make a search easier and more successful. Excellent tutorials exist on the PubMed site to teach you the basics of a good search. Also, several helpful review articles have been written that address PubMed tools and features. [30-32]

The query box in PubMed (Fig. 10.1) allows keyword, Medical Subject Heading (MESH) and natural language (Google-type) entries. Search terms may be entered alone or connected by search operators, such as "AND" or "OR". The goal of the search is to find specific citations on the topic described by the search terms.

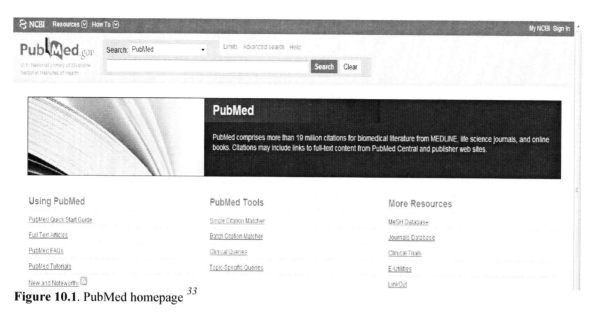

Figure 10.1. PubMed homepage [33]

PubMed citations include the author, title, journal, publication date and PubMed identification number (PMID) as shown in Fig. 10.2 and 65% of the citations include an author abstract. PubMed does not search the full-text of cited articles.

Randomised controlled trial of integrated care to reduce disability from chronic low back pain in working and private life
Lambeek LC, van Mechelen W, Knol DL, Loisel P, Anema JR.
BMJ. 2010 Mar 16;340:c1035. doi: 10.1136/bmj.c1035.
PMID: 20234040 [PubMed - indexed for MEDLINE]   Free PMC Article   Free text
Related citations

Figure 10.2. Medline citation (courtesy National Library of Medicine)

Medical Subject Heading (MeSH). Journal articles are categorized by NLM indexers in order to facilitate searching. Articles are saved under two or more subject headings using a structured vocabulary called MeSH. Understanding what these terms are and how they can refine a search is an important first step in harnessing the power of PubMed. As you can imagine, terms such as low back pain could be labeled lumbar pain, osteoarthritis of the lumbar spine, etc. It will improve your search significantly, if you search with the preferred term, so take a moment to look at MeSH. You can access MeSH in the drop down menu in the search window or by choosing the MeSH Database in the menu on the left.

Figure 10.3 shows how the term "low back pain" is organized in MeSH. The MeSH entry shows a definition of the term and its synonyms and displays a set of subheadings with which to narrow a search on low back pain. You can even restrict your search to the term as a "Major Topic."

Figure 10.3. MESH term display (courtesy National Library of Medicine)

Figure 10.4 illustrates a search for sinusitis in MeSH. Different types of sinusitis are listed. Under each term "entry terms" are provided as above (Figure 10.3).

Figure 10.4. MESH term search

At the bottom of each MeSH entry is the categorical display or "MeSH Tree" as shown in Fig. 10.5. Searching the term, sinusitis, includes all the specific types listed under it, broadening the search. Conversely, reviewing sinusitis in MeSh allows you to discover the specific type of sinusitis available sot that you may search the one that fits your query the best. MeSH is valuable in broadening or narrowing a search query.

- All MeSH Categories
  - Diseases Category
    - Respiratory Tract Diseases
      - Nose Diseases
        - Paranasal Sinus Diseases
          - **Sinusitis**
            - Ethmoid Sinusitis
            - Frontal Sinusitis
            - Maxillary Sinusitis
            - Sphenoid Sinusitis

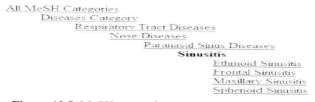

Figure 10.5. MeSH categories

If you are struggling with your search terms in PubMed and not finding what you need, you may want to check your search terms in MeSH to see if the term is accepted by PubMed. Searching with the correct term can make all the difference. As is the case with Google searching, learning to use filters such as MeSH will result in more successful retrieval of information.

PubMed Limits Option allows a search to be narrowed by date, age of subjects, gender, humans or animals, language, publication types, topics and field tags.

- You can also search for full text and free full text articles and abstracts. (Keep in mind that most articles before 1975 did not contain abstracts.)
- You can search by author or journal name
- Searchable main publication types include Clinical Trial, Editorial, Letter, Meta-Analysis, Practice Guideline, Randomized Controlled Trial, and Review
- Searchable topic subsets include AIDS, Bioethics, Cancer, Complementary Medicine, History of Medicine, Systematic Reviews and Toxicology
- Field tags. You can stipulate whether you want the search term in the title or body of the article. Multiple other choices are listed as well

Entering a Search in PubMed

PubMed is based on an architecture that uses indexed concepts (MESH Headings) and Boolean logic to retrieve information. Search questions should be analyzed and broken down into concepts that are described using MESH Headings or textwords. These search terms are then joined together by AND to retrieve articles that contain both concepts, or joined together by OR to retrieve articles that contain either concept. The Boolean operators should be capitalized. To search for articles on sinusitis caused by bacteria search bacterial infections AND sinusitis.

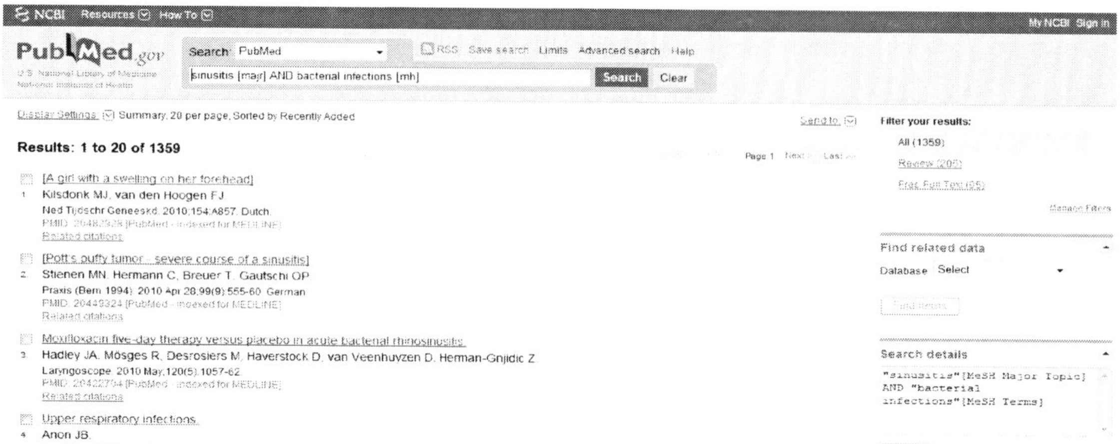

Figure 10.6 Combining MeSH terms with Boolean operators

Although the search box in PubMed looks very much like Google, the words entered in the search box are processed based on concept searching rather than natural language searching. Recognizing that most searchers are accustomed to Google searching PubMed is developing a natural language search engine that works together with the concept search engine to retrieve articles.

Selecting limits

Once the concept search has been entered you may limit the search with the search parameters. We are now going to limit the sinusitis search. In addition to searching for articles where sinusitis is the main topic (sinusitis [MAJR]) we will limit the search by age (Adult: 19-44), humans, Core clinical journals (all of which are in English), added to PubMed in past 5 years and those with links to free full text. We could have also selected clinical trial, random controlled trial or review or checked the box for all four.

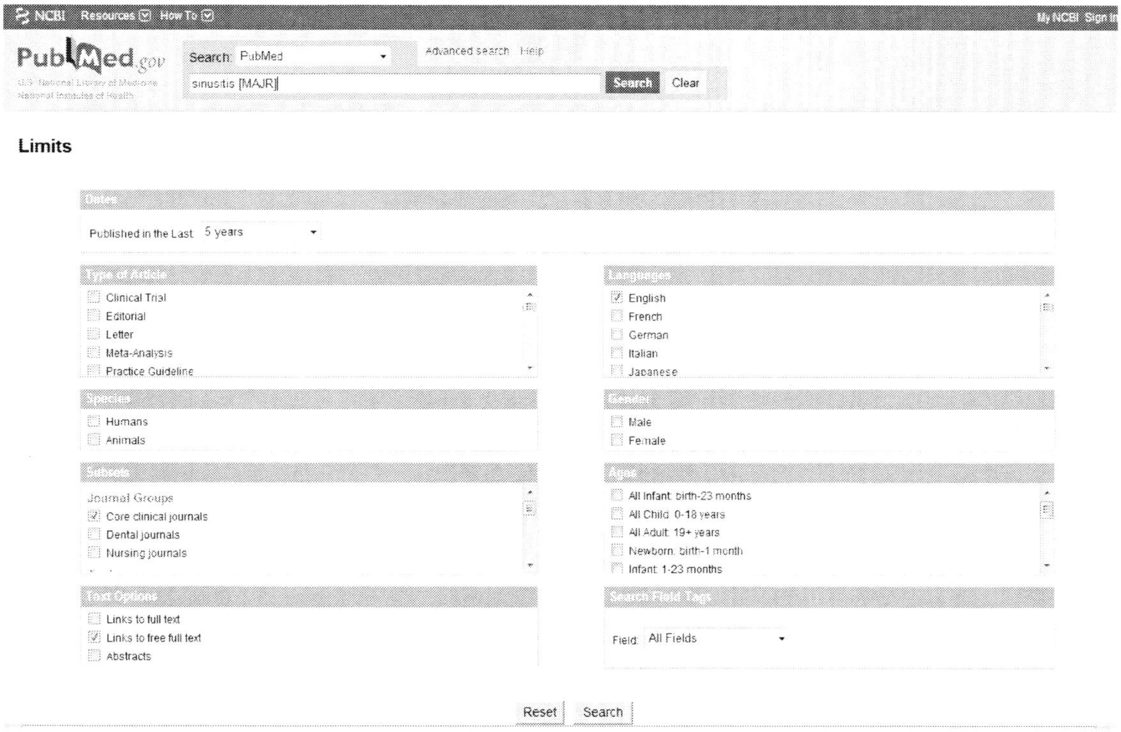

Figure 10.7. Selecting multiple search limits (courtesy National Library of Medicine)

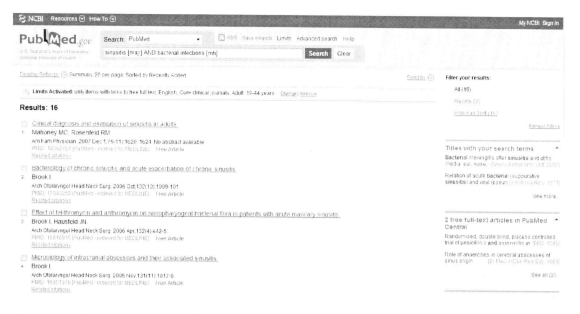

Figure 10.8. Search for sinusitis with multiple limits (courtesy National Library of Medicine)

Our search with limits has greatly reduced the number of returned citations and improved the quality (Fig. 10.8). Requesting free full text articles also reduces the search considerably. Changing the search to any abstract, instead of free full text articles, returns 79 citations.

- Note that the most current articles are listed first.
- Many articles are associated with an abstract that summarizes the article (Fig.10.9)
- You must go to the full text article for more detail

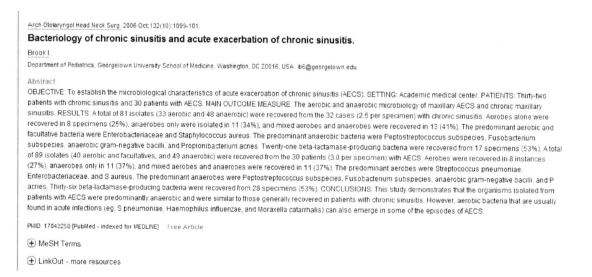

Figure 10.9. Example of an abstract (courtesy National Library of Medicine)

Other options

Using the Advanced Search Screen

As you search, PubMed records your search statements. To review your previous searches and to combine search statements go to Advanced Search. Combine the statements by clicking search statement numbers and choosing the appropriate Boolean operator from the menu.

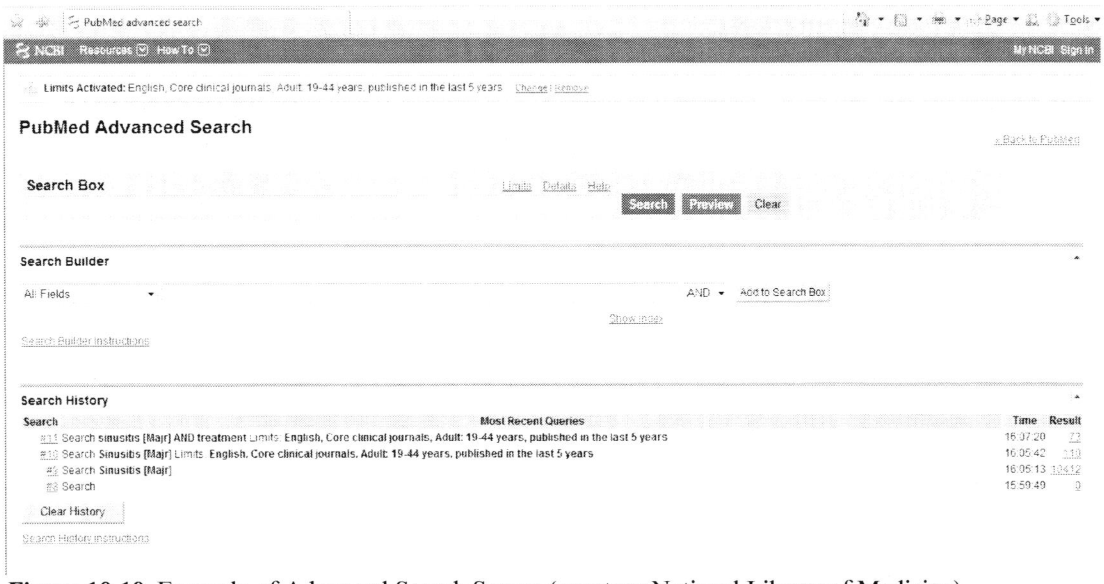

Figure 10.10. Example of Advanced Search Screen (courtesy National Library of Medicine)

To see how the search you entered was executed by PubMed click on **Details.** On the Advanced Search Screen at the bottom of the page you will find navigation links for access to other PubMed modules.

Single Citation Matcher

When you are trying to locate a specific article and you only have fragments of the citation you may find Single Citation Matcher helpful. Type the information you know into the form to find the article of interest. You can search by author, journal, date, volume, issue, page or title words.

Clinical Queries

Clinical Queries provides another way to search for articles reporting the results of randomized controlled trials with the use of built-in filters. The research methodology behind the filters was at McMaster University and the filters are built into the *Clinical Queries* interface, so you can search for randomized controlled trials by etiology (cause), diagnosis, therapy, prognosis or search for clinical prediction guides. Searches can be modified to be either broad/sensitive manner or narrow/specific.

Systematic reviews, a type of review that critically appraises multiple random controlled trials to give conclusions more strength (covered in more detail in the chapter on evidence based medicine), can be searched from the Clinical Queries page.

Display Options

Search results display automatically defaults to summary view, but you can also select abstract, Medline, XML and others
- Under Show tab you can elect to show up to 200 citations per page
- Under Sort by tab you can sort by author, journal or published date

Options for Saving Your Results

- **Clipboard** found under the *Send to* link allows you store up to 200 citations for up to 8 hours
- **E-mail** lets you send your selected results to a colleague or yourself

- **Text** lets you save the results as a text file. (Sometimes useful for bringing the citations into MS Word.)
- **File** lets you put the results into a format that is suitable for a bibliographic software program
- **Order** allows you to send the citation to an affiliated library under the Loansome Doc program for document delivery
- **My NCBI,** located on upper right hand corner of the main screen, provides a valuable storage area for searches and collections of articles you have retrieved allowing you to
  - Save searches (otherwise gone in 8 hrs)
  - Set up e-mail alerts so you are notified when new articles are published on your topic of interest
  - Display links to online full-text of articles (LinkOut)
  - Choose filters that group search results
  - Registration is required for this free service

Other Features

Related articles and links

- To the right of each article you will see a hyperlink to similar or related articles. Select Links and it can link you to:
  - PubMed books
  - PubMed Central (www.pubmedcentral.nih.gov) journal articles links to articles in free and full text
  - Link Out—links to external resources such as OVID or MDConsult to which the library with which you are affiliated subscribes
  - Patient information from MedlinePlus
- PubMed® for Handhelds website offers several search options for MEDLINE® with the web browser of any mobile device. (Figure 10.11)
- PubMed PICO search, part of PubMed for Handhelds (Fig. 10.12) —aids in the construction of a well thought out question prior to initiating a search. This tool it divides the question into sections defining the patient or problem, intervention, comparison and outcome (P.I.C.O). The URL or web address could be a desktop icon shortcut or a program on your handheld for fast searches.[34]
  - **(P)** atient or problem—how do you describe the patient group you are interested in? Elderly? Gender?
  - **(I)** ntervention, prognostic factor or exposure-- Drug? Lab test? Tobacco?
  - **(C)** omparison—with another drug or placebo?
  - **(O)** utcome—what are you trying to measure? Mortality? Reduced heart attacks?

PubMed® for Handhelds website [35]

National Library of Medicine
The world's largest medical library

PubMed for Handhelds
- PICO search
  Patient, Intervention. Comparison, Outcome
- *ask*MEDLINE
  free-text, natural language search
- MEDLINE/PubMed
  Search MEDLINE/PubMed
  Read Journal Abstracts
- Disease Associations
  Search case reports for disease associations

Feedback
Disclaimer

Figure 10.11. Example of PubMed® for Handhelds (courtesy National Library of Medicine)

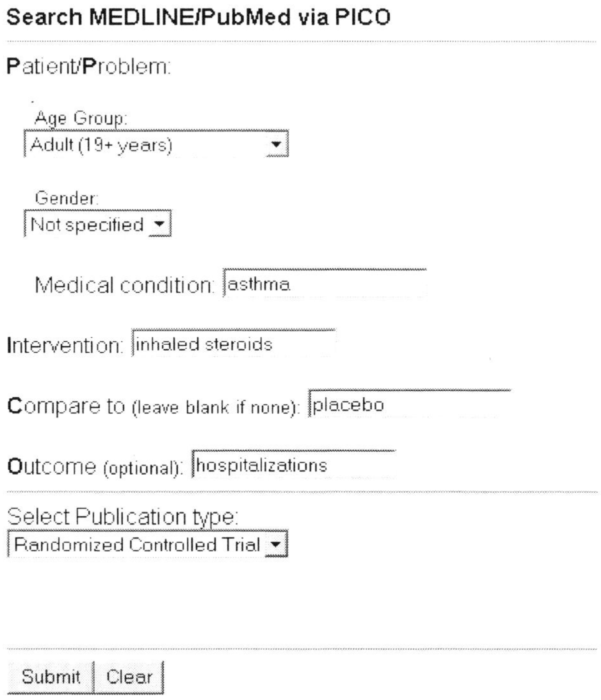

Search MEDLINE/PubMed via PICO

Patient/Problem:

Age Group:
Adult (19+ years)

Gender:
Not specified

Medical condition: asthma

Intervention: inhaled steroids

Compare to (leave blank if none): placebo

Outcome (optional): hospitalizations

Select Publication type:
Randomized Controlled Trial

Submit | Clear

Figure 10.12. PICO search (courtesy National Library of Medicine)

- PubMed Central—hosts multiple free full text articles. Unfortunately, many are located in minor journals of recent vintage. They are also more weighted towards a bioinformatics search.

Third Party PubMed tools

The National Library of Medicine (NLM) makes its database of citations available to the public for searching, and it also makes its data available through an application programming interface (API). The API allows interested users to write programs that mine the MEDLINE database in new ways. Several applications designed to optimize MEDLINE searching are available and others are emerging continually in an effort to exploit the MEDLINE data and may more accessible to the user.[36] Below are some examples of these third party PubMed tools that are noteworthy.

PubMed PubReMiner.[37] Medical Subject Heading searching gives power to a PubMed searching. PubReMiner allows the searcher to enter keywords or PMIDs related to a query and then analyzes the relevant PubMed citations and their indexing to develop a list of terms with which to expand the search. PubReMiner is available directly from the website or through a web browser plug-in that is available for Mozilla Firefox or Internet Explorer 7.0.

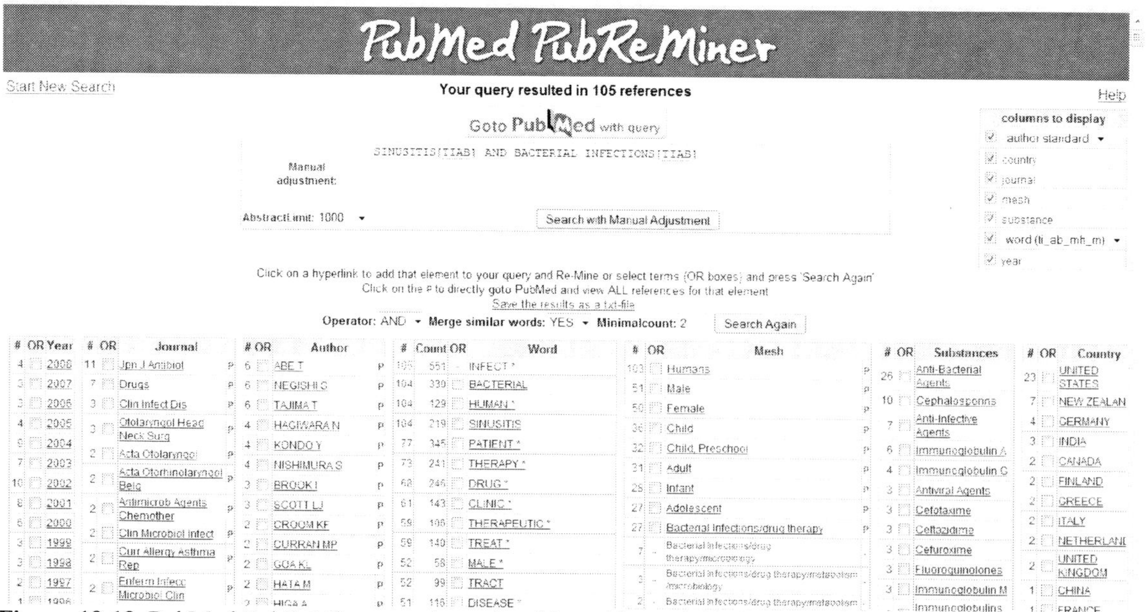

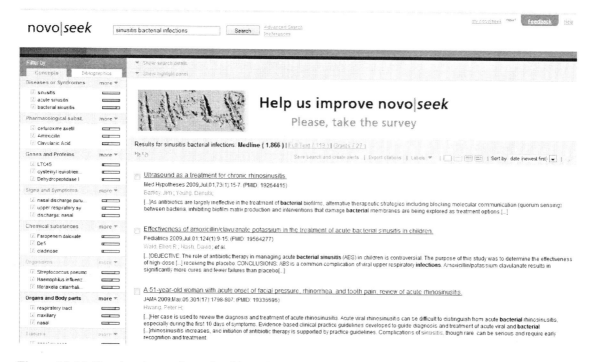

Figure 10.13. PubMed PubReMiner search on sinusitis and bacterial infections in title and abstract

Novo/seek[38] indexes the medical literature with text mining tools that identify key biomedical terms. Novo|seek indexes the biomedical literature and US grants for over 75 institutions including the US National Institutes of Health (NIH), the National Science Foundation, US Department of Agriculture and the Centers for Disease Control and Prevention (CDCP) with a text mining technology that enables the identification of key terms. Novo|seek is able to retrieve every document where a term is mentioned regardless of synonym.

Novo|seek extracts precise information, retrieves key biomedical concepts, filters results, reviews key information derived from thousands of documents, identifies key research concepts by author and links to relevant external chemical and biological information.

Figure 10.14. Novo| seek search on sinusitis and bacterial infections

PubMed EX. PubMed EX is a browser extension for Mozilla Firefox and Internet Explorer that marks up PubMed search results with additional information derived from data mining. PubMed EX provides background information that allows searchers to focus on key concepts in the retrieved abstracts.[39]

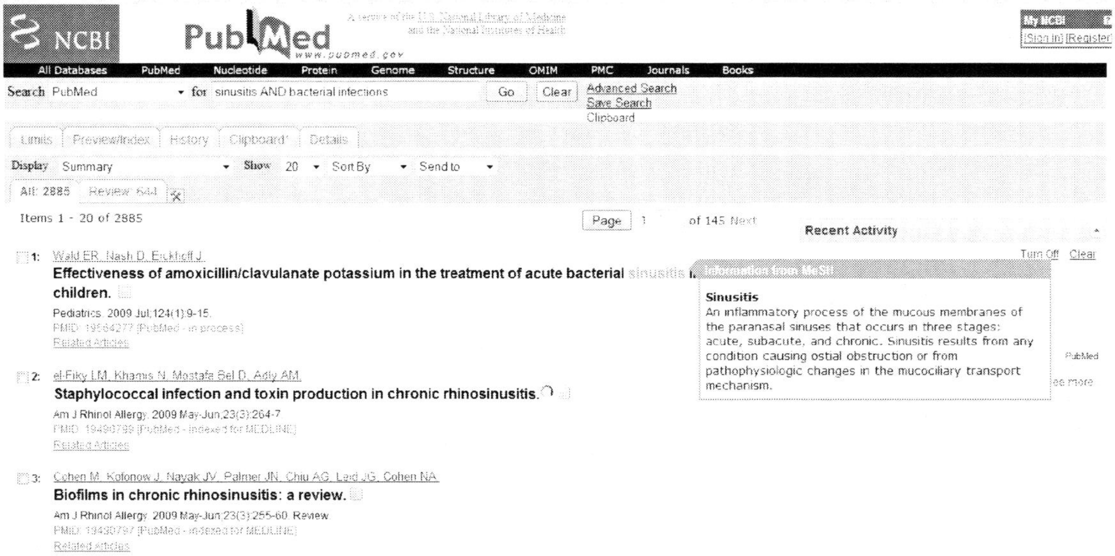

Figure 10.15. PubMed search on sinusitis and bacterial infections using a browser with the PubMed EX add-on

Gopubmed is a semantic search engine for searching PubMed enabling searchers to find articles easier and faster. Simply enter keywords or MeSH headings into the search box and the search engine will display the frequency of relevant terms with which to formulate your search. In Figure 10.16 the types of sinusitis as defined by MeSH is displayed on the left of the gopubmed screen. [40]

Figure 10.16. GoPubMed Basic Search on sinusitis and bacterial infections

Quertle

- Is a semantic search engine that helps users:
    o Find true relationships between terms, rather that simple co-occurrences
    o Use Power Terms to find appropriate categories of terms

o Use easily employed filters to limit a search by year or publication type [41]

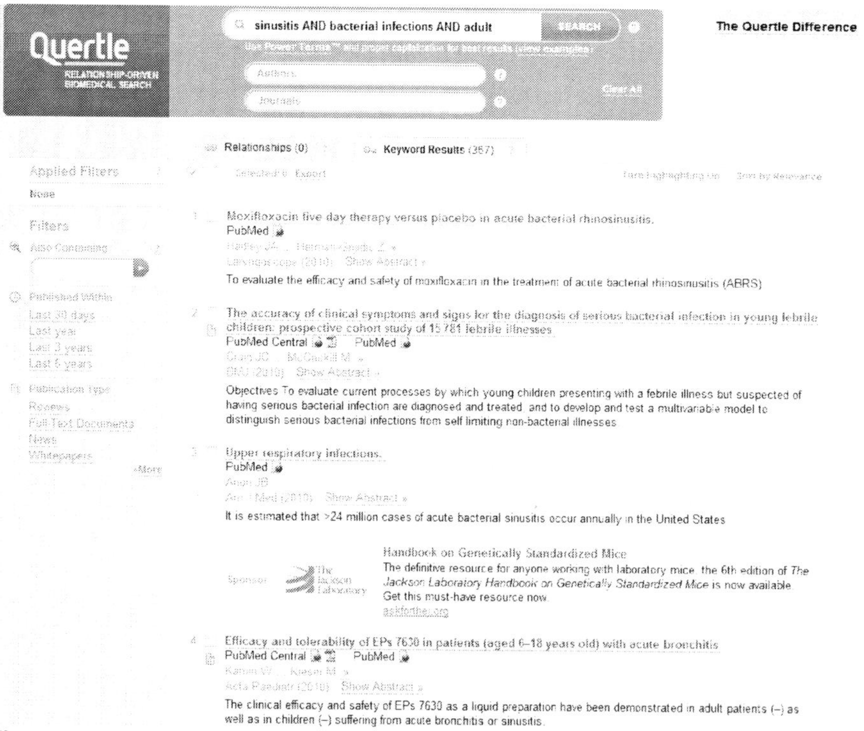

Figure 10.17. Quertle Search on sinusitis and bacterial infections in adults

NLM Mobile

NLM Mobile includes several programs for Palm devices and Pocket PCs some of which can be used from the desktop These programs include AIDSinfo PDA Tools, WISER (Wireless System for First Responders, PubMed for Handhelds and the NCBI Bookshelf.[42]

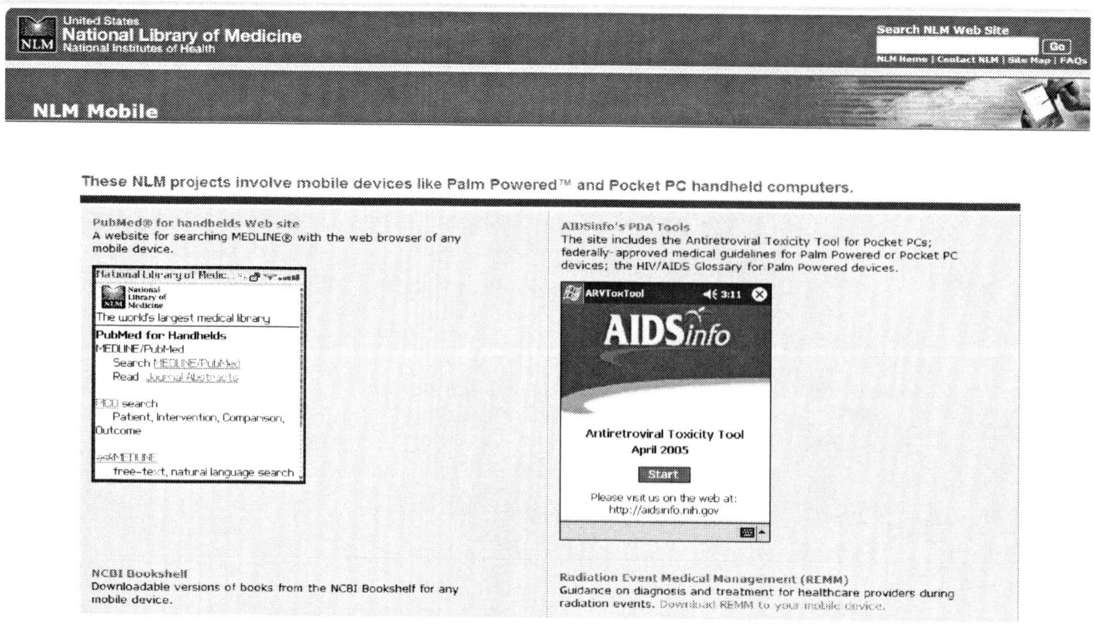

Figure 10.18. NLM Mobile (courtesy National Library of Medicine)

Key Points

- Search engines exist that can provide rapid high quality medical searches
- Google has become a defacto initial medical search engine for many
- New search engines and meta-search engines continue to appear on the scene
- PubMed searches are important for formal searches of the medical literature
- All searches benefit from appropriate filters
- PubMed searching is enhanced by third party PubMed tools

Conclusion

At this time, Google is the premier search engine for non-medical and perhaps medical searches. With proper filtering and experience, Google can be used with significant success. Today the average person can search for answers to a variety of medical questions. Although this may produce some "cyber-hypochondria" in a minority of searchers, it is likely to produce better- informed patients in the majority. Better studies are needed to compare Google with other search engines and PubMed for quality and speed of retrieval. Familiarity with PubMed and its new features is important for healthcare workers who need to conduct formal searches of the medical literature. With knowledge and experience a PubMed search can result in relevant results in a timely fashion.

References

1. Al-Ubaydli. Using search engines to find online medical information   PLOS Medicine 2005 http://medicine.plosjournals.org/archive/1549-1676/2/9/pdf/10.1371_journal.pmed.0020228-S.pdf (Accessed 3 July 2010)
2. Goldman D. Search wars: WoldframAlpha joins the battle. CNN Money.com. http://money.cnn.com/2009/05/27/technology/search_engines (Accessed 3 July 2010)
3. Google search basics: more search help http://www.google.com/support/bin/static.py?page=searchguides.html&ctx=advanced&hl=en (Accessed 3 July 2010)
4. Google.pedia. The ultimate Google resource. Michael Miller. Que publishing. 2007
5. Friedman, Thomas. The World is Flat. Farrar, Straus and Giroux. New York. 2006
6. The Anatomy of a large scale hypertextual web search engine http://infolab.stanford.edu/pub/papers/google.pdf (Accessed 3 July 2010)
7. Wikipedia: Google http://en.wikipedia.org/wiki/Google (Accessed 12 July 2009)
8. Giglia E. To Google or not to Google, that is the question Eur J Phys Rehabil Med 2008; 44:221-7
9. Steinbrook R. Searching for the Right Search—Reaching the Medical Literature NEJM 2006;354:4-7
10. Meats E et al. Using the Turning Research into Practice (TRIP) database: how do clinicians really search? JMLA 2007; 95(2): 156-163
11. Tang H, Ng JHK. Googling for a diagnosis-use of Google as a diagnostic aid: Internet study BMJ Nov 11 2006. http://www.bmj.com/cgi/reprint/333/7579/1143 (Accessed 5 July 2010)
12. Correspondence. And a Diagnostic Test was Performed. NEJM 2005;353:2089-2090

13. Turner MJ. Accidental epipen injection into a digit—the value of a Google search Ann R Coll Surg Engl.2004;86:218-9

14. Google Health Gains Partners. March 3 2010. http://news.cnet.com/8301-27083_3-10462961-247.html

15. Google search basics: basic search help http://www.google.com/help/basics.html (Accessed 3 July 2010)

16. Google Health www.google.com/health Google Health Topics https://www.google.com/health/ref/index.html (Accessed 3 July 2010)

17. Google Book Search http://books.google.com/ (Accessed 3 July 2010)

18. Butler D. Science searches shift up a gear as Google Starts Scholar engine Nature 2004 Nov 25;432(7016):423

19. Jacso P. Google Scholar revisited. Online Information Review 2008; 32 (1): 102-114.

20. Giustini D How Google is changing medicine BMJ. 2005 December 24; 331(7531): 1487–1488

21. Hazing AW, Vander Wal R. Google Scholar: the democratization of citation analysis. http://www.harzing.com/dlownload/gsdemo.pdf (Accessed 25 July 2010)

22. Sure, Google Scholar is ideal for some things The Search Principle blog http://blogs.ubc.ca/dean/2010/05/sure-google-scholar-is-ideal-for-some-things/ (Accessed 25 July 2010)

23. Brandon J. Outside the box: search the web with power PC Today Dec 2004: 39-41

24. Bing http://bing.com (Accessed 3 July 2010)

25. Clusty http://www.clusty.com/ (Accessed 3 July 2010)

26. Dogpile www.dogpile.com (Accessed 3 July 2010)

27. Omni Medical Search www.omnimedicalsearch.com (Accessed 3 July 2010)

28. NML Databases http://www.nlm.nih.gov/databases/ (Accessed 4 July 2010)

29. Fact Sheet MEDLINE http://www.nlm.nih.gov/pubs/factsheets/medline.html (Accessed 3 July 2010)

30. Ebbert JO, Dupras DM, Erwin PJ. Searching the Medical Literature Using PubMed: A tutorial Mayo Clin Proc. 2003; 78:87-91

31. Sood A, Erwin PJ, Ebbert JO. Using Advanced Search Tools on PubMed for Citation Retrieval Mayo Clin Proc. 2004; 79:1295-1300

32. Haynes RB, Wilczynski N. Finding the gold in MEDLINE: Clinical Queries ACP Journal Club Jan/Feb 2005;142: A8-A9

33. Entrez PubMed http://www.pubmed.gov/ (Accessed 3 July 2010)

34. NLM PICO http://askmedline.nlm.nih.gov/ask/pico.php (Accessed 3 July 2010)

35. PubMed for Handhelds http://pubmedhh.nlm.nih.gov/nlmd/ (Accessed 3 July 2010)

36. Rothman D. Archive for 3rd Party PubMed/MEDLINE Tools http://davidrothman.net/category/technology/3rd-party-pubmedmedline-tools/ (Accessed 25 July 2010)

37. PubReMiner http://hgserver2.amc.nl/cgi-bin/miner/miner2.cgi (Accessed 3 July 2010)

38. novo/seek http://www.novoseek.com/Welcome.action (Accessed 3 July 2010)

39. PubMed EX http://bws.iis.sinica.edu.tw/PubMed-EX/index.html (Accessed 3 July 2010)

40. gopubmed http://www.gopubmed.com (Accessed 25 July 2010)

41. Quertle http://www.quertle.info/v2/ (Accessed 25 July 2010)

42. NLM Mobile http://www.nlm.nih.gov/mobile/ (Accessed 3 July 2010)

11

Mobile Technology

ROBERT E HOYT
ROBERT W CRUZ

Learning Objectives

After reading this chapter the reader should be able to:

- Describe the history of medical mobile technology and the evolution from personal digital assistants to smartphones
- List the essential features of a medical smartphone
- Compare and contrast the medical software programs most helpful for clinicians
- Identify the limitations of handheld technology

The Washington Post
New Tool in the MD's Bag: A Smartphone

By Sindya N. Bhanoo
Special to The Washington Post
Tuesday, May 19, 2009

In the first edition of our textbook this chapter discussed personal digital assistants (PDAs) in the field of medicine. With the advent of "smartphones" (PDA phones, BlackBerries, iPhones, G-phones, etc), we feel mobile technology is a broader and more appropriate terminology. Handheld technology is also an acceptable alternative term. Mobile technology is a logical transitional step from the personal computer. With improving speed, memory, wireless connectivity and shrinking form factor (size and shape), users desire a mobile platform for their information and applications as well as phone capability, e-mail and access to the Internet. Although mobile technology is not necessarily part of a Medical Informatics curriculum, widespread popularity of this technology in medicine makes it worth mentioning. Some are arguing that we have entered the mHealth (mobile health) era in which smartphones will play a much larger role. We will discuss mHealth in the last chapter on emerging trends.

History

The history of personal digital assistants is quite recent. In the early 1990s the Apple Newton appeared with the hope that this new technology would appeal to the average user. This monochrome PDA weighed .9 lbs, measured 7.25 x 4.5 x .75 inches, had 150 K of SRAM, a processor speed of only 20 MHz, short battery life and a cost of $700.[1] It obviously did not succeed because it was too big, heavy, slow and costly for the average consumer.

The next handheld product to catch the public's attention was the Palm Pilot 1000, invented by Jeff Hawkins in 1994 and released in 1996.[2] It was smaller, less expensive and had 128 K of memory. Synchronizing with the personal computer was a one-step operation. One could argue that the PDA did not become popular with the medical profession until the "*killer application*"

Epocrates was released in 1999.[3] First, there was the excitement of knowing that drug facts could be retrieved much more rapidly with the PDA compared to the Physician Desk Reference (PDR) and secondly, the program was free. The PDA was also a platform to store all medical "pearls" rather than stuffing notes into the pockets of a white coat.

Other companies got on the bandwagon to produce PDAs. Hewlett-Packard produced the iPaq, while Dell and Sony have since abandoned the PDA market. Microsoft's Windows operating system (OS) made major inroads into what initially was mainly Palm territory for the medical profession. Symbian and Linux are also operating systems that can be used in the PDA but are not prevalent in the United States. BlackBerry by Research in Motion is popular for e-mail functionality and now has multiple medical software programs available. Newer platforms for smartphones and mobile devices such as iOS and Android have appeared and will be covered later in this chapter.[4]

According to a 2006 study in the Journal of General Internal Medicine, 50 percent of practicing physicians and about 60-70 percent of medical students used PDAs on a regular basis.[5] Many medical schools fully support handheld technology as part of their educational programs. The shift from PDAs to smartphones has been rapid, occurring over only 6-8 years. Manhattan Research reported in 2009 sixty four percent of physicians used smartphones, compared to 30% in 2001. They predicted 81% would use smartphones by 2012.[6]

There is no industry-wide definition of a smartphone. Some define it as having an operating system that can support the execution of third party applications and others define it as simply having more functionality than conventional cell phones. For the purpose of this chapter we will use the term smartphone and include only those that have operating systems capable of hosting medical software. With the evolution of cloud computing, more and more medical programs will be hosted in the cloud and not on the device. Therefore, one could eventually state that a smartphone is one that is Internet capable. There are likely to be further convergence of handheld technologies, as slate devices such as the Apple iPad blur the distinction between smartphones and laptop/tablet computers. It is too early to know if the iPad will be used by physicians making rounds or evaluating patients in an exam room. There is a paucity of evidence to prove this type of mobile technology is superior to the standard desktop computer. [7]

Later in this chapter we will discuss each operating system with an example of a smartphone available in 2010 for each platform.

Medical Software Programs

The Palm PDA was the first mobile technology to have medical software available and they became the leader for about the next 5 years. Soon thereafter Windows CE and Windows Mobile-based technology offered similar programs. In the past 3-5 years other smartphones such as BlackBerry and the iPhone have appeared that have essentially out-paced Palm and Windows for medical software due to the fact that their devices are more popular. In part because of Apple's choice to partner with developers for the iOS application distribution channel, by 2009 more than 100,000 apps were available for download. Approximately 2500 health-related programs are available from the Apple iTunes App store. In 2010 there is an extensive inventory of medical software, both freeware and fee-based for anyone in the medical field. Software programs can be downloaded from a variety of web sites but all medical programs for the iPhone/iTouch or iPad must be downloaded from the Apple iTunes App Store. An advantage of purchasing through iTunes is that it lists when the program was last updated, program version, size, language, vendor, requirements and rating/comments. [8]

The following section lists major categories of popular medical software with specific examples.

Medical Software Categories

Calculators

Handheld technology is ideal for medical calculators and the following are just a sample of those currently available for the average clinician:

1. **StatCoder™** currently supports iOS that includes the iPhone, iPod Touch and iPad.
    a. STAT E&M Coder helps determine the evaluation and management level for a visit. It is available as a free Lite edition or fully functional application at $49.99.
    b. STAT Depression Screener calculates the likelihood of depression and is free
    c. Cardiovascular (CV) screening:
        i. Screening for heart attack prevention and education (Shape) for men ages 45-75 and women ages 55-75. A free application.
        ii. Stat Reynolds Risk Score that determines CV risk based on cholesterol, C-reactive protein, family history, systolic blood pressure and hemoglobin A1c levels. A free application.
        iii. Framingham Risk Score determines the 10 year risk of CV disease based on age, gender, cholesterol and systolic blood pressure. Also determines heart/vascular age. A free application. Figure 11.1 demonstrates the 10 year risk of cardiovascular disease and heart age score for a male smoker in the age range 40-44 with several other risk factors.
    d. NHLBI Guidelines for Asthma is a free program based on national clinical practice guidelines
    e. STAT Growth Charts plots typical growth charts for children. Lite version is free and fully functional version is $5.99
    f. STAT ICD-9 Coder is available as a Lite version for free or full version for $29.99
    g. STAT BMI Obesity Chart determines if a patient is overweight or obese. Costs $4.99
    h. STAT Immunizations is a program for adults and children that costs $4.99. The lite version that covers only adults is free.
    i. STAT Cardiac Clearance is based on a 2007 national guideline to determine what type of evaluation is necessary for a patient facing non-cardiac surgery. A free application.
    j. STAT Insulin DM2 is a program that discusses many of the issues regarding the use of insulin in type 2 diabetes. A free application. [9]

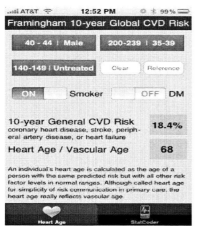

Figure 11.1. Framingham Risk Score for iPhone (courtesy iTunes)

2. **MedCalc.** A very popular free program for the Palm, Windows Mobile and iOS platforms. They are creating new programs only for the iPhone and offer over 125 commonly used formulas. The platform is open source so it can be modified. It is also available in English, French, Spanish and German. You can customize the calculator views to make programs very user-friendly. Program includes IV infusion rate calculators that should improve medication safety.[10] Figure 11.2 shows a typical Medcalc menu on the iPhone

Figure 11.2 MedCalc for the iPhone (courtesy MedCalc)

3. **Archimedes.** A free program by Skyscape for Palm, Palm WebOS, iPhone, BlackBerry, Windows Pocket PC and Windows Mobile operating systems that is similar to Medcalc.

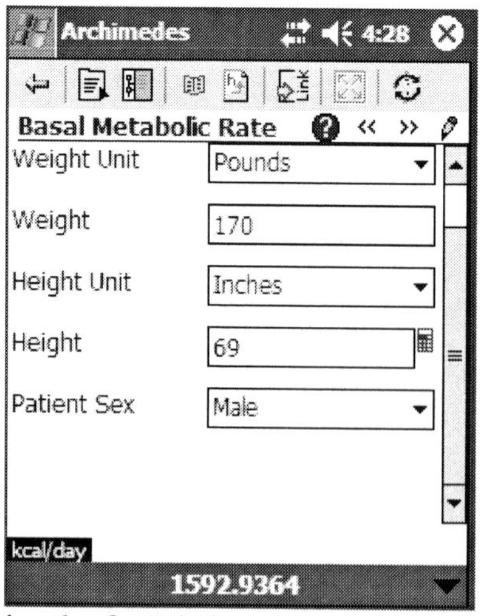

Figure 11.3. Archimedes for Windows Pocket PC (courtesy Skyscape)

They also offer a fee-based program with an extra 75 calculators for $24.95. In addition, they offer a free online calculator.[11] Figure 11.3 shows Archimedes for the Windows platform.

4. **ABG Pro.** A free Palm OS program that interprets arterial blood gases. Figure 11.4 shows a typical screen

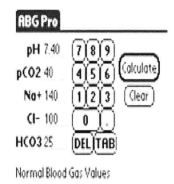

Figure 11.4 ABG Pro calculator (courtesy Stacworks)

5. **Evidence Based Medicine (EBM) Calculator.** A free program available for both the Palm OS and Windows OS systems. It calculates relative risk reduction (RRR), absolute risk reduction (ARR) and the number needed to treat (NNT) for randomized controlled trials (RCTs).[13] We will discuss this in more detail in the chapter on EBM.

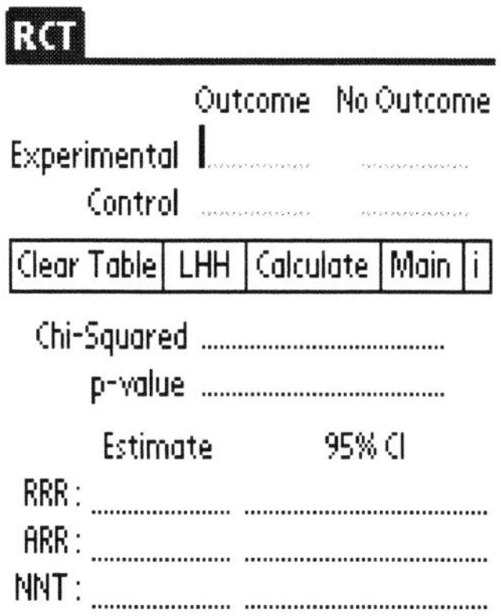

Figure 11.5 Randomized Controlled Trial calculator (courtesy CEBM)

6. **Epocrates.** Although this application is noted for its drug look-up capabilities, it also comes with commonly used formulas for easy calculation. This application will be discussed in detail under drug look up programs. Available for Palm, Windows Mobile, Android, BlackBerry and iOS smartphones. [3]

7. **ICU Math Medical** calculator for the adult ICU uses 85 medical equations, including pulmonary, cardiology, BNP-CHF nomogram, pharmacokinetic dosing, renal, electrolyte, chemistry, nutrition, TPN, peri-operative risk, biostatistics, ACLS, Apache II, unit conversions and rules of thumb. Available only for the Palm platform. [14]

8. **QxMD** offers free content for the iOS, BlackBerry and Android devices. They offer bundled calculators for the iPhone and for the other platforms they offer separate Cardiology, Nephrology, Gastroenterology and Hematology calculators as well as a pregnancy wheel.[15]

Textbooks for Mobile Technology

Most medical textbooks are also available for a mobile platform. They can be stored on internal memory or on external memory (usually secure digital card). There are too many textbooks to highlight so we will instead point readers to common web sites where they can be downloaded. While e-books and e-readers are on the rise, some users may not like navigating and reading a textbook on a smartphone due to the small screen. Certain textbooks such as the 5-Minute Clinical Consultant (5MCC) are better suited to the small screen associated with handheld technology. Figure 11.6 demonstrates a typical screen shot from the 5MCC. A typical disease topic will be organized under basic, diagnosis, treatment, ongoing care, miscellaneous, references and author. In addition to disease content, they offer additional sections such as algorithms, signs and symptoms, calculators, ICD-9 look up and SNOMED look up.

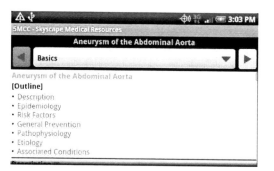

Figure 11.6 5MCC Typical screen shot for abdominal aortic aneurysm on an Android smartphone

One of the most popular sites for e-textbooks for mobile technology is Skyscape.[11] They offer almost 400 titles covering primary care and multiple specialties for all of the major smartphones. MobiPocket is another company that sells medical e-books (1700+) that require a special reader downloaded to the smartphone. E-books reader software is available for Palm, BlackBerry, Symbian and Windows Mobile devices.[16]

Drug Look-Up Programs

Like calculators, drug look up programs are ideal for mobile technology. That is why they are the most popular category of medical software.

1. **Epocrates.** The most dominant drug program has been Epocrates, because it was the first available (1999), was intuitive, innovative and inexpensive. They claim 900,000 healthcare users in 2010. The program is organized in a manner that is logical to most healthcare workers. After searching for a drug, the viewer has the choice to go to adult or pediatric dosing, black box warnings, contraindications, drug interactions, adverse reactions, safety/monitoring, pharmacology, manufacturer/price and pill pictures. Figure 11.7 displays a typical screenshot of Epocrates for the iPhone.

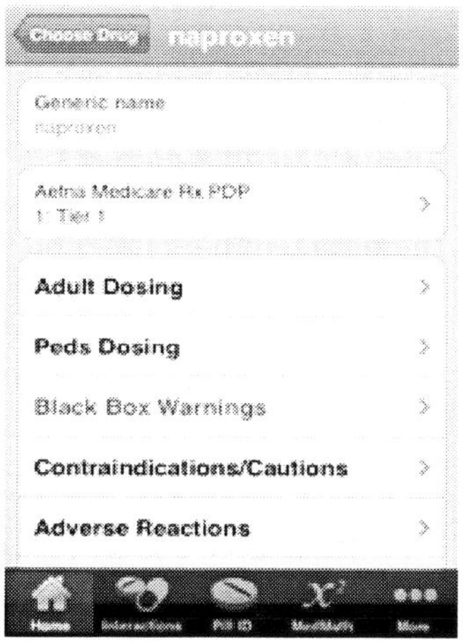

Figure 11.7 Epocrates program on the iPhone (courtesy Epocrates)

The software also has the following standard features:

a. The Medicare part D formulary: important with new prescription benefits and e-prescribing
b. Continuing medical information (CME): users can fulfill their state-specific requirements with this option
c. MedTools: a set of free medical calculators
d. Doc Alerts: medical alerts and news downloaded to your smartphone
e. Local formulary hosting: available at an additional cost
f. Epocrates Patient Snapshot: a 2009 program that allows a patient to designate a physician to access their Google Health profile. A physician must sign up for the program and obtain an ID. A patient can then designate an Epocrates-using physician to access their profile
g. Epocrates OTC drugs: a new mid-2009 offering that is available with their premium software programs only

They claim to have an EHR available for smartphones by the fall 2010. As of mid-2010, Epocrates is available for Palm, Windows mobile, BlackBerry, Android and *i*OS as the following programs:

i. **Epocrates Rx.** A free program with more limited drug content than fee-based programs. Special discounted generic drugs from Target and Kmart are now available as separate formularies.
ii. **Epocrates RxPro.** Provides the drug program plus an infectious disease (ID) program. Program covers 3,300 brand/generic drugs and 600 alternative drugs for $99/1 year or $169/2 years.
iii. **Epocrates Essentials.** Includes a drug program, an infectious disease (ID) program, a lab test reference, symptom assessment tool and disease monographs for $159/1 year or $269/2 years.
iv. **Epocrates Essentials Deluxe.** Includes the Essentials package plus a medical dictionary (100,000 terms) and a coding tool (ICD-9 and CPT) for $199 for one year or $299/2 years. [3]

According to a 2002 study at the Brigham and Woman's Hospital regarding the use of Epocrates: 82% thought it helped inform patients; 63% thought it reduced adverse drug events and 50% thought it reduced about one medication error per week.[17] A 2006 study of 3600 users of Epocrates showed that the majority accessed this program more than 6 times daily and about 40% used it for more than 50% of patient encounters.[18]

2. **Tarascon Pharmacopeia.** It is simple to use and similar to the popular pocket book version. It is available for Palm, BlackBerry, Windows and iPhone smartphones. Cost is $39.95.[19]

3. **Medscape Mobile 2.0.** This free program covers 6000 brand name, generic and over-the-counter (OTC) and herbal drugs for iOS devices with plans to be available for the BlackBerry in 2010. Standard features include a drug-drug, drug-herbal and drug-supplement interaction engines. It also includes medical news, CME, physician, pharmacy and hospital directories and peer-reviewed disease content (3200 topics plus images and videos).[20]

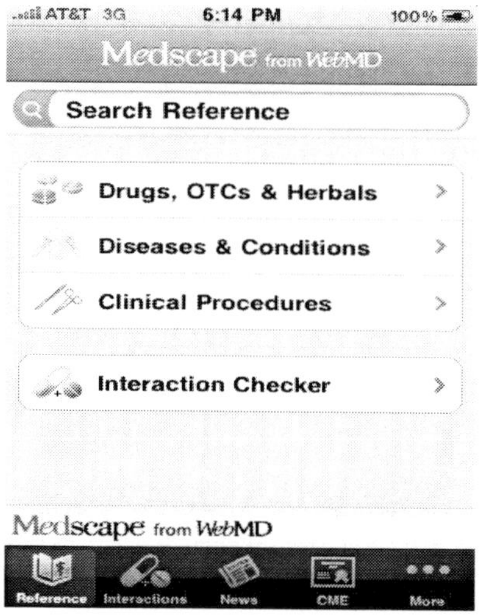

Figure 11.8 Medscape Mobile for iPhone (courtesy Medscape)

4. **Thomson Clinical Xpert.** This program is available free for Palm or Windows Mobile. It has a drug database, drug-drug interactions, toxicology, lab test resource, disease databases, calculators and medical news.[21]

5. **Gold Standard/Elsevier Clinical Pharmacology Mobile.** Basic module has features similar to the other programs mentioned. Companion modules are: drug iDentifier, IV drug compatibility and formulary management. Currently available for Palm or Windows Mobile. Basic module costs $199; addition of companion modules adds $39-$139.[22]

6. **Drug e-Books.** Skyscape has 67 drug-related e-textbooks on their web site. Drug Handbook has all of the basic features already discussed and available for multiple operating systems. Prices range from $4-$60. It is important to point out that unlike the previous choices the e-books do not "autoupdate" to keep content current.[11]

Infectious Disease Programs

After calculators and drug look-up programs, the next most popular downloadable category for the smartphone is Infectious Disease (ID) programs. Like drug programs they provide quick answers to straightforward questions. According to one study, infectious disease programs for the handheld were found to be a good reference for a majority of general internal medicine hospital admissions with infectious disease conditions.[23] These programs may not be able to answer highly complicated questions, however. For an excellent review of the first three programs below, we refer you to an article by Miller.[24]

Epocrates Infectious Disease Treatment Guide. This program can only be purchased as part of a fee-based Epocrates program and not part of the free program. [3]

Sanford Guide to Antimicrobial Therapy 2010. The electronic version of the highly popular handbook is updated yearly. It is available for Palm, Windows and BlackBerry smartphones with an annual cost of $34.95.[25]

Johns Hopkins Antibiotic Guide. The Editor-in-Chief is the well-respected John Bartlett MD. The program is available as a desktop or smartphone program with auto-updates. The program was upgraded in September 2007 to include sections on diagnosis, drugs, pathogens, management and vaccines. The program costs $24.95 and is available for the Palm, Windows Pocket PC, Android, BlackBerry, Symbian and iOS devices. [11]

IDdx is an infectious disease clinical decision support for the Palm, BlackBerry and Pocket PC platforms. It includes 275 infectious diseases, 119 signs and symptoms, 39 epidemiologic factors, covering 16 regions of the world. It is based on Microsoft Access so queries are easy. Cost is $29.95 [26]

Database Programs

Because most handheld operating systems store, retrieve and calculate data rapidly, a database program can be a valuable addition. Several excellent mobile database programs exist, but only one program will be discussed because it is simple, yet robust with scalable medical databases.

HanDBase. This program creates a database on your desktop that you can send to a PDA or smartphone. Simply set up the fields with the characteristics you want in the desktop program and synchronize that to the smartphone. The database is created and saved on both your PC and your handheld. Figure 11.9 shows a simple antibiogram (a database that shows antibiotic sensitivity to common bacteria). You can create small databases for inventories, lists of all of your usernames and passwords and so forth. It's easy to add a password to keep your information secure. The software is available for Palm, Windows, iOS, Symbian and BlackBerry devices. HanDBase is a relational database and not a flat file, so one database can be related to another database. The professional version comes with "Forms" that provides an attractive "front end" user interface (UI) for data entry. An example of forms is seen in figure 11.10. Over 2000 free databases are already created and available on the website; all you have to do is fill in the data. As an example, you can find a patient tracking or wine inventory database. Databases can also synchronize to Microsoft Excel and Access. The cost for the professional version is $39.[27]

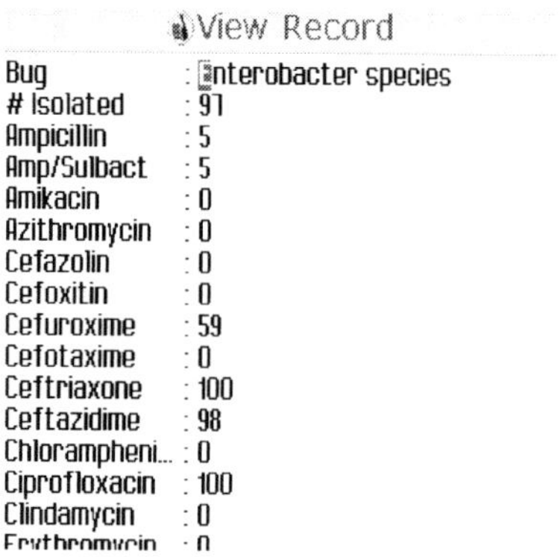

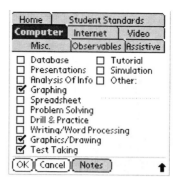

Figure 11.9 Simple database

Figure 11.10 Form interface

Document Readers

PDAs and smartphones will not automatically read every document type. While searching commercial medical software web sites you will occasionally see the comment that you need a document reader. There are numerous good products on the market but only two will be discussed.

1. **iSilo.** With this program you can read documents that you download or create. Using iSilo (and iSiloX), one of the authors converted a very lengthy guideline to prevent blood clots (DVTs) in hospitalized patients to the handheld format. Because the program was written with html you can easily hyperlink one page to another. High text compression with this program results in a much smaller file size. After a few screen taps, you can find the answer. iSilo is available for Palm, Windows, iOS, BlackBerry, Android and Symbian devices. The program will allow rich text, images and tables to be added. The cost of this reader is $20 and that includes frequent upgrades.[28] iSiloX is a free desktop program that converts html documents, text documents and images so they can be seen on your handheld (in the iSilo document reader). The DVT program below (figures 11.11-11.13) was written in html and converted with iSiloX.[29]

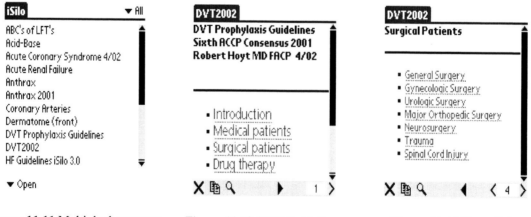

Figure 11.11 Multiple documents   Figure 11.12 DVT Guidelines   Figure 11.13 Hyperlinks

2. **MobiPocket Reader.** This free program for Palm, Windows mobile, BlackBerry and Symbian devices allows the reader to read e-books and access approximately 1700+ fee-based medical e-books on their mobile platform. With the free MobiPocket Creator program

you can create a simple e-book or database for download to your favorite mobile device. E-books can be created by conversion of Word documents, simple text and pdfs.[16]

Patient Tracking

Very few clinicians are willing to manually input patient data into their PDA or smartphone on a regular basis. It sounds impressive to say that you have all of your patient data on your smartphone but the reality is, this is very time consuming. As mentioned, HanDbase has several free databases that could be used in this manner. Another program called Patient Tracker offers a mobile (Palm and Windows CE PDAs) solution for $30.[30] Another product is Patient Keeper that has substantially modified its product line to match "Meaningful Use" criteria discussed in the chapter on electronic health records. This web based solution has a suite of clinical and financial modules that can be accessed by any smartphone with a web browser and include modules such as order entry, e-prescribing, clinical notes, results, medication histories and physician portal. [31]

E-prescribing

Electronic prescribing can be part of an electronic health record or be a standalone application that uses a mobile technology platform such as a smartphone. E-prescribing will be discussed in detail in the chapter on e-prescribing.

Forms on the Mobile Platform

Mobile forms are offered by a variety of vendors, generally for specific operating systems because client software needs to be loaded on the device. With the trend towards "Software as a Service", you can now find programs that are associated with the form on the smartphone but the data stored, collated and analyzed in the cloud.

Pendragon Forms. With this program you can create customizable surveys. As an example, home health nurses could record the status of a patient and synchronize the information to the office PC at the end of the day. Data is uploaded as an Excel file that saves time and money by going paperless. Data can also be synchronized to Microsoft Access or a SQL server. The company also offers the ability to synchronize to a remote server so workers from multiple areas can send information to a head office. Cost for an individual license of version 5.1 is $299 and is available for Palm and Window Pocket PC operating systems. A two week free trial is available.[32]

Go Canvas. This fee-based forms tool is available for the Android, iOS, BlackBerry and Windows Mobile Oss. A user can download a pre-designed form from a forms library covering 15 different industries to the smartphone. Alternately, a user can create a unique new form using web tools. The data is uploaded to a remote server where reports can be generated as pdfs, Excel files or XML documents. Bar codes and images can be part of a newly created form. Figure 11.14 shows a screen shot of a diabetic tracking form used on a HTC smartphone. [33]

Other software

There are simply too many smartphone medical or healthcare-related software programs and vendors to list. An April 2010 search for the word "medical" at Handango returned 3523 medical software programs for the iPhone, 538 for Windows Mobile, 421 for BlackBerry Storm, 32 for the Palm Pre and 17 for Android.[34] Applications have been written for every member of the healthcare team, as well as for patients. A variety of medical dictionaries, encyclopedias and e-books are available in addition to tools to access PubMed for literature searches. Table 11.1 lists medical software program vendors and helpful smartphone general information web sites.

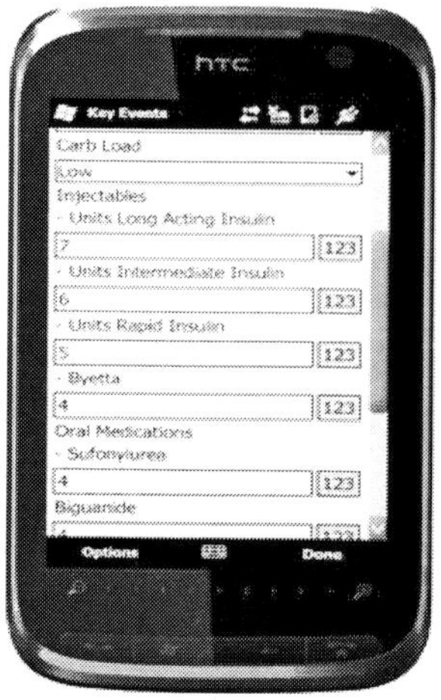

Figure 11.14 Diabetic tracking form (Courtesy Go Canvas)

Table 11.1 Medical software for smartphones (A = Android, BB = BlackBerry, IP = iPhone/iTouch, P = Palm, PPC = Pocket PC, WM = Windows Mobile, S = Symbian)

Website	URL	Platforms supported	Cost
Electronic Preventative Services Selector	http://epss.ahrq.gov/PDA/index.jsp	BB, P, WM, IP	Free
Engadget Mobile	www.engadgetmobile.com	A,BB, P, IP, PPC, WM, S	Free
Handango	www.handango.com	A, BB, IP, P, WM, S	$
iTunes App Store	www.apple.com/itunes	IP	$, Free
Lexi-Comp	www.lexi.com	A, BB, IP, P, PPC, WM	$
MeisterMed	www.meistermed.com	A, BB, IP, PPC, WM, S	$, Free
MemoWare	www.memoware.com	A, BB, IP, PPC, WM, S	Free
Mobihealthnews	http://mobihealthnews.com	A, BB, IP, PPC, WM, S	Free
Mobile MerckMedicus	www.merckmedicus.com	A, BB, P, IP PPC, WM	Free
Palm Doc Chronicles	www.palmdoc.net	A, BB, IP, PPC, WM, S	Free
PDA Medisoft	www.pdamedisoft.com	A, BB, P, PPC, S, WM	$
Pepid	www.pepid.com	A, BB, IP, P, PPC, WM	$
Pocketgear	www.pocketgear.com	BB, P, PPC, S, WM	$
Shots 2010	www.immunizationed.org	P, PPC (web access for IP and BB)	Free
Unbound Medicine	www.unboundmedicine.com	A, BB, IP, P, PPC, WM	$
USBMIS	www.USBMIS.com	BB, P, PPC	$, Free

Smartphones

"I have always wished that my computer would be as easy to use as my telephone. My wish has come true. I no longer know how to use my phone"
Bjarne Stroustrup

We previously defined smartphones as having an operating system allowing for medical programs to be installed. Additional capabilities include phone service, e-mail, Internet access, calendars, contact lists, task lists, cameras and video capability. Synchronization to a computer can be by Bluetooth, WiFi or a USB connection. Touch screens and speech recognition have made data entry easier, compared to a stylus. Internal memory is no longer an issue with smartphones because most have slots for mini SD cards, available in the 1-32 GB range. Physicians who may have carried a pager, cell phone and PDA can have a single multi-purpose device to receive routine phone calls, text messages or voice mails.[35] Moreover, with much faster Internet access we can anticipate more interest in using smartphones to e-prescribe, access online resources, access EHRs and many more functions. NextGen and Allscripts, for example, offer access to their EHRs via the iPhone.

Since December 2005 most carriers offered the benefits of faster 3G networks, with 4G speeds being offered in some US markets. Currently AT&T and T-Mobile use HSDPA standard for their networks, while Sprint and Verizon use EV-DO Rev.A. These services offer a download speed of about 1.5 Mbps. In the 4G space Mobile WiMax and LTE offer much faster upload and download speeds.[36] For additional information about 4G networks, see the chapter on architectures of information systems.

More web sites are producing mobile versions of their web sites to accommodate the smaller screen size of most smartphones. Web sites can detect the browser version requesting a page from the server, and can redirect the browser to a mobile-only page or use style sheets made specifically for the mobile platform to display the content in a more compact and easier to browse manner. For further reading on the various platforms and technical specifications of each smartphone we refer you to this resource.[37]

We will discuss six smartphone operating systems (OSs) and highlight their similarities and differences. Table 11.2 will give an example of a popular smartphone in each operating system category:

1. **Android OS.** This operating system was developed by Google, leveraging a modified Linux kernel with its major modules released as open source and offered in smartphones by T-Mobile, AT&T, Sprint and Verizon in 2010. This OS executes applications built on the software development kit (SDK) available to developers worldwide to create new medical and non-medical applications. In mid-2010 Google released App Inventor that would allow non-programmers to create apps. The first Android device, the T-Mobile G1 has a touch screen, trackball and QWERTY keyboard. As anticipated, this operating system is seamlessly integrated with Google G-mail, Calendar, Maps and Chrome Lite browser. The Android Market claimed 50,000 applications in the early 2010 time frame. Google released the HTC Nexus One in early 2010 that is a phone built by HTC and offered by T-Mobile and AT&T. While it has several excellent features such as a very fast 1 GHz processor and 5 megapixel camera it only stores applications in internal memory and not in its secure digital card. The HTC Incredible was released in April 2010 with an 8 megapixel camera and other features similar to the Nexus One and offered by Verizon. The HTC Evo was released in June 2010 with a larger screen, 4G

support, and more memory. Software is available through multiple vendors and the Android Market. [38]

2. **BlackBerry OS**. BlackBerry has been known for its excellent platform to receive and send e-mail, but since late 2007 they have become a more valuable platform for the medical profession because of expanding medical software. In mid-2008 the BlackBerry 9000 (Bold) was released that included a faster HSDPA network, a faster processor at 624 MHz, more memory at 1GB, GPS, WiFi and a sharper screen. The BlackBerry 8900 (Curve) was also recently released with many similar features to the BOLD but with a sharper screen and ability to place phone calls via WiFi. Later in 2008 the touch screen BlackBerry 9530 (Storm) was released and its features are outlined in Table 11.2. Like the iPhone, an application store was opened in early 2009 for the BlackBerry. In early 2010 the BlackBerry App Store listed 30 fitness programs, 42 health & diet, 21 healthcare services and 99 medical guides. Enterprise Business (WLAN) solutions are available to incorporate nurse call functions, lab data, decision support, alerts, charge capture, e-prescribing, bar coding and RFID. Enterprise solutions allow for group e-mails to be pushed to clients. BlackBerry now offers the BlackBerry Enterprise Server Express software free on their web site to help create an enterprise solution. [39]

3. **iOS**. Without a doubt, the operating system whose supported devices have received the most buzz in the past few years has been iOS. In its first 3 months, Apple sold about 1 million iPhones. It was the first touch screen smartphone that was easy to navigate and robust in web applications. The web experience is enhanced by its Safari browser and large screen. Within a year most medical software companies had created popular solutions for the iPhone. Some would argue that the BlackBerry platform is for those who use e-mail extensively and the iPhone for those who spend a lot of time on the Web. The iPod Touch has become another Apple platform for medical programs because, in essence, it has the iPhone capabilities without the phone feature and is less expensive. The Apple App Store continues to be a major drawing card for this platform and more and more developers create innovative software. By early 2010 Apple claimed 150,000 applications, with approximately 2500 being health-related. Significant improvements occurred with an OS3 upgrade that includes: a system wide search, keyboard in landscape mode, cut and paste capability, multimedia messaging and voice control of contacts and music. The newer iPhone 3G S was released in June 2009 and included these upgrades plus: built in compass, video recording capability, faster application launching, memory increase to 32 GB, better 3-D graphics, better camera and voice memos. A new iPhone 4 was released in June 2010. Details about the iPad can be found in the last chapter on emerging trends. One significant drawback of the iPhone is the fact that the battery cannot be replaced by the user and must be sent off for replacement for about $80 plus shipping. [40]

4. **Palm Garnet OS**. Palm was the first operating system offered on early PDAs and as a result this platform probably has more medical software programs than any other. Because its OS is considered outdated and sales of Palm PDAs faltered, they have migrated to the newer Palm Pre and Palm Pixi we will discuss in the next section. Palm has a simple, intuitive and reliable OS and this explains the initial success of the Palm Treo phone. [2]

5. **Web OS**. In an effort to stay competitive in the smartphone arena, Palm developed WebOS, derived from Linux, for the smartphone platform and released in June 2009. This platform has a touch screen as well as a QWERTY slide out keyboard. It does not

offer additional memory with a memory card but offers a unique wireless charging with electromagnetic induction. The Pre is able to automatically consolidate contact, e-mail and calendar information from multiple sources to include Google, Outlook and Facebook. You can connect with friends via text messages, instant messaging, e-mail, phone calls or Facebook. WebOS allows multitasking; new pages like a deck of cards will appear so that e-mail and example calendar functions will be accessible at the same time. Palm Pixi was released in 2009 that was thinner than the Pre and also offered by Sprint and later on the AT&T network. A user must have a program known as "Classic" (an emulator), in order to view older Palm software programs.[2] In April 2010 Palm was sold to Hewlett-Packard and it is too early to know how this will impact this platform.

6. **Symbian OS.** This operating system is prominent in Europe but not the United States. As of 2010, there are limited medical software programs for this OS.[41] In 2008, The Symbian Foundation was created to promote the use of the new open source Symbian operating system. [42] Clearly, this effort was intended to compete with open source Android and the myriad of applications being developed for all smartphone operating systems. The largest Symbian smartphone platform is Nokia and an example of a Nokia smartphone is presented in Table 11.2.

Table 11.2 Smartphone operating system platforms and features

Specs:	HTC Incredible	BlackBerry Storm	iPhone 3 GS	Palm Pre Plus	Nokia Nuron 5230	Samsung Saga
Provider	Verizon	Verizon	AT&T	Sprint Verizon	AT&T T-Mobile	Verizon
OS	Android 2.1	BB	iOS	Web OS	Symbian	Windows Mobile 6.1
Screen size (Inches)	3.7	3.2	3.5	3.1	3.2	2.6
Screen Resolution	800 x 480	360 x 480	320 x 480	320 x 480	640 x 360	320 x 320
Camera (Megapixels)	8	3.2	3	3	2	2
Expandable Memory	MicroSD	MicroSD	No	No	MicroSD	MicroSD
WiFi	Yes	Yes	Yes	Yes	No	Yes
Bluetooth	Yes	Yes	Yes	Yes	Yes	Yes
Network	GSM, UMTS	GSM, CDMA, UMTS	GSM, UMTS	CDMA	GSM, UMTS	GSM, CDMA
Internal Memory	8GB	8 GB	16/32 GB	16 GB	70 MB	112 MB
Processor Speed	1 GHz	500 MHz	600 MHz	600 MHz	381 MHz	400 MHz
Touch Screen	Yes	Yes	Yes	Yes	Yes	Yes
Price	$199	$199	$199-$299	$49.99	$69.99	$24.99

7. **Windows Mobile OS.** Before the emergence of BlackBerry and the iPhone the Windows operating system was clearly competitive with Palm. One of the attractions was the ability of this OS to integrate well with standard Microsoft Office applications, Microsoft Outlook and a mobile version of Internet Explorer. Pocket PC is actually the hardware specification for the Microsoft handheld computer, whereas Windows Mobile is the operating system. At the time that Windows Mobile 6 was announced Microsoft stopped using the name Pocket PC. Devices without a phone are called Windows Mobile Classic devices, whereas devices with a phone and touch screen are called Windows Professional devices. Those devices without a touch screen but with a phone are called Windows Mobile Standard devices. The most recent version is Mobile 6.5, but the Windows Phone 7 series operating system should be out later in 2010. Microsoft created My Phone which is a free service to back up phone information to the cloud for Mobile 6 and higher OSs. Like Android, BlackBerry and the iPhone Microsoft has created a Windows Marketplace for Mobile to encourage developers to create attractive applications for their OS.[43-44] Examples of phones that use this OS are Motorola Q9h, Samsung SCH-i760, HTC Touch Pro, Motorola MC7090 and Palm Treo Pro.

Limitations of mobile technology

Smartphones were not initially intended to replace the PC or laptop, in spite of their impressive evolution. Nevertheless, as computers such as netbooks get smaller and smartphones improve access speed to the Internet and user interface options, there is some convergence. Widespread broadband wireless will also change the utility of smartphones as rapid access to the Internet becomes a reality. Regardless, limitations currently exist for mobile technology:

- Slow inputting: typical keyboards are too small and inputting with styluses is too slow. Touchscreens are popular but have a learning curve all their own due to the lack of true haptic feedback on key press. Given the great recent advances in voice recognition performance, it seems likely that this may help solve some of the inputting issues [45-47]
- Small screen size: traditionally, handheld computers had tiny screens. Phones such as the iPhone, EVO and HTC Incredible have larger screens to make viewing of web pages and documents more reasonable. Newer large screen smartphones also utilize multitouch gesturing which can allow for more rapid navigation through multiple pages
- Security: as smartphones are used to access enterprise or patient level data, better security will need to be standard. Encryption and anti-viral programs for the mobile platform exist. [47-48] Programs such as Mobile Defense allow you to locate, lock, backup and wipe your lost smartphone. It is available free as a public beta download, for the Android OS.[49] iOS provides similar functionality through Apple's MobileMe service For other platforms, WaveSecure provides similar functionality at $19.90 per year [50]
- Expense: not everyone is willing to pay for a data plan and other bells and whistles associated with smartphones
- Adoption: is probably no longer a major issue if 64% of physicians now embrace this technology.[6] It seems likely that most professionals will be using smartphones for professional and personal reasons in the near future
- Lack of Proof: there is a paucity of data concerning the impact of PDAs or smartphones on physician efficiency, quality of care delivered, improved patient safety or return on investment. More research is needed [51]

Key Points

- Handheld technology has moved quickly from personal digital assistants (PDAs) to smartphones

- Multiple medical software programs are available for mobile platforms that are free, shareware or fee-based

- Drug look-up programs and calculators are very fast and intuitive and therefore essential to the average clinician

- The technology is improving rapidly such that a majority of clinicians now own and use a smartphone

Conclusion

Mobile technology continues to improve and gain popularity in the medical profession at an amazing pace. Smartphones are being used for storing medical information, telephonic communication, patient monitoring and clinical decision support. In the not too distant future, they will likely be used commonly for geo-location and connectivity to electronic health records and other hospital networks. Smartphones have replaced most PDAs as processer speed, memory, network access and multimedia features continue to improve. Interest in smartphones will continue to increase due to more medical and non-medical applications developed, as well as evolving 4G networks. We believe that voice recognition has improved to the degree that it may become a prominent means of inputting for the smartphone in the not too distant future. Competition among the various operating systems is intense, driving functionality up and cost down. However, mobile technology has definite limitations and further research is needed to determine their actual impact and place in the armetarium of most physicians. An excellent review *How Smartphones Are Changing Health Care for Consumers and Providers by the California HealthCare Foundation* appeared in 2010 that addresses the current and future state of smartphones and mobile technology.[52]

References

1. Everymac.com http://www.everymac.com/systems/apple/messagepad/stats/newton_mp_omp.html (Accessed February 1 2006)
2. Palm http://www.palm.com/us/ (Accessed April 10 2010)
3. Epocrates www.epocrates.com (Accessed April 10 2010)
4. PDA Medisoft http://www.pdamedisoft.com/BlackBerry (Accessed February 2 2007)
5. Kho A, Henderson LE, Dressler DD. Use of Handheld computers in medical education. A Systematic Review. J Gen Int Med 2006;21:531-537
6. Manhattan Research. Taking the Pulse v9.0 www.manhattanresearch.com April 22 2009 (Accessed April 23 2009)
7. Anderson P, Lindgaard AM, Prgomet M et al. Is selection of hardware device related to clinical task? A multi-method study of mobile and fixed computer use by doctors and nurses on hospital wards. J Med Internet Res 2009;11(3):e32

8. Apple iTunes App Store. www.apple.com/itunes (Accessed April 10 2010)

9. StatCoder www.statcoder.com (Accessed April 1 2010)

10. MedCalc www.med-ia.ch/medcalc/download/html (Accessed April 1 2010)

11. Skyscape www.skyscape.com (Accessed April 1 2010)

12. ABG Pro http://www.stacworks.com/index.html (April 2 2010)

13. Center for Evidence Based Medicine http://www.cebm.utoronto.ca/practise/ca/statscal/ (Accessed January 2 2009)

14. FreewarePalm http://www.freewarepalm.com/medical/icumath.shtml (Accessed June 1 2009)

15. QXMD www.qxmd.com (Accessed April 10 2010)

16. MobiPocket www.mobipocket.com (Accessed June 1 2009)

17. Rothschild JM et al Clinician use of a palmtop drug reference guide JAMIA 2002;9:223-229

18. Rothschild JM, Fang E, Liu V, et al. Use and Perceived Benefits of Handheld Computer-based Clinical References JAMIA 2006 2006;13:619-626

19. Tarascon http://www.tarascon.com/home.php?cat=6 (Accessed April 10 2010)

20. Medscape Mobile www.medscape.com (Accessed April 5 2010)

21. Thomson Clinical Xpert http://www.pdr.net/pda-medical/ (Accessed April 10 2010)

22. Gold Standard. http://www.clinicalpharmacologyonhand.com/marketing/about_cpoh.html (Accessed April 101 2010)

23. Burdette SD, Herchline TE and Richardson WS Killing bugs at the bedside: a prospective hospital survey of how frequently personal digital assistants provide expert recommendations in the treatment of infectious diseases Ann Clin Micro and Antim. 2004;3 or http://www.ann-clinmicrob.com/content/3/1/22 (Accessed January 1 2008)

24. Miller SM, Beattie MM and Butt AA Personal Digital Assistant Infectious Diseases Applications for Health Care Professionals Clin Inf Dis 2003;36:1018-1029

25. The Sanford Guide www.sanfordguide.com (Accessed April 10 2010)

26. IDdx http://www.usbmis.com/iddx/# (Accessed April 14 2010)

27. HanDbase www.ddhsoftware.com (Accessed April 8 2010)

28. iSilo www.isilo.com (Accessed April 8 2010)

29. iSiloX www.isiloX.com (Accessed April 8 2010)

30. Patient Tracker http://www.patienttracker.com/product_patienttracker.shtml (Accessed April 9 2010)

31. Patient Keeper http://www.patientkeeper.com/ (Accessed April 10 2010)

32. Pendragon forms www.pendragon-software.com (Accessed April 10 2010)

33. Go Canvas. www.gocanvas.com (Accessed July 1 2010)

34. Handango www.handango.com (Accessed April 10 2010)

35. Burdette, SD, Herchline TE, Dehler R. Practicing Medicine in a Technological Age: Using Smartphones in Clinical Practice. Surfing the web. CID 2008;47:117-22

36. Tips & Tricks . Mobile Technology. Laptop Magazine August 2008

37. 2010 Smartphone Product Comparisons http://cell-phones.toptenreviews.com/smartphones/ (Accessed April 10 2010)

38. Google Nexus One www.google.com/phone (Accessed April 11 2010)

39. BlackBerry. www.blackberry.com (Accessed April 10 2010)

40. iPhone www.apple.com/iphone (Accessed April 10 2010)

41. Symbian http://reviews.cnet.com/45210-11309_7-6624304-5.html (Accessed May 25 2009)

42. Symbian Foundation www.symbian.org (April 10 2010)

43. Windows Mobile www.wikipedia.com (Accessed May 26 2009)

44. Windows Mobile www.microsoft.com/windowsmobile/6-1/default.mspx (Accessed May 26 2009)

45. VoiceIT http://promo.palmgear.com/voiceit/ (Accessed January 1 2008)
46. Rottman, R. The role of speech on ultra-smart smartphones. http://www.24100.net/2009/03/the-role-of-speech-on-ultra-smart-smartphones/ (Accessed June 5 2009)
47. Ultimaco http://americas.utimaco.com/products/pda/ (Accessed January 1 2008)
48. Avast http://www.avast.com/eng/download-avast-pda.html (January 1 2008)
49. Mobile Defense www.mobiledefense.com (Accessed April 29 2010)
50. WaveSecure www.wavesecure.com (Accessed April 29 2010)
51. Prgomet, M, Georgiou A, Westbrook J. The impact of mobile handheld technology on hospital physicians' work practices and patient care: a systematic review. J Am Med Infor Assoc 2009;16:792-801
52. Sarasohn-Kahn J. How Smartphones Are Changing Health Care for Consumers and Providers. California HealthCare Foundation. www.chcf.org April 2010 (Accessed April 10 2010)

Evidence Based Medicine

ROBERT E HOYT
M HASSAN MURAD

Learning Objectives:

After reading this chapter the reader should be able to:

- State the definition and origin of evidence based medicine
- Define the benefits and limitations of evidence based medicine
- Describe the evidence pyramid and levels of evidence
- State the process of using evidence based medicine to answer a medical question
- Compare and contrast the most important online and PDA/smartphone evidence based medicine resources

> *"The great tragedy of Science- the slaying of a beautiful hypothesis by an ugly fact"*
> Thomas Huxley (1825-1875)

Some might ask why Evidence Based Medicine (EBM) is included in a textbook on Medical Informatics. The reason is that medical performance is based on quality and quality is based on the best available evidence. Clearly, information technology has the potential to improve decision making through online medical resources, electronic clinical practice guidelines, electronic health records (EHRs) with decision support, online literature searches, statistical analysis and online continuing medical education (CME). This chapter is devoted to finding the best available evidence. Although one could argue that EBM is a buzz word like quality, in reality it means that clinicians should seek and apply the highest level of evidence available. According to the Center for Evidence Based Medicine, EBM can be defined as:

> *"the conscientious, explicit and judicious use of current best evidence in making decisions about the care of individual patients"*[1]

In *Crossing the Quality Chasm,* the Institute of Medicine (IOM) states:

> *"Patients should receive care based on the best available scientific knowledge. Care should not vary illogically from clinician to clinician or from place to place"*[2]

What the IOM is saying, is that every effort should be made to find the best answers and that these answers should be standardized and shared among clinicians. Such standardization implies that clinical practice should be consistent with the best available evidence that would apply to the majority of patients. This is easier said than done because so many clinicians are independent practitioners with little allegiance to any one healthcare organization. It is true that many questions cannot be answered by current evidence so clinicians may have to turn to subject experts. It is also true that the medical profession lacks the time and the tools to seek the best evidence. More than 1,800 citations are added to MEDLINE every day, making it impossible for a practicing clinician to stay up-to-date with the medical literature, not to mention that interpreting this evidence requires certain expertise and knowledge that not every clinician has. One does not have to look very far to see how evidence changes recommendations, e.g. bed rest is no longer recommended for low back pain [3] or following a spinal tap (lumbar puncture); routine activity is recommended instead. [4]

Until these older recommendations were challenged with high quality randomized controlled trials the medical profession had to rely on expert opinion, best guess or limited research studies.

Three pioneers are closely linked to the development of EBM. Gordon Guyatt coined the term EBM in 1991 in the American College of Physician (ACP) Journal Club. [5] The initial focus of EBM was on clinical epidemiology, methodology and detection of bias. This created the first fundamental principle of EBM: not all evidence is equal; there is a hierarchy of evidence that exists. In the mid-1990's, it was realized that patients' values and preferences are essential in the process of decision making, and addressing these values has become the second fundamental principle of EBM, after the hierarchy of evidence. Archie Cochrane, a British epidemiologist, was another early proponent of EBM. Cochrane Centers and the International Cochrane Collaboration were named after him as a tribute to his early work. The Cochrane Collaboration consists of review groups, centers, fields, methods groups and a consumer network. Review groups, located in 13 countries, look at randomized controlled trials. As of 2005 they have completed about 2000+ systematic reviews, even though there have been 300,000 randomized controlled trials published. [6-7] The rigorous reviews are performed by volunteers, so efforts are slow. David Sackett is another EBM pioneer who has been hugely influential at The Centre for Evidence Based Medicine in Oxford, England and at McMaster University, Ontario, Canada. EBM has also been fostered at McMaster University by Brian Haynes who is the Chairman of the Department of Clinical Epidemiology and Biostatistics and the editor of the American College of Physician's (ACP) Journal Club. Although EBM is popular in the United Kingdom and Canada it has received mixed reviews in the United States. The primary criticisms are that EBM tends to be a very labor intensive process and in spite of the effort, frequently no answer is found.

The first randomized controlled trial was published in 1948. [8] For the first time subjects who received a drug were compared with similar subjects who would receive another drug or placebo and the outcomes were evaluated. Subsequently, studies became "double blinded" meaning that both the investigators and the subjects did not know whether they received an active medication or a placebo. Until the 1980s evidence was summarized in review articles written by experts. However, in the early 1990s, systematic reviews and meta-analyses became known as a better and more rigorous way to summarize the evidence and the preferred way to present the best available evidence to clinicians and policy makers. Since the late 1980s more emphasis has been placed on improved study design and true patient outcomes research. It is no longer adequate to show that a drug reduces blood pressure or cholesterol; it should demonstrate an improvement in patient-important outcomes such as reduced strokes or heart attacks. [9]

Why is EBM Important?

Learning EBM is like climbing a mountain to gain a better view. You might not make it to the top and find the perfect answer but you will undoubtedly have a better vantage point than those who choose to stay at sea level. Reasons for studying EBM resources and tools include:

- Current methods of keeping medically or educationally up to date do not work
- Translation of research into practice is often very slow
- Lack of time and the volume of published material result in information overload
- The pharmaceutical industry bombards clinicians and patients everyday; often with misleading or biased information
- Much of what we consider as the "standard of care" in every day practice has yet to be challenged and could be wrong

Without proper EBM training we will not be able to appraise the best information resulting in poor clinical guidelines and wasted resources.

How have we traditionally gained medical knowledge?

Continuing Medical Education (CME). Traditional CME is desired by many clinicians but the evidence shows it to be highly ineffective and does not lead to changes in practice. In general, busy clinicians are looking for a non-stressful evening away from their practice or hospital with food and drink provided.[10-11] Much of CME is provided free by pharmaceutical companies with their inherent biases. Better educational methods must be developed. A recent study demonstrated that online CME was at least comparable, if not superior to traditional CME.[12]

Clinical Practice Guidelines (CPGs). This will be covered in more detail in the next chapter. Unfortunately, just publishing CPGs does not in and of itself change how medicine is practiced and the quality of CPGs is often variable and inconsistent.

Expert advice. Experts often approach a patient in a significantly different way compared to primary care clinicians because they deal with a highly selective patient population. Patients are often referred to specialists because they are not doing well and have failed treatment. For that reason, expert opinion needs to be evaluated with the knowledge that their recommendations may not be relevant to a primary care population. Expert opinion therefore should complement and not replace EBM.

Reading. It is clear that most clinicians are unable to keep up with medical journals published in their specialty. Most clinicians can only devote a few hours each week to reading. All too often information comes from pharmaceutical representatives visiting the office. Moreover, recent studies may contradict similar prior studies, leaving clinicians confused as to the best course.

What are the normal EBM steps towards answering a question?

The following are the typical steps a clinician might take to answer a patient-related question:
1. You see a patient and generate a question
2. You next formulate a well constructed question. Here is the PICO method, developed by the National Library of Medicine, to formulate a question:
   a) **Patient or problem:** how do you describe the patient group you are interested in? Elderly? Gender? Diabetic?
   b) **Intervention:** what is being introduced, a new drug or test?
   c) **Comparison:** with another drug or placebo?
   d) **Outcome:** what are you trying to measure? Mortality? Hospitalizations? A web based PICO tool has been created by the National Library of Medicine to search Medline.[13] This tool can be placed as a short cut on any computer.
   e) It has been recently suggested to add a T to PICO (i.e., PICOT) to indicate the **T**ype of study that would best answer the PICO question.
3. Seek the best evidence for that question via an EBM resource or PubMed.
4. Appraise that evidence using tools mentioned in this chapter.
5. Apply the evidence to your patient considering patient's values, preferences, and circumstances.[14]

Evidence appraisal:

When evaluating evidence, one needs to assess its validity, results and applicability.

Validity means: is the study believable? If apparent biases or errors in selecting patients, measuring outcomes, conducting the study, or analysis are present, then the study is less valid.

Results should be assessed in terms of the magnitude of treatment effect and precision (narrower confidence intervals or statistically significant results indicate higher precision).

Applicability, also called external validity, indicates that the results reported in the study can be generalized to the patients of interest.

The most common types of clinical questions:

1. Therapy question. This is the most common area for medical questions and the only one we will discuss in this chapter
2. Prognosis question
3. Diagnosis question
4. Harm question
5. Cost question

The Evidence Pyramid

The pyramid in figure 12.1 represents the different types of medical studies and their relative ranking. The starting point for research is often animal studies and the pinnacle of evidence is the meta-analysis of randomized trials. With each step up the pyramid our evidence is of higher quality associated with fewer articles published.[15] Although systematic reviews and meta-analyses are the most rigorous means to evaluate a medical question, they are expensive, labor intensive, and their inferences are limited by the quality of the evidence of the original studies.

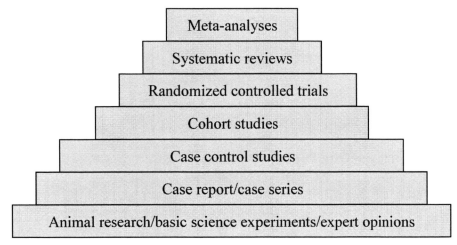

Figure 12.1. The Evidence Pyramid

Case reports/case series. Consist of collections of reports on the treatment of individual patients without control groups, therefore they have much less scientific significance.

Case control studies. Study patients with a specific condition (retrospective or after the fact) and compare with people who do not. These types of studies are often less reliable than randomized controlled trials and cohort studies because showing a statistical relationship does not mean that one factor necessarily caused the other.

Cohort studies. Evaluate (prospectively or followed over time) and follow patients who have a specific exposure or receive a particular treatment over time and compare them with another group that is similar but has not been affected by the exposure being studied. Cohort studies are not as reliable as randomized controlled studies, since the two groups may differ in ways other than the variable under study.

Randomized controlled trials (RCTs). Subjects are randomly assigned to a treatment or a control group that received placebo or no treatment. The randomization assures to a great extent that patients in the two groups are balanced in both known and unknown prognostic factors, and that the only difference between the two groups is the intervention being studied. RCTs are often "double blinded" meaning that both the investigators and the subjects do not know whether they received an active medication or a placebo. This assures that patients and clinicians are less likely to become biased during the conduct of a trial, and the randomization effect remains protected throughout the trial. RCTs are considered the gold standard design to test therapeutic interventions.

Systematic reviews. Defined as protocol-driven comprehensive reproducible searches that aim at answering a focused question; thus, multiple RCTs are evaluated to answer a specific question. Extensive literature searches are conducted (usually by several different researchers to reduce selection bias of references) to identify studies with sound methodology; a very time consuming process. The benefit is that multiple RCTs are analyzed, not just one study.

Meta-analyses. Defined as the quantitative summary of systematic reviews that take the systematic review a step further by using statistical techniques to combine the results of several studies as if they were one large single study.[15] Meta-analyses offer two advantages compared to individual studies. First, they include a larger number of events, leading to more precise (i.e., statistically significant) findings. Second, their results apply to a wider range of patients because the inclusion criteria of systematic reviews are inclusive of criteria of all the included studies.

We will be dealing exclusively with therapy questions so note that randomized controlled trials are the suggested study of choice.[16]

Table 12.1 Suggested studies for questions asked

Type of Question	Suggested Best Type of Study
Therapy	RCT > cohort > case control > case series
Diagnosis	Prospective, blind comparison to a gold standard
Harm	RCT + cohort > case control > case series
Prognosis	Cohort study > case control > case series
Cost	Economic analysis and modeling

Studies that don't randomize patients or introduce a therapy along with a control group are referred to as observational studies (case control, case series and cohort). Most studies that have been reported on health information technology (HIT) are observational studies. This is important because cause and effect are difficult to prove, compared to a RCT. As an example, randomizing physicians to electronic prescribing (vs. paper prescribing) is difficult to implement and often impractical. In an observational study, physicians who volunteer to try electronic prescribing are likely "early adopters" and not representative of average physicians, which would skew the results. Alternate methods of randomization are feasible and desired, however. For example, "cluster randomization" would be a practical methodology in this situation. Here, several clinics or hospitals can be randomized as a whole practice to electronic prescribing whereas other clinics or hospitals can be randomized to paper prescribing.

Evidence of harm should be derived from both RCTs and cohort study designs. Cohort studies have certain advantages over RCTs when it comes to assessing harm: larger sample size, longer follow up duration, and more permissive inclusion criteria that allow a wide range of patients representing a real world utilization of the intervention to be included in the study.

Levels of Evidence (LOE)

Several methods have been suggested to grade the quality of evidence, which on occasion, can be confusing. The most up-to-date and acceptable framework is the GRADE (Grading of Recommendations, Assessment, Development and Evaluation).[17] The following is a description of the levels of evidence in this framework:

Level 1: High quality evidence (usually derived from consistent and methodologically sound RCTs)

Level 2: Moderate quality evidence (usually derived from inconsistent or less methodologically sound RCTs; or exceptionally strong observational evidence)

Level 3: Low quality evidence (usually derived from observational studies)

Level 4: Very low quality evidence (usually derived from flawed observational studies, indirect evidence or expert opinion

In this framework, RCTs start with a level 1 and observational studies start with a level 3; both can be upgraded or downgraded if they met certain criteria based on their methodology and applicability.

Risk Measures and Terminology

Overall, therapy trials are the most common area of research and ask questions such as, is drug A better than drug B or placebo? In order to determine what the true effect of a study is, it is important to understand the concept of risk reduction and the number needed to treat. These concepts are used in studies that have dichotomous outcomes (i.e., only 2 possible answers such as dead or alive, improved or not improved); which are more commonly utilized outcomes. We will define these concepts and then present an example for illustration.

Risk is defined as the rate of events during a specific period of time. It is calculated by dividing the number of patients suffering events by the total number of patients at risk for events.

Odds are defined as the ratio of the number of patients with events to the number of patients without events.

Notice that *Odds=1/(1+risk)*

Consider this example; Amazingstatin is a drug that lowers cholesterol. If we treat a 100 patients with this drug and 5 of them suffer a heart attack over a period of 12 months, the risk of having a heart attack in the treated group would be 5/100= 0.050 (or 5%). The odds of having a heart attack would be 5/95= 0.052. In the control group, if we treat 100 patients with placebo and 7 suffer heart attacks, the risk in this group is 7/100=0.070 or 7% and the odds are 7/93=0.075.
Notice that the risk in the experimental group is called experimental event rate (EER) and the risk in the control group is called control event rate (CER).

To compare risk in 2 groups, we use the following terms:

Relative Risk (RR) is the ratio of 2 risks as defined above. Thus, it is the ratio of the event rate of the outcome in the experimental group (EER) to the event rate in the control group (CER). *RR = EER/CER.*

Relative Risk Reduction (RRR) is the difference between the experimental event rate (EER) and the control event rate (CER), expressed as a percentage of the control event rate.

RRR = (EER-CER)/CER.

Absolute Risk Reduction (ARR) is the difference between the EER and the CER.

ARR = EER-CER

Number Needed to Treat (NNT) is the number of patients who have to receive the intervention to prevent one adverse outcome.[18]

NNT = 1/ARR (or 100/ARR, if ARR is expressed as a percentage instead of a fraction)

Odds ratio (OR) is the ratio of odds (instead of risk) of the outcome occurring in the intervention group to the odds of the outcome in the control group.

Consider the example of Amazingstatin:

On Amazingstatin, 5% (EER) of patients have a heart attack after 12 months of treatment. On placebo 7% (CER) of patients have a heart attack over 12 months

RR = 5% /7% = 0.71

RRR = (7% - 5%) /7% = 29%

ARR = 7% - 5% = 2%

NNT = 100/2 = 50

As we calculated above, the odds for the intervention and control group respectively are 0.052 and 0.075; the odds ratio (OR) = 0.52/0.075 = 0.69

Comments: RR and OR are very similar concepts and as long as the event rate is low, their results are almost identical. These results show that this drug cuts the risk of heart attacks by 29% (almost by a third), which seems like an impressive effect. However, the absolute reduction in risk is only 2% and we need to treat 50 patients to prevent one adverse event. Although this NNT may be acceptable, using RRR seems to exaggerate our impression of risk reduction compared with ARR. Most of what we see written in the medical literature and the lay press will quote the RRR. Unfortunately, very few studies offer NNT data, but it is very easy to calculate if you know the ARR specific to your patient. Nuovo et al. noted that NNT data was infrequently reported by five of the top medical journals in spite of being recommended. [19] In another interesting article, Lacy and co-authors studied the willingness of US and UK physicians to treat a medical condition based on the way data was presented. Ironically, the data was actually the same but presented in three different ways. Table 12.2 suggests that US physicians may need more training in EBM. [20]

Table 12.2 Physician's Likelihood of Prescribing Medication Based on How Research Data is Presented

Physicians From	Relative Risk Reduction (RRR)	Absolute Risk Reduction (ARR)	Number Needed To Treat (NNT)
United States	54%	4%	10%
United Kingdom	24%	11%	22%

Examples of using RRR, ARR and NNT. A full page article appeared in a December 2005 Washington Post newspaper touting the almost 50% reduction of strokes by a cholesterol lowering drug. This presented an opportunity to take a look at how drug companies usually advertise the benefits of their drugs. Firstly, in small print, you note that patients have to be diabetic with one other risk factor for heart disease to see benefit. Secondly, there are no references. The statistics are derived from the CARDS Study published in the Lancet in Aug 2004.[21] Stroke was reported to occur in 2.8% in patients on a placebo and 1.5% in patients taking the drug Lipitor. The NNT is therefore 100/1.3 or 77. So, you had to treat 77 patients for an average of 3.9 years (the average length of the trial) to prevent one stroke. This doesn't sound as good as "cuts the risk by nearly half". Now armed with these EBM tools, look further the next time you read about a miraculous drug effect.

Number needed to harm (NNH) is calculated similarly to the NNT. If, for example, Amazingstatin was associated with intestinal bleeding in 6% of patients compared to 3% on placebo, the NNH is calculated by dividing the ARR (%) into 100. For our example the calculation is 100/.03 = 33. In other words, the treatment of 33 patients with Amazingstatin for one year resulted, on average, in one case of intestinal bleeding as a result of the treatment. Unlike NNT, the higher the NNH, the better.

Cost of preventing an event (COPE). Many people reviewing a medical article would want to know what the cost of the intervention is. A simple formula exists that sheds some light on the cost: COPE = NNT x number of years treated x 365 days x the daily cost of the treatment. Using our example of Amazingstatin = 40 x 1 x 365 x $2 or $29,200 to treat 40 patients for one year to prevent one heart attack. Now you can compare COPE scores with other similar treatments.[22]

Limitations of the Medical Literature and EBM

Because evidence is based on information published in the medical literature, it is important to point out some of the limitations researchers and clinicians must deal with on a regular basis:
- There is a low yield of clinically useful articles in general [23]
- Conclusions from randomized drug trials tend to be more positive if they are from for-profit organizations [24]
- Up to 16% of well publicized articles are contradicted in subsequent studies [25]
- Peer reviewers are "unpaid, anonymous and unaccountable" so it is often not known who reviewed an article and how rigorous the review was [26]
- Many medical studies are poorly designed [27]
  - The recruitment process was not described [28]
  - Inadequate power (size) to make accurate conclusions. In other words, not enough subjects were studied [29]
  - Studies with negative results (i.e., results that are not statistically significant) are not always published or take more time to be published, resulting in "publication bias". In an effort to prevent this type of bias the American Medical Association advocates mandatory registration of all clinical trials in public registries. Also, the International Committee of Medical Journal Editors requires registration as a condition to publish in one of their journals. However, they do not require publishing the results in the registry at this time. Registries could be a data warehouse for future mining and some of the well known registries include:
    - ClinicalTrials.gov
    - WHO International Clinical Trials Registry
    - Global Trial Bank of the American Medical Informatics Association
    - Trial Bank Project of the University of California, San Francisco [30]

In spite of the fact that EBM is considered a highly academic process towards gaining medical truth, numerous problems exist:
- Different evidence rating systems by various medical organizations

- Different conclusions by experts evaluating the same study
- Time intensive exercise to evaluate existing evidence
- Systematic reviews are limited in the topics reviewed (3000 in the Cochrane database) and are time intensive to complete (6-24 months). Often the conclusion is that current evidence is weak and further high quality studies are necessary
- Randomized controlled trials are expensive. Drug companies tend to fund only studies that help a current non-generic drug they would like to promote
- Results may not be applicable to every patient population
- Some view EBM as "cookbook medicine" [31]
- There is not good evidence that teaching EBM changes behavior [32]

Recent advancements in EBM

The field of EBM continues to evolve. Methodologists continue to identify opportunities to improve our understanding and interpretation of research findings. Two studies published in 2010 help refine our knowledge base:

- Trials are often stopped early when extreme benefits are noted in the intervention group. The rationale for stopping enrollments of participants is that it is "unethical" to continue randomizing patients to the placebo arm because we are depriving them from the benefits of the intervention. However, it was found that stopping trials early for benefit exaggerates treatment effect by more than 30%; simply because we are stopping the trial at a point of extreme benefit that is clearly made extreme by chance. Such exaggeration leads to the wrong conclusions by patients and physicians embarking on comparing the pros and cons of a treatment and also leads to the wrong decisions by policymakers. In fact, stopping early may be unethical from a societal and individual point of view [33]
- The second recent advancement in methodology relates to the finding that authors who have financial affiliation with the industry are 3 times more likely to make statements that are favorable to the sponsored interventions. It is very plausible that this bias is subconscious and unintentional; nevertheless, as readers of the literature, we should recognize the potential and implications of this bias [34]

Other Approaches

EBM has had both strong advocates and skeptics since its inception. One of its strongest proponents, Dr David Sackett published his experience with an "Evidence Cart" on inpatient rounds in 1998. The cart contained numerous EBM references but was so bulky that it could not be taken into patient rooms.[35] Since that article, multiple, more convenient EBM solutions exist. While there are those EBM advocates who would suggest we use solely EBM resources, many others feel that EBM "may have set standards that are untenable for practicing physicians".[36-37]

Dr. Frank Davidoff believes that most clinicians are too busy to perform literature searches for the best evidence. He believes that we need "Informationists", who are experts at retrieving information.[38] To date, only clinical medical librarians (CMLs) have the formal training to take on this role. At large academic centers CMLs join the medical team on inpatient rounds and attach pertinent and filtered articles to the chart. As an example, Vanderbilt's Eskind Library has a Clinical Informatics Consult Service.[39-40] The obvious drawback is that CMLs are only available at large medical centers and are unlikely to research outpatient questions.

According to Slawson and Shaughnessy you must become an "information master" to sort through the "information jungle". They define the usefulness of medical information as:

Usefulness = $\dfrac{Validity \times Relevance}{Work}$

Only the clinician can determine if the article is relevant to his/her patient population and if the work to retrieve the information is worthwhile. Slawson and Shaughnessy also developed the notion of looking for "patient oriented evidence that matters" (POEM) and not "disease oriented evidence that matters" (DOEM). POEMS look at mortality, morbidity and quality of life whereas DOEMS tend to look at laboratory or experimental results. They point out that it is more important to know that a drug reduces heart attacks or deaths from heart attacks (POEM), rather than just reducing cholesterol levels (DOEM). [41] This school of thought also recommends that you not read medical articles blindly each week but should instead learn how to search for patient specific answers using EBM resources. [42] This also implies that you are highly motivated to pursue an answer, have adequate time and have the appropriate training.

EBM Resources

There are many first-rate online medical resources that provide EBM type answers. They are all well referenced, current and written by subject experts. Several include the level of evidence (LOE). These resources can be classified as **filtered** (an expert has appraised and selected the best evidence, e.g., Up-to-date) or **unfiltered** (non selected evidence, e.g., PubMed). For the EBM purist, the following are considered traditional or classic EBM resources:

- Clinical Evidence [43]
  - British Medical Journal product with two issues per year
  - New drug safety alert section
  - New "latest research" results section
  - Evidence is oriented towards patient outcomes (POEMS)
  - Very evidence based with single page summaries and links to national guidelines
  - Available in paperback (Concise), CD-ROM, online or PDA format

- Cochrane Library [44]
  - Database of systematic reviews. Each review answers a clinical question
  - Database of review abstracts of effectiveness (DARE)
  - Controlled Trials Register
  - Methodology reviews and register
  - Fee-based

- Cochrane Reviews [45]
  - Part of the Cochrane Collaboration
  - Reviews can be accessed for a fee but abstracts are free. A search for low back pain, as an example, returned 44 reviews (abstracts)

- EvidenceUpdates [46]
  - Since 2002 BMJ Updates has been filtering all of the major medical literature. Articles are not posted until they has been reviewed for newsworthiness and relevance; not strict EBM guidelines
  - You can go to their site and do a search or you can choose to have article abstracts e-mailed to you on a regular basis
  - These same updates are available through www.Medscape.com

- ACP Journal Club [47]

- o Bimonthly journal that can be accessed from OVID or free if a member of the American College of Physicians (ACP)
  - o Over 100 journals are reviewed but very few articles make the cut: in 1992 only 13% of articles from the NEJM made the Journal Club, all other journals were much lower

- Practical Pointers for Primary Care [48]

  - o Free online review of articles from the New England Journal of Medicine, Journal of the American Medical Journal, British Medical Journal, the Lancet, the Annals of Internal Medicine and the Archives of Internal Medicine
  - o Program can be accessed via the web or monthly reports e-mailed to those who subscribe
  - o Editor dissects the study and makes summary comments that are very helpful to the average reader

- Evidence-Based On-Call [49]

  - o User friendly site intended for quick look-ups for clinicians on call
  - o Has multiple critically appraised topics (CATs) that point out the most important clinical pearls, with level of evidence

- PDA EBM Resources

  - o Centre for EBM www.cebm.utoronto.ca
  - o Duke Medical Center Library www.mclibrary.duke.edu/training/pdaformat/
  - o EBM 2 go http://www.ebm2go.com

- Others

  - o TRIP Database [50]
  - o OVID has the ability to search the Cochrane Database of Systematic Reviews, DARE, ACP Journal Club and Cochrane Controlled Trials Register at the same time. Also includes Evidence-Based Medicine Reviews [51]
  - o SUMSearch. Free site that searches Medline, National Guideline Clearing House and DARE [52]
  - o Bandolier. Free online EBM journal; used mainly by primary care docs in England. Provides simple summaries with NNTs. Resource also includes multiple monographs on EBM that are easy to read and understand [53]
  - o Denison Memorial Library: Evidence Based Medicine Resources with extensive links located on the same page [54]

Key Points

- Evidence Based Medicine (EBM) is the academic pursuit of the best available answer to a clinical question

- The two fundamental principles of EBM are: 1) a hierarchy of evidence exists (i.e., not all evidence is equal), and 2) evidence alone is insufficient for medical decision making. It should rather be complemented by patient's values, preferences and circumstances

- Medical Informatics will hopefully improve medical quality, which is primarily based on EBM

- There are multiple limitations of both EBM and the medical literature

- The average clinician should have a basic understanding of EBM and know how to find answers using EBM resources

Conclusion

Knowledge of evidence based medicine is important if you are involved with patient care, quality of care issues or research. Rapid access to a variety of online EBM resources has changed how we practice medicine. In spite of its shortcomings, an evidence based approach helps healthcare workers find the best possible answers. Busy clinicians are likely to choose commercial high quality resources, while academic clinicians are likely to select true EBM resources. Ultimately, EBM tools and resources will be integrated into all electronic health records as clinical decision tools.

Acknowledgement: we would like to thank Dr. Brian Haynes and Dr. Ramón Puchades for their contributions to this chapter

References

1. Evidence Based Medicine: What it is, what it isn't. http://www.cebm.net/ebm_is_isnt.asp (Accessed September 3 2005)
2. Crossing the Quality Chasm: A new health system for the 21th century (2001) The National Academies Press http://www.nap.edu/books/0309072808/html/
3. MedlinePlus http://www.nlm.nih.gov/medlineplus/ency/article/003108.htm (Accessed September 3 2006)
4. Teece I, Crawford I. Bed rest after spinal puncture. BMJ http://emj.bmjjournals.com/cgi/content/full/19/5/432 (Accessed Aug 24 2006)
5. Guyatt GH. Evidence-based medicine. ACP J Club 1991;114:A16
6. Evidence Based Medicine. Wikipedia. http://en.wikipedia.org/wiki/Evidence_based-medicine (Accessed September 5 2005)
7. Levin A. The Cochrane Collaboration Ann of Int Med 2001;135:309-312
8. Medical Research Council. Streptomycin treatment of pulmonary tuberculosis. BMJ 1948;2:769-82
9. Gandhi GY, Murad MH, Fujiyoshi A, et al. Patient-important outcomes in registered diabetes trials. JAMA. Jun 4 2008;299 (21):2543-2549.

10. Davis DA et al. Changing physician performance. A systematic review of the effect of continuing medical education strategies. JAMA 1995; 274: 700-1.

11. Sibley JC. A randomized trial of continuing medical education. N Engl J Med 1982; 306: 511-5.

12. Fordis M et al. Comparison of the Instructional Efficacy of Internet-Based CME with Live Interactive CME Workshops. JAMA 2005;294:1043-1051

13. National Library of Medicine PICO http://askmedline.nlm.nih.gov/ask/pico.php (Accessed September 7 2005)

14. Centre for Evidence Based Medicine http://www.cebm.net/learning_ebm.asp (Accessed September 7 2007)

15. Haynes RB. Of studies, syntheses, synopses and systems: the "4S evolution of services for finding the best evidence". ACP J Club 2001;134: A11-13

16. The well built clinical question. University of North Carolina Library http://www.hsl.unc.edu/Services/Tutorials/EBM/Supplements/QuestionSupplement.htm (Accessed September 20 2005)

17. Guyatt GH, Oxman AD, Vist G, Kunz R, Falck-Ytter Y, Alonso-Coello P, Schünemann HJ. The GRADE Working Group. Rating quality of evidence and strength of recommendations GRADE: an emerging consensus on rating quality of evidence and strength of recommendations. BMJ 2008;336:924-926

18. Henley E. Understanding the Risks of Medical Interventions Fam Pract Man May 2000;59-60

19. Nouvo J, Melnikow J, Chang D. Reporting the Number Needed to Treat and Absolute Risk Reduction in Randomized Controlled Trials JAMA 2002;287:2813-2814

20. Lacy CR et al. Impact of Presentation of Research Results on Likelihood of Prescribing Medications to Patients with Left Ventricular Dysfunction. Am J Card 2001;87:203-207

21. Collaborative Atorvastatin Diabetes Study (CARDS) Lancet 2004;364:685-96

22. Maharaj R. Adding cost to number needed to treat: the COPE statistic. Evidence Based Medicine 2007;12:101-102

23. Haynes RB. Where's the Meat in Clinical Journals? ACP Journal Club Nov/Dec 1993: A-22-23

24. Als-Neilsen B, Chen W, Gluud C, Kjaergard LL. Association of Funding and Conclusions in Randomized Drug Trials. JAMA 2003; 290:921-928

25. Ioannidis JPA. Contradicted and Initially Stronger Effects in Highly Cited Clinical Research JAMA 2005;294:218-228

26. Kranish M. Flaws are found in validating medical studies The Boston Globe August 15 2005 http://www.boston.com/news/nation/articles/2005/08/15/flaws_are_found_in_validating_medical_studies/ (Accessed June 12 2007)

27. Altman DG. Poor Quality Medical Research: What can journals do? JAMA 2002;287:2765-2767

28. Gross CP et al. Reporting the Recruitment Process in Clinical Trials: Who are these Patients and how did they get there? Ann of Int Med 2002;137:10-16

29. Moher D, Dulberg CS, Wells GA. Statistical Power, sample size and their reporting in randomized controlled trials JAMA 1994;22:1220-1224

30. Evidence Based Medicine. Clinfowiki. www.informatics-review.com/wiki/index.php/EBM (Accessed June 19 2007)

31. Straus SE, McAlister FA Evidence Based Medicine: a commentary on common criticisms Can Med Assoc J 2000;163:837-841

32. Dobbie AE et al. What Evidence Supports Teaching Evidence Based Medicine? Acad Med 2000;75:1184-1185
33. Bassler RD, Briel M, Murad MH et al. Stopping Randomized Trials Early for Benefiit and Estimation of Treatment Effects: Systematic Review and Meta-Regression Analysis. JAMA 2010;303 (12):1180-7
34. Wang AT, McCoy CP, Murad MH. Association Between Affiliation and Position on Cardiovascular Risk with Rosiglitazone: Cross Sectional Systematic Review. BMJ 2010. March 18:340.c1344. doi:10.1136/bmj.c1344 (Accessed April 10 2010)
35. Sackett DL, Staus SE. Finding and Applying Evidence During Clinical Rounds: The "Evidence Cart" JAMA 1998;280:1336-1338
36. Grandage K et al. When less is more: a practical approach to searching for evidence-based answers. J Med Libr Assoc 90(3) July 2002
37. Schilling LM et al. Resident's Patient Specific Clinical Questions: Opportunities for Evidence Based Learning Acad Med 2005;80:51-56
38. Davidoff F, Florance V.The Informationist: A New Health Profession? Ann of Int Med 2000;132:996-999
39. Giuse NB et al. Clinical medical librarianship: the Vanderbilt experience Bull Med Libr Assoc 1998;86:412-416
40. Westberg EE, Randolph AM. The Basis for Using the Internet to Support the Information Needs of Primary Care JAMIA 1999;6:6-25
41. Slawson DC, Shaughnessy AF, Bennett JH. Becoming a Medical Information Master: Feeling Good About Not Knowing Everything J of Fam Pract 1994;38:505-513
42. Shaughnessy AF, Slawson DC and Bennett JH. Becoming an Information Master: A Guidebook to the Medical Information Jungle J of Fam Pract 1994;39:489-499
43. Clinical Evidence www.clinicalevidence.com (Accessed January 2 2008)
44. Cochrane Library http://www3.interscience.wiley.com/cgi-bin/mrwhome/106568753/HELP_Cochrane.html (Accessed January 2 2008)
45. Cochrane Review http://www.cochrane.org/reviews/index.htm (Accessed September 18 2007)
46. EvidenceUpdates http://plus.mcmaster.ca/evidenceupdates (Accessed August 2 2009)
47. ACP Journal Club. http://plus.mcmaster.ca/acpjc (Accessed August 2 2009)
48. Practical Pointers for Primary Care www.practicalpointers.org (Accessed January 26 2008)
49. Trip Database www.tripdatabase.com (Accessed January 2 2008)
50. Evidence-Based On-call. www.eboncall.org (Accessed February 29 2008)
51. OVID http://gateway.ovid.com (Accessed January 2 2008)
52. SUMSearch http://sumsearch.uthscsa.edu (Accessed January 2 2008)
53. Bandolier www.jr2.ox.ac.uk/bandolier (Accessed January 2 2008)
54. Denison Memorial Library http://denison.uchsc.edu/evidence_based.html (Accessed January 2 2008)

Clinical Practice Guidelines

ROBERT E. HOYT
M. HASSAN MURAD

Learning Objectives

After reading this chapter the reader should be able to:

- Define the utility of clinical practice guidelines
- Describe the interrelationship between clinical practice guidelines, evidence based medicine, electronic health records and pay for performance
- Define the processes required to write and implement a clinical practice guideline
- Compare and contrast the potential benefits and obstacles of clinical practice guidelines
- Describe clinical practice guidelines in an electronic format
- List the most significant clinical practice guideline resources

The Institute of Medicine in 1990 defined clinical practice guidelines (CPGs) as:

"systematically developed statements to assist practitioner and patient decisions about health care for specific clinical circumstances" [1]

Clinical practice guidelines (CPGs) take the very best evidence based medical information and formulate a game plan to treat a specific disease or condition. If one considers evidence as a continuum that starts by data generated from a single study, appraised and synthesized in a systematic review, CPGs would represent the next logical step in which evidence is transformed into a recommendation. Many medical organizations use CPGs with the intent to improve quality of care, patient safety and/or reduce costs. Information technology assists CPGs by expediting the search for the best evidence and linking the results to EHRs, PDAs and smartphones for easy access. Two areas in which CPGs may be potentially beneficial include disease management and pay for performance, covered in other chapters. CPGs are important for several reasons such as 83% of Medicare beneficiaries have at least one chronic condition and 68% of Medicare's budget is devoted to the 23% who have 5 or more chronic conditions. [2] There is some evidence that guidelines that address multiple comorbidities (concurrent chronic diseases) actually do work. As an example, in one study of diabetics, there was a 50% decrease in cardiovascular and microvascular complications with intensive treatment of multiple risk factors. [3]

In spite of evidence to suggest benefit, several studies have shown poor CPG compliance by patients and physicians. The well publicized 2003 RAND study in the New England Journal of Medicine demonstrated that *"overall, patients received 54% of recommended care"*. [4-5] In another study of guidelines at a major teaching hospital there was overuse of statin therapy (cholesterol lowering drugs). Overuse occurred in 69% of primary prevention (to prevent a disease) and 47% of secondary prevention (to prevent disease recurrence or progression), compared to national recommendations. [6]

It should be emphasized that creating or importing a guideline is the easy part because hundreds have already been created by a variety of national organizations. Implementing CPGs and achieving buy-in by all healthcare workers, particularly physicians, is the hard part.

How are CPGs Developed?

Ideally, the process starts with a panel of content and methodology experts commissioned by a professional organization. As an example, if the guideline is about preventing venous thrombosis and pulmonary embolism, multi-disciplinary content experts would be pulmonologists, hematologists, pharmacists and hospitalists.

Methodology experts are experts in evidence based medicine, epidemiology, statistics, cost analysis, etc. The panel refines the questions, usually in PICO format, that was discussed in the previous chapter. A systematic literature search and evidence synthesis takes place. Evidence is graded and recommendations are negotiated. Panel members have their own biases and conflicts of interest that should be declared to CPG users. Voting is often needed to build consensus since disagreement is a natural phenomenon in this context.

The Strength of Recommendations

Guideline panels usually associate their recommendations by a grading that describes how confident they are in their statement. Ideally, panels should separately describe their confidence in the evidence (the quality of evidence, described in previous chapter) from the strength of the recommendation. The reason for this separation is that there are factors other than evidence that may affect the strength of recommendation. These factors are: 1) how closely balanced are the benefits and harms of the recommended intervention, 2) patients' values and preferences, and 3) resource allocation.

For example, even if there is very high quality evidence from randomized trials showing that warfarin (a blood thinner) decreases the risk of stroke in some patients, the panel may issue a weak recommendation considering that the harms associated with this medicine are substantial. Similarly, if high quality evidence suggests that a treatment is very beneficial, but this treatment is very expensive and only available in very few large academic centers in the US, the panel may issue a weak recommendation because this treatment is not easily available or accessible.

Application to Individuals

A physician should consider a strong recommendation to be applicable to all patients who are able to receive it. Therefore, physicians should spend his/her time and effort on explaining to patients how to use the recommended intervention and integrate it in their daily routine.

On the other hand, a weak recommendation may only apply to certain patients. Physicians should spend more time discussing pros and cons of the intervention with patients, use risk calculators and tools designed to stratify patients' risk to better determine the balance of harms and benefit for the individual. Weak recommendations are the optimal condition to use decision aids, which are available in written, videographic and electronic formats and may help in the decision-making process by increasing knowledge acquisition by patients and reduce their anxiety and decisional conflicts.

Appraisal and Validity of Guidelines

There are several tools suggested to appraise CPGs and determine their validity. These tools assess the process of conducting CPGs, the quality and rigor of the recommendations and the clarity of their presentation. The following list includes some of the attributes that guidelines users (clinicians, patients, policy makers) should seek to determine if a particular CPG is valid and has acceptable quality:

- Evidence based, preferably linked to systematic reviews of the literature
- Considers all relevant patients groups and management options
- Considers patient-important outcomes (as opposed to surrogate outcomes)
- Updated frequently
- Clarity and transparency in describing the process of CPGs development (e.g., voting, etc.)
- Clarity and transparency in describing the conflicts of interests of the guideline panel
- Addresses patients' values and preferences
- Level of evidence and strength of recommendation are given
- Simple summary or algorithm that is easy to understand
- Available in multiple formats (print, online, PDA, etc.) and in multiple locations
- Compatibility with existing practices
- Simplifies, not complicates decision making [7]

Barriers to CPGs

Attempts to standardize medicine by applying evidence based medicine and clinical practice guidelines has been surprisingly difficult due to multiple barriers:

- Practice setting: inadequate incentives, inadequate time and fear of liability. A 2003 study estimated that it would require 7.4 hours/working day just to comply with all of the US Preventive Services Task Force recommendations for the average clinician's practice! [8]
- Contrary opinions: local experts do not always agree with CPG or clinicians hear different messages from drug detail representatives
- Sparse data: there are several medical areas in which the evidence is of lower quality or sparse. Guideline panels in these areas would heavily depend on their expertise and should issue weak recommendations (e.g. suggestions) or no recommendations if they did not reach a consensus. These areas are problematic to patients and physicians and are clearly not ready for quality improvement projects or pay for performance incentives. For years, diabetologists advocated tight glycemic control of patients with type 2 diabetes; however, it turned out from results of recent large randomized trials that this strategy does not result in improved outcomes [9]
- We need more information about why clinicians don't follow CPGs. Persell et al. reported in a 2010 study that 94% of the time when clinicians chose an exception to the CPG it was appropriate. Three percent were inappropriate and 3% were unclear [10]
- Knowledge and attitudes: there is a lack of confidence to either not perform a test (malpractice concern) or to order a new treatment (don't know enough yet). Information overload is always a problem [11]
- CPGs can be too long, impractical or confusing. One study of Family Physicians stated CPGs should be no longer than 2 pages.[12-13] Most national CPGs are 50-150 pages long and don't always include a summary of recommendations
- Where and how do you post CPGs? What should be the format?
- Less buy-in if data reported is not local since physicians tend to respond to data reported from their hospital or clinic
- No uniform level of evidence (LOE) rating system
- Too many CPGs posted on the National Guideline Clearinghouse. For instance, a non-filtered search in January 2010 by one author for "diabetes" yielded 535 diabetes-related CPGs. The detailed search option helps filter the search but is not very user friendly [14]

- Lack of available local champions to promote CPGs
- Excessive influence by drug companies: A survey of 192 authors of 44 CPGs in the 1991-1999 time frame showed:
    - o 87% had some tie to drug companies
    - o 58% received financial support
    - o 59% represented drugs mentioned in the CPG
    - o 55% of respondents with ties to drug companies said they did not believe they had to disclose involvement [15]
- National Guidelines are not necessarily of high quality. A 2009 review of CPGs from the American Heart Association and the American College of Cardiology (1984-Sept 2008) concluded that many of the recommendations were based on a lower level of evidence or expert opinion, not high quality studies [16]
- No patient input. At this point patients are not normally involved in any aspect of CPGs, even though they receive recommendations based on CPGs. In an interesting 2008 study, patients who received an electronic message about guidelines experienced a 12.8% increase in compliance. This study utilized claims data as well as a robust rules engine to analyze patient data. Patients received alerts (usually mail) about the need for screening, diagnostic and monitoring tests. The most common alerts were for adding a cholesterol lowering drug, screening women over age 65 for osteoporosis, doing eye exams in diabetics, adding an ACE inhibitor drug for diabetes and testing diabetics for urine microalbumin.[17] It makes good sense that patients should be knowledgeable about national recommendations and should have these guidelines written in plain language and available in multiple formats. Also, because many patients are highly "connected" they could receive messages via cell phones, social networking software, etc to improve monitoring and treatment

Where should Hospitals or Individuals Start? Examples of Starting Points:

- High cost conditions: heart failure
- High volume conditions: diabetes
- Preventable admissions: asthma
- There is variation in care compared to national recommendations: deep vein thrombophlebitis (DVT) prevention
- High litigation areas: failure to diagnose or treat
- Patient safety areas: intravenous (IV) drug monitoring

The Strategy

- Leadership support is crucial
- Use process improvement tools such as the Plan-Do-Study-Act (PDSA) model
- Identify gaps in knowledge between national recommendations and local practice
- Locate a guideline champion who is a well respected clinician expert.[18] A champion acts as an advocate for implementation based on his/her support of a new guideline
- Other potential team members
    - o Clinician selection based on the nature of the CPG
    - o Administrative or support staff
    - o Quality Management staff
- Develop action plans
- Educate all staff involved with CPGs, not just clinicians
- Pilot implementation

- Provide frequent feedback to clinicians and other staff regarding results
- Consider using the checklist for reporting clinical practice guidelines developed by the 2002 Conference on Guideline Standardization (COGS) [19]

Examples of Clinical Practice Guidelines

The CPG in figure 13.1 was written for the treatment of uncomplicated bladder infections (cystitis) in women. The goal was to use less expensive antibiotics and treat for fewer days. The protocol or algorithm can be administered by a triage nurse when a patient telephones or walks in. Figure 13.2 demonstrates that the use of the first line drug (sulfa family) increased after the start of the CPG, whereas second line drug use decreased. Success of this program was based on educating all members of the healthcare team and reporting the results at medical staff meetings and other venues. It was also aided by an easy to follow guideline and full support by the nursing staff

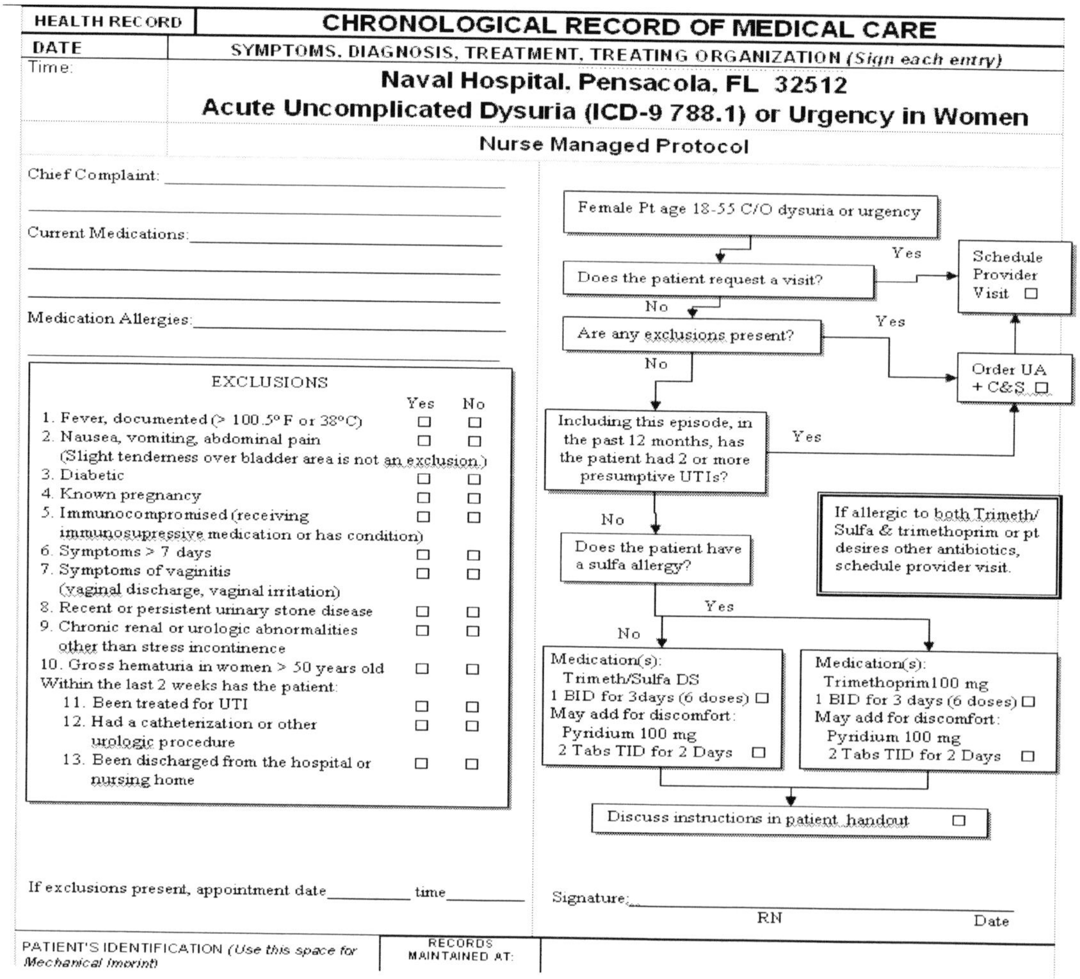
HEALTH RECORD | CHRONOLOGICAL RECORD OF MEDICAL CARE

DATE | SYMPTOMS, DIAGNOSIS, TREATMENT, TREATING ORGANIZATION *(Sign each entry)*

Time:

Naval Hospital, Pensacola, FL 32512

Acute Uncomplicated Dysuria (ICD-9 788.1) or Urgency in Women

Nurse Managed Protocol

Chief Complaint: ______

Current Medications: ______

Medication Allergies: ______

EXCLUSIONS

	Yes	No
1. Fever, documented (> 100.5° F or 38°C)	☐	☐
2. Nausea, vomiting, abdominal pain (Slight tenderness over bladder area is not an exclusion)	☐	☐
3. Diabetic	☐	☐
4. Known pregnancy	☐	☐
5. Immunocompromised (receiving immunosupressive medication or has condition)	☐	☐
6. Symptoms > 7 days	☐	☐
7. Symptoms of vaginitis (vaginal discharge, vaginal irritation)	☐	☐
8. Recent or persistent urinary stone disease	☐	☐
9. Chronic renal or urologic abnormalities other than stress incontinence	☐	☐
10. Gross hematuria in women > 50 years old	☐	☐
Within the last 2 weeks has the patient:		
11. Been treated for UTI	☐	☐
12. Had a catheterization or other urologic procedure	☐	☐
13. Been discharged from the hospital or nursing home	☐	☐

If exclusions present, appointment date______ time______

Signature: ______ RN Date

PATIENT'S IDENTIFICATION *(Use this space for Mechanical Imprint)* | RECORDS MAINTAINED AT:

Figure 13.1 CPG for uncomplicated dysuria or urgency in women

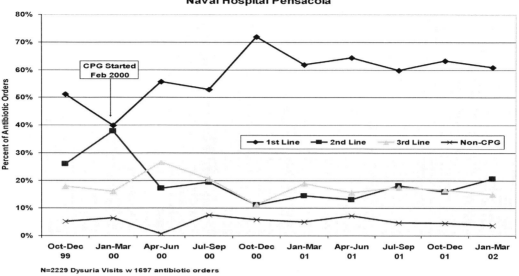

Figure 13.2 Results of CPG implementation (Courtesy Naval Hospital Pensacola)

Clinical Practice Guidelines in electronic format

CPGs have been traditionally paper based and often accompanied by a flow diagram or algorithm. With time, more are being created in an electronic format and posted on the Internet or Intranet for easy access. Zielstorff outlined the issues, obstacles and future prospects of online practice guidelines in a 1998 review.[20] What has changed since then is the ability to integrate CPGs with electronic health records and PDAs/Smartphones.

CPGs on PDAs/Smartphones: These mobile platforms function well in this area as each step in an algorithm is simply a tap or touch of the screen. In addition to commercially hosted PDA/Smartphone programs, CPGs can be created on the PDA or smartphone using iSiloX or MobiPocket Creator (see chapter on Mobile Technology). In figures 13.3 and 13.4 programs are shown that are based on national guidelines for cardiac risk and cardiac clearance. Figure 13.3 depicts a calculator that determines the 10 year risk of heart disease based on serum cholesterol and other risk factors. A cardiac clearance program determines whether a patient needs further cardiac testing prior to an operation (figure 13.4) [20] Many excellent guidelines for the PDA/Smartphone exist that will be listed later in this chapter.

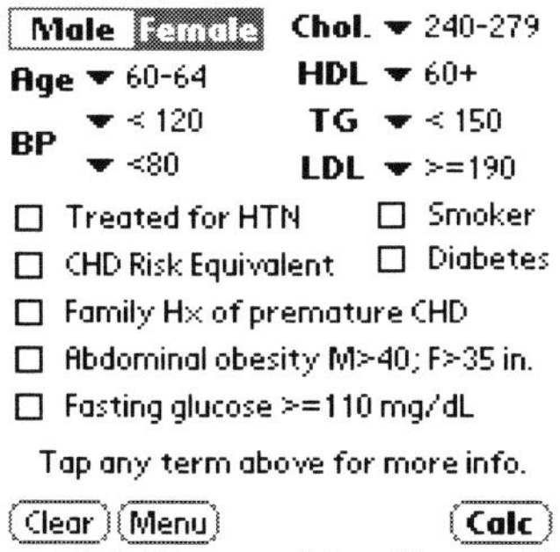

Figure 13.3 10 year risk of heart disease

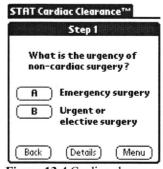

Figure 13.4 Cardiac clearance

Risk Calculators: Many of these are available on a mobile platform and are also available online. While these are not CPGs exactly, they are based on population studies and are felt to be part of EBM and can give direction to the clinician. As an example, some experts feel that aspirin has little benefit in preventing a heart attack unless your 10 year risk of one exceeds 20%. The following is a short list of some of the more popular online calculators:

- ATP III Cardiac risk calculator: estimates the 10 year risk of a heart attack or death based on your cholesterol, age, gender, etc. http://hp2010.nhlbihin.net/atpiii/calculator.asp?usertype=prof
- FRAX fracture risk calculator: estimates the 10 year risk of a hip or other fracture based on all of the common risk factors for osteoporosis. Takes into account a patient's bone mineral density score, gender and ethnicity. http://www.shef.ac.uk/FRAX/
- GAIL breast cancer risk assessment tool: estimates a patient's risk of breast cancer, again, based on known and accepted risk factors. http://www.cancer.gov/bcrisktool/
- Stroke risk calculator: based on the Framingham study it predicts 10 year risk of a stroke based on known risk factors. http://www.stroke-education.com/calc/risk_calc.do
- Risk of stroke or death for new onset atrial fibrillation: also based on the Framingham study, it calculates 5 year risk of stroke or death. http://www.zunis.org/FHS%20Afib%20Risk%20Calculator.htm

HTML CPGs: Another simple technique to post a CPG in a user-friendly format is to write it as a Word document but save it as a web page (html). Hyperlinking to other pages or simply "book marking" to information located further down on the same page makes navigation much easier and reduces the size of the CPG. This allows a CPG to be written on a single document that is easy to e-mail or post. The CPG in figure 13.5 is an example of the 2004 American College of Chest Physicians (ACCP) Guideline to prevent venous thromboembolism (blood clots in legs and lungs). Select a hyperlink such as General Surgery and it takes you to another section with specific recommendations, thus making the document compact and easy to navigate. It is also easy to update or correct.

EHR CPGs: Although a minority of electronic health records have embedded CPGs, there is definite interest in providing local or national CPGs at the point of care. CPGs embedded in the EHR are clearly a form of decision support. They can be linked to the diagnosis or the order entry process. In addition, they can be standalone resources available by clicking, for example, an "info-button". Clinical decision support provides treatment reminders for disease states that may include the use of more cost effective drugs. Institutions such as Vanderbilt University have integrated more than 750 CPGs into their EHR by linking the CPGs to ICD-9 codes.[21] The results of embedded CPGs appears to be mixed. In a study by Durieux using computerized decision support reminders, orthopedic surgeons showed improved compliance to guidelines to prevent deep vein thrombophlebitis.[22] On the other hand, three studies by Tierney, failed to demonstrate improved compliance to guidelines using computer reminders for hypertension, heart disease and asthma.[23-25] Clinical decision support, to include order sets is discussed in more detail in the chapters on electronic health records and patient safety.

PREVENTION OF VENOUS THROMBOEMBOLISM **Naval Hospital Pensacola September 2004**

General Surgery Gynecologic Surgery Urologic Surgery Orthopedic Surgery Neurosurgery and Trauma
Medical Conditions Reference Risk Assessment

General recommendations:

Every hospital should develop a written policy or other formal strategy for preventing thromboembolic complications, especially for high-risk patients.

We do **not** recommend prophylactic use of aspirin as a sole prevention strategy, because the measures listed in the tables below are more efficacious (**grade 1A**).

We recommend anticoagulant prophylaxis or therapy should be used with **caution** in all cases involving either spinal puncture or placement of epidural catheters for regional anesthesia or analgesia (**grade 1C+**).

Level of Evidence:

Grade of recommendation	Implications
1A	Strong. Applies to most patients in most circumstances
1B	Strong. Likely to apply to most patients
1C+	Strong. Can apply to most patients in most circumstances
1C	Intermediate. May change when stronger evidence is available

Glossary:

ES=elastic stockings **IPC**=intermittent pneumatic compression
LDUH=low dose unfractionated heparin **LMWH**=low molecular weight heparin

Figure 13.5 DVT prevention CPG (Courtesy Naval Hospital Pensacola)

There are other ways to use electronic tools to promulgate CPGs. In an interesting paper by Javitt, primary care clinicians were sent reminders on outpatient treatment guidelines based only on claims data. Outliers were located by using a rules engine (Care Engine) to compare a patient's care with national guidelines. They were able to show a decrease in hospitalizations and cost as a result of alerts that notified physicians by phone, fax or letter. This demonstrates one additional means of changing physician behavior using CPGs and information technology not linked to the electronic health record.[26] Critics might argue that claims data is not as accurate, robust or current as actual clinical results.

Software is now available (EBM Connect) that can compute compliance with guidelines automatically using administrative data. The program translates guidelines from text to algorithms for 20 disease conditions and therefore would be much more efficient than chart reviews. Keep in mind it will tell you if, for example, LDL cholesterol was ordered, not the actual results.[27]

CPG Resources

Web based CPGs

- National Guideline Clearinghouse. This program is an initiative of the Department of Health and Human Services and is the largest and most comprehensive of all CPG resources. Features offered:
  - Includes about 2200 guidelines
  - There is extensive search engine filtering i.e. you can search by year, language, gender, specialty, level of evidence, etc
  - Abstracts are available as well as links to full text guidelines where available
  - CPG comparison tool
  - Forum for discussion of guidelines
  - Annotated bibliography
  - They link to 17 international CPG resource sites[28]

- National Institute for Health and Clinical Excellence (NICE)
    - Service of the British National Health Service
    - Approximately 100 CPGs are posted and dated
    - A user-friendly short summary is available as well as a lengthy guideline, both in downloadable pdf format
    - Podcasts are available [29]

- Colorado Clinical Guidelines Collaborative
    - Free downloads available for Colorado physicians and members of CCGC
    - They currently have 14 CPGs available
    - Guidelines are in easy to read tables, written in a pdf format
    - References, resources and patient handouts are available [30]

PDA/Smartphone based CPGs

- National Guideline Clearinghouse
    - CPGs for the Palm OS PDA. Word documents are available to download for the Pocket PC OS. Document reader (Apprisor) is necessary for Palm OS [28]
- American Diabetes Association
    - For Palm OS
    - Includes the 2009 treatment CPG for diabetes [31]
- Sites that require free Apprisor software for downloads: [32]
    - American Heart Association
        - For Palm and Pocket PC OS
        - Covers: heart failure, angioplasty/stents, ventricular arrhythmias, stable and unstable angina, myocardial infarction, bypass surgery, atrial fibrillation, echocardiography, pacemakers, stroke and other preventative guidelines
    - American College of Chest Physicians
        - Covers: lung cancer, emphysema, ventilator related pneumonia, weaning from the ventilator, blood clots, cough and pulmonary rehabilitation
    - Colorado Clinical Guideline Collaborative
        - Covers colorectal cancer screening, diabetes and major depression
    - National Heart, Lung and Blood Institute
        - Covers the JNC VII guide for the treatment of hypertension and the national guidelines for the treatment of cholesterol
    - American College of Physicians
        - Covers: emphysema treatment, ICD-9 codes, bioterrorism and chemical terrorism
    - American Academy of Family Physicians
        - Covers their recommendations for preventive testing
- Centers for Disease Control (CDC)
    - Palm OS only
    - Guideline is for the treatment of sexually transmitted diseases [33]

Key Points

- Clinical Practice Guidelines (CPGs), based on evidence based medicine, are the roadmap to standardize medical care

- CPGs are valuable for chronic disease management or as a means to measure quality of care

- CPGs can be part of electronic health records and order sets

- CPGs can be installed on PDAs or smartphones

- CPGs can be digital or paper based

- Hundreds of CPGs are readily available for download from many excellent sites

Conclusion

The jury is out regarding the impact of CPGs on physician behavior or patient outcomes. Busy clinicians are slow to accept new information, including CPGs. Whether embedding CPGs into EHRs will result in significant changes in behavior that will consistently result in improved quality, patient safety or cost savings remains to be seen. It is also unknown if linking CPGs to better reimbursement (pay for performance) will result in a higher level of acceptance. While we are determining how to optimally improve healthcare with CPGs, most authorities agree that CPGs need to be concise, practical and accessible at the point of care. Every attempt should be made to make them electronic and integrated into the workflow of clinicians. Moreover, further research is needed to determine if patients consistently benefit from having CPGs customized for patients as well as electronic alerting mechanisms.

References

1. Institute of Medicine (1990). Clinical Practice Guidelines: Directions for a New Program. Field MJ and Lohr KN (eds). Washington DC. National Academy Press. Page 38
2. O'Connor P. Adding Value to Evidence Based Clinical Guidelines JAMA 2005;294:741-743
3. Gaede P. Multifactorial intervention and cardiovascular disease in patients with type 2 diabetes NEJM 2003;348:383-393
4. McGlynn E . Quality of Health Care Delivered to Adults in the US RAND Health Study NEJM Jun 26 2003
5. Crossing the Quality Chasm: A new Health System for the 21th century 2001. IOM. http://darwin.nap.edu/books/0309072808/html/227.html (Accessed March 5 2006)
6. Abookire SA, Karson AS, Fiskio J, Bates DW. Use and monitoring of "statin" lipid-lowering drugs compared with guidelines Arch Int Med 2001;161:2626-7
7. Oxman A, Flottorp S. An overview of strategies to promote implementation of evidence based health care. In: Silagy C, Haines A, eds Evidence based practice in primary care, 2nd ed. London: BMJ books 2001
8. Yarnall KSH, Pollak KL, Østbye T et al. Primary Care: Is There Enough Time for Prevention? Am J Pub Health 2003;93 (4):635-641

9. Montori VM, Fernandez-Balsells M. Glycemic control in type 2 diabetes: time for an evidence-based about face? Ann Intern Med 2009;150 (11):803-808

10. Persell SD, Dolan NC, Friesema EM et al. Frequency of Inappropriate Medical Exceptions to Quality Measures. Ann Intern Med 2010;152:225-231

11. Grol R, Grimshaw J. From Best evidence to best practice: effective implementation of change in patient's care Lancet 2003;362:1225-30

12. Wolff M, Bower DJ, Marabella AM, Casanova JE. US Family Physicians experiences with practice guidelines. Fam Med 1998;30:117-121

13. Zielstorff RD. Online Practice Guidelines JAMIA 1998;5:227-236

14. National Guideline Clearinghouse www.guideline.gov (Accessed January 3 2009)

15. Choudry NK et al. Relationships between authors of clinical practice guidelines and the pharmaceutical industry JAMA 2002;287:612-7

16. Tricoci P, Allen JM, Kramer JM et al. Scientific Evidence Underlying the ACC/AHA Clinical Practice Guidelines JAMA 2009;301(8):831-841

17. Rosenberg SN, Shnaiden TL, Wegh AA et al. Supporting the Patient's Role in Guideline Compliance: A Controlled Study. Am J Manag Care 2008;14 (11):737-744

18. Stross JK. The educationally influential physician – Journal of Continuing Education Health Professionals 1996; 16: 167-172)

19. Shiffman RN, Shekelle P, Overhage JM et al. Standardized Reporting of Clinical Practice Guidelines: A Proposal form the Conference on Guideline Standardization. Ann Intern Med 2003;139:493-498

20. Zielstorff, RD. Online Practice Guidelines. Issues, Obstacles, and Future Prospects. JAMIA 1998;5:227-236

21. Cardiac Clearance www.statcoder.com (Accessed March 2 2009)

22. Giuse N et al. Evolution of a Mature Clinical Informationist Model JAIMA 2005;12:249-255

23. Durieux P et al. A Clinical Decision Support System for Prevention of Venous Thromboembolism: Effect on Physician Behavior JAMA 2000;283:2816-2821

24. Tierney WM et al. Effects of Computerized Guidelines for Managing Heart Disease in Primary Care J Gen Int Med 2003;18:967-976

25. Murray et al. Failure of computerized treatment suggestions to improve health outcomes of outpatients with uncomplicated hypertension: results of a randomized controlled trial Pharmacotherapy 2004;3:324-37

26. Tierney et al. Can Computer Generated Evidence Based Care Suggestions Enhance Evidence Based Management of Asthma and Chronic Obstructive Pulmonary Disease? A Randomized Controlled Trial Health Serv Res 2005;40:477-97

27. Javitt JC et al. Using a Claims Data Based Sentinel System to Improve Compliance with Clinical Guidelines: Results of a Randomized Prospective Study Amer J of Man Care 2005;11:93-102

28. Welch, PW et al. Electronic Health Records in Four Community Physician Practices: Impact on Quality and Cost of Care. JAMIA 2007;14:320-328

29. National Guideline Clearing House www.guideline.gov (Accessed January 1 2000)

30. National Institute for Health and Clinical Excellence www.nice.org.uk (Accessed June 18 2009)

31. Colorado Clinical Guidelines Collaborative www.coloradoguidelines.org (Accessed June 18 2009)

32. ADA DiabetesPro http://professional.diabetes.org/CPR_Search.aspx (Accessed September 29 2009)

33. Apprisor http://www.apprisor.com/dlselect.cfm (Accessed December 29 2007)

34. CDC PDA download http://www.cdcnpin.org/scripts/STD/pda.asp (Accessed December 30 2007)

Disease Management and Disease Registries

ROBERT E HOYT

Learning Objectives

After reading this chapter the reader should be able to:

- Define the role of disease management in chronic disease
- Describe the need for rapid retrieval of patient and population statistics to manage patients with chronic diseases
- Compare and contrast the various disease registry formats including those that integrate with electronic health records
- Describe the interrelationships between disease registries, evidence based medicine and pay for performance

Background

According to Epstein, Disease Management (DM) is:

"a systematic population based approach to identify persons at risk, intervene with a specific program of care and measure clinical and other outcomes " [1]

Disease Management (DM) Programs are important, as pointed out by the Institute of Medicine because:

"The current delivery system responds primarily to acute and urgent health care problems. Those with chronic conditions are better served by a systematic approach that emphasizes self management, care planning with a multidisciplinary team and ongoing assessment and follow up" [2]

DM is generally considered part of Population Health and is divided into the following categories:

Disease management: focuses on specific diseases (e.g. diabetes)
Lifestyle management: focuses on personal risk factors (e.g. smoking)
Demand management: focuses on improved utilization (e.g. emergency room usage)
Condition management: focuses on temporary conditions (e.g. pregnancy)

Disease Management is discussed under Medical Informatics because it is dependent on information technology for several processes:

- Automated data collection and analysis
- Clinical Practice Guidelines (CPGs) that are web based or embedded into the electronic health records (EHRs)
- Automated disease registries
- Telemonitoring of patients at home

- Patient tracking
- Web based portals
- Networks and health information exchanges to connect multiple healthcare workers on the DM team

In the future, CPGs, EBM programs and DMPs will be embedded or linked within all EHRs. Quality reports will be generated from electronic health records and health information organizations and shared with pay for performance programs, researchers, insurers and healthcare organizations.

DMPs were created in part because health maintenance organizations (HMOs) wanted to control the rising cost of chronic diseases. DMPs were established in the 1980s at Group Health of Puget Sound and Lovelace Health System in New Mexico and now are part of many large health care organizations. As an example, in a survey of over 1000 healthcare organizations, disease registries were established with the following frequencies: diabetes (40.3%), asthma (31.2 %), heart failure (34.8 %) and depression (15.7 %).[3]

Disease management is interrelated to many other topics and chapters in this book, as depicted in Figure 14.1.

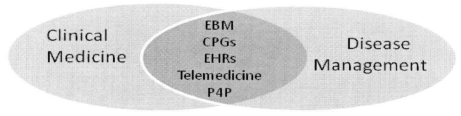

Figure 14.1. Interrelationships between disease management, evidence based medicine (EBM), clinical medicine, pay for performance (P4P), electronic health records (EHRs), clinical practice guidelines (CPGs) and telemedicine

New attention may be paid to DMPs if pay for performance (P4P) (discussed in next chapter) becomes a standard quality improvement strategy. Chronic diseases affect about 20% of the general population yet account for 75% of health care spending. By the year 2030, 20% of the US population will be 65 or older. Chronic diseases are more likely to affect lower income populations who have limited access to medical care and limited insurance coverage. Figure 14.2 shows the predicted prevalence of chronic disease by year.[4]

The most common chronic diseases to be managed are heart failure, diabetes and asthma due to high prevalence and cost. Following close behind are obesity, hypertension, chronic renal failure and chronic obstructive lung disease (COPD).

Chronic Disease by Year

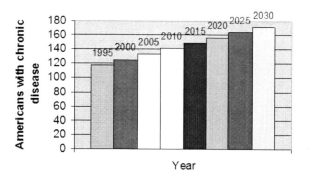

Figure 14.2. Predicted chronic disease prevalence (millions) by year

Disease Management programs involve multiple players:

- Quality Improvement Organizations (QIOs)
- State and Federal Governments (Medicare and Medicaid)
- Pharmacy organizations
- Pharmaceutical companies
- Hospital Systems, including information technology
- Physicians and their office staff
- Employers
- Insurers
- Independent vendors; including EHR vendors
- Health Maintenance Organizations (HMOs)
- Health Information Organizations (HIOs)

The integration of multiple players is best demonstrated by the classic *Chronic Care Model* created by Dr. E.Wagner and the Macoll Institute for Healthcare Innovation. His model incorporates community resources, healthcare systems, information technology, patient participation and a disease management team.[5]

The usual processes involved in Disease Management are:

- Identification of a problem and a target population
- Comparison of local to national data (how do we compare to others?)
- Review of existing clinical practice guidelines to see if they can be used or modified
- Evaluation of patient self-management education
- Evaluation of process and outcomes measurements
- Feedback to clinicians and other hospital workers
- Emphasize systems and populations, not individuals
- Coordination among multiple services and agencies

Prior to initiating a DM Program the following questions should be addressed:

- What is the goal? Decrease diabetic complications? Decrease trips to the emergency room by asthmatics?
- What population will be studied? Rural and urban? Insured and uninsured?
- Is the problem high volume or high cost or both?

- Are there preventable complications such as hospital admissions?
- How prevalent is the disease? Common enough to create a DM program?
- Are there practice variations among different medical groups?
- Are payers (insurance companies) interested?
- Does a guideline already exist or will a new CPG need to be created?
- Is the treatment feasible or practical?
- Are outcomes clearly defined, measurable and meaningful?
- Do information systems already exist for the program? Data retrieval is easier if systems are already in place

The goal of all DMPs is to improve patient outcomes: clinical, behavioral, cost, patient functional status and quality of life.

Disease Management Programs

Centers for Medicare and Medicaid's (CMS) position. Medicare and Medicaid costs account for about one third of national health expenditures so those programs are constantly looking for ways to improve quality and reduce costs. In 2009 Medicare expenditures were more than $400 billion or 20% of all healthcare expenditures. A quote from the CMS web site: *"About 14 percent of Medicare beneficiaries have congestive heart failure but they account for 43 percent of Medicare spending. About 18 percent of Medicare beneficiaries have diabetes, yet they account for 32 percent of Medicare spending. By better managing and coordinating the care of these beneficiaries, the new Medicare initiatives will help reduce health risks, improve quality of life, and provide savings to the program and the beneficiaries".* [6]

CMS has created 10 pilot programs to see if disease management can save the government money over a three year period (phase I). The Chronic Care Improvement Program (part of the Medicare Modernization Act of 2003) is now known as the Medical Health Support Program. Companies involved will not get paid for disease management unless they can show a total savings of 5% compared to a control group. Companies that can demonstrate improved outcomes are asked to participate in phase II and will likely tackle diabetes or heart failure. The companies selected were: American Healthways, XL Health, Health Dialog Services, LifeMasters and McKesson Health Solutions. All participants will need robust information technology to succeed. As of April 2010 there has been no final report. [7]

Meaningful Use and Disease Management. There is the expectation by the federal government that health IT, in particular EHRs, will result in better management and reporting of chronic diseases. Furthermore, health information organizations should also help with directing EHR reports to appropriate agencies. In early 2010 it appears that most EHRs are not capable of sending robust reports and government organizations such as Medicare/Medicaid are not ready to receive an avalanche of quality reports. To improve population health/disease management reporting, the Office of the National Coordinator released a free open source (Apache 2.0 license) population health reporting tool (popHealth) in early 2010. The goal of this tool is to allow for easier submission of quality reports to public health organizations. In addition, it will allow clinicians to create new ad hoc reports and perform their own population health analyses. Importantly, this tool integrates with EHRs because it complies with multiple data standards (CCD and CCR) and integrates with open source CONNECT, discussed in the chapters on health information exchange and data standards. The popHealth application runs on JRuby or the Ruby programming language atop the Java Virtual Machine (JVM). Figure 14.3 shows how this application would run within the network of the user and generate quality reports. Figure 14.4 shows a screenshot of a typical quality report.

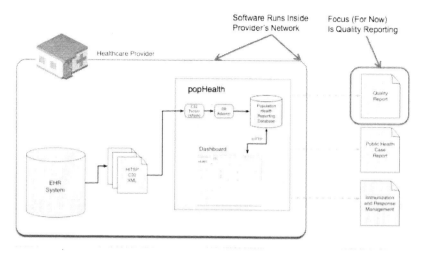

Figure 14.3 popHealth application (courtesy Project popHealth)

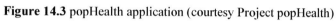

Figure 14.4 Patient dashboard (courtesy Project popHealth)

On the left is a disease and condition menu demonstrating overall patient compliance with goals such as LDL cholesterol under 100. On the right the user can select gender or age to analyze the data further. [8]

The Patient-Centered Medical Home (PCMH) This model is intended to be an improved model for healthcare delivery, to include chronic disease management. It is based on the relationship between the patient and their primary care physician (PCP). It is up to the PCP and his/her team to manage and coordinate chronic diseases with the goal of keeping the patient healthy and at home. Although the concept has been around since 1967, it was promoted by major Medical Associations in 2007. Since then the concept has been embraced by private insurers [9], Medicare [10] and the Department of Defense.[11] Part of the concept for PCMH is technology support using disease registries, EHRs, personal health records, e-prescribing, patient portals, health information exchanges and telehomecare. In this model, practices would have to handle more walk-ins and same-day appointments. For a review of the topic and more detail we refer you to a 2008 article by Rosenthal. [12] We also refer you to the Patient-Centered Primary Care Collaborative that published a 2010 monograph "Meaningful Connections" that outlines how health IT supports this model. [13] Bates and Bitton review the role of HIT in PCMHs in a 2010 article in Health Affairs and maintain that EHRs are pivotal for this model but frequently lack the desired functionality. [14]

What do we know from current DM programs and the medical literature?

- A study in the Journal of the American Medical Association (JAMA) demonstrated that 32 of 39 interventions showed improvement in at least one process or outcome measurement for diabetic patients; 18 of 27 studies involving three chronic conditions also demonstrated lower health care costs and/or lower utilization of services [15]
- A comprehensive DM program for African-American diabetics showed large reductions in amputations, hospitalizations, emergency room visits and missed work days with an aggressive foot care program [16]
- HealthPartners Optimal Diabetes Care Impact: Program noted 400 fewer cases of retinopathy (eye damage) each year; 120 fewer amputations each year and 40-80 fewer myocardial infarctions (heart attacks) per year [17]
- A systematic review/meta-analysis of DM programs for heart failure concluded that programs are effective in reducing admissions in elderly patients [18]
- A DM program for myocardial infarctions reduced readmissions, emergency room visits and insurance claims [19]
- A study of almost 800 chronically ill veterans using a web based interactive disease dialogue telemedicine strategy at home was able to show a reduction in emergency room visits (40%), a reduction in hospital admissions (63%), a reduction in hospital bed days (60%), a reduction in nursing home admissions (64%) and a reduction in nursing home bed days (88%). Medication compliance improved as did compliance with national guidelines [20]
- Grant et al. studied the effect of a specific diabetic web portal/personal health record that was integrated with an EHR. Although participants were more likely to have medications changed, their diabetic, blood pressure and cholesterol control was not better than a similar group of patients who had access to a standard web portal. One of the lessons learned was that patient participation in this trial was only 5% of their diabetic population. Also, poorly controlled diabetics were less likely to enroll in such a study [21]
- Peikes et al. reviewed 15 disease management programs (Medicare Coordinated Care Demonstration) funded by Medicare. They studied 18,000 patients to determine if care

coordination by nurses would improve chronic disease care or decrease costs. Only 2 of the 12 largest programs showed any statistically significant effects on hospital admissions. Expenditures were 8-41% higher in the intervention groups compared to controls. None of the programs generated net savings. They subsequently terminated all but 2 of the programs. They concluded that care coordinators (nurses) must interact with patients in person and not rely on telephones and technology. Also, coordinators must collaborate with the primary care clinicians to be successful [22]

- Nephrologists (kidney specialists) working for Kaiser Permanente in Hawaii wanted to improve the number of referrals from generalists so they could intervene earlier for chronic kidney disease. Because they all used the same electronic health record, they were able to monitor kidney function in the entire population of 214,000 patients. Access to lab results, clinical notes and secure messaging allowed the specialists to contact the generalists with advice and schedule consultations with themselves rather than waiting for the generalists. The end result was the decrease in late referrals from 32% to 12%. This was a good example of using a disease registry to improve population health. Rather than rely on a computerized clinical decision support, the specialists provided the decision support. Actual patient outcomes such as whether kidney dialysis was delayed due to the specialists intervening early were not included. They outlined the key features of the EHR-based electronic population management database:
  - Access to comprehensive, current patient information
  - Database permitted risk stratification
  - Ability to annotate records to improve communication
  - Seamless integration of new data into the longitudinal record
  - Electronic messaging between specialists and generalists
  - Electronic alerts for deteriorating lab results
  - Generation of population level statistics
  - Ability to flag patient records by status [23]

- In another study from Kaiser Permanente they used their EHR to collect information on 650,000 individuals from 2002-2007. The EHR allowed them to easily note who had had a bone mineral density test (DEXA), who had a fracture and what meds the patients were on. Armed with this information they were able to show that hip fractures decreased 38%, DEXA testing increased 263% over the five years and the number of people on anti-osteoporosis drugs increased 153%. Again, population health is much easier with computable information obtained from robust EHR systems [24]

In spite of some encouraging reports like those cited above, there are problems with the quality of the studies published thus far, such as lack of randomization or lack of a control group. In addition, many studies do not convincingly prove a reasonable return on investment.

Disease Registries

Definition of an electronic disease registry:

> "*A software application for capturing, managing and providing access to condition specific information for a list of patients to support organized clinical care* " [25]

Registries are tools that disease management programs use to track patients with chronic diseases, such as diabetes. As a result of this data DM programs can remind clinicians, nurses and patients to get lab work done and keep appointments. In addition, they can aggregate data to show, for example, the average hemoglobin A1c levels (blood test to measure blood sugar control) of a single clinician or an entire clinic that could be useful for "pay for performance" (next chapter) programs. Disease registries are available in several data inputting formats:

- **Manual**: data manually inputted onto paper or a computer database or spreadsheet or into a web based program
- **Automatic**: data automatically inputted into standalone software or web based site using client-server software and integrated with, for example, a laboratory result program using LOINC and HL7 standards
- **Automated and Integrated**: data input, retrieval, tracking and graphing are all automatic and part of an electronic health record or health information organization. This is the least common scenario currently but is felt to have great potential in DMP and pay for performance programs

Potential disease registry limitations were summarized in a 2010 monograph by the Robert Woods Johnson Foundation:

- Standardizing data elements among disparate disease registries
- Uniform method for patient identification
- Assistance in linking registries with EHRs
- Standardizing methodologies for statistical analysis
- Ensuring high clinician participation
- Guaranteeing registry sustainability
- Clinical and administrative (claims) data should be combined in a registry [26]

Other limitations include: manual data inputting, need for accurate coding, need for frequent updating and need for additional staff to maintain a registry. Disease management team members need to meet on a regular basis to discuss data reports and analysis. With time it is anticipated that more team functions will be automated to streamline workflow. The ultimate solution is to have all clinicians use an electronic health record with an integrated disease management and disease registry application. In this way all fields are automatically populated with patient data, to include lab results, etc (figure 14.5). As discussed in chapter two, stage I Meaningful Use criteria included the requirement to generate patient lists lists for specific conditions to use for quality improvement, reduction of disparities and outreach. It also required outpatient quality reports and the ability to send reminders to patients for preventive care. [27]

Approximately 50 disease registries exist that are free or fee-based. Cost is usually $500-$600 annually per user for commercial registries. In general, free public registries have less functionality than commercial registries. For an excellent in-depth review of 16 registries see *Chronic Disease Registries: A Product Review by the California HealthCare Foundation.*[28] They also review the IT tools used for chronic disease management. Five California foundations have combined resources to support a $4.5 million project known as "Tools for Quality" to test disease registries for the low income and underserved populations in their state. They have recruited 33 clinics thus far that will be paid on average about $40,000 to acquire and maintain disease registries.[29] The California HealthCare Foundation also has a 2008 monograph that compares electronic health records with disease management systems but does not offer specific examples or vendors. [30] A paper by Khan et al. discusses the current and future status of diabetic registries that has implications with other diseases.[31]

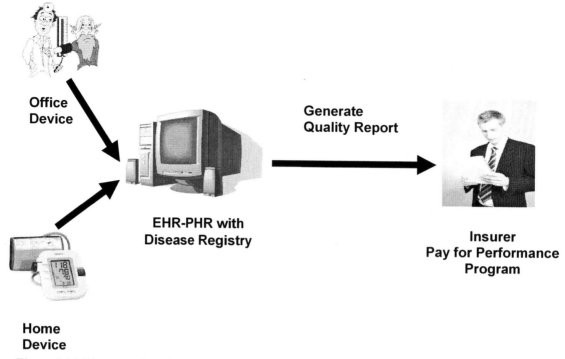

Figure 14.5 Disease registry integrated with EHR to generate reports for pay for performance programs

Disease Registry Examples

Chronic Disease Electronic Management Systems (CDEMS). This popular program is Microsoft Access-based and tracks diabetes and adult preventive health. The program is customizable and includes lab reminders for clinicians. The reports generated are also customizable and users have access to a web forum to discuss issues. A free add-on program inputs data automatically from several laboratory information systems (Quest, Labcorp, Dynacare and PAML). Shortcomings include the need to manually input data and access is limited to ten concurrent users.[32]

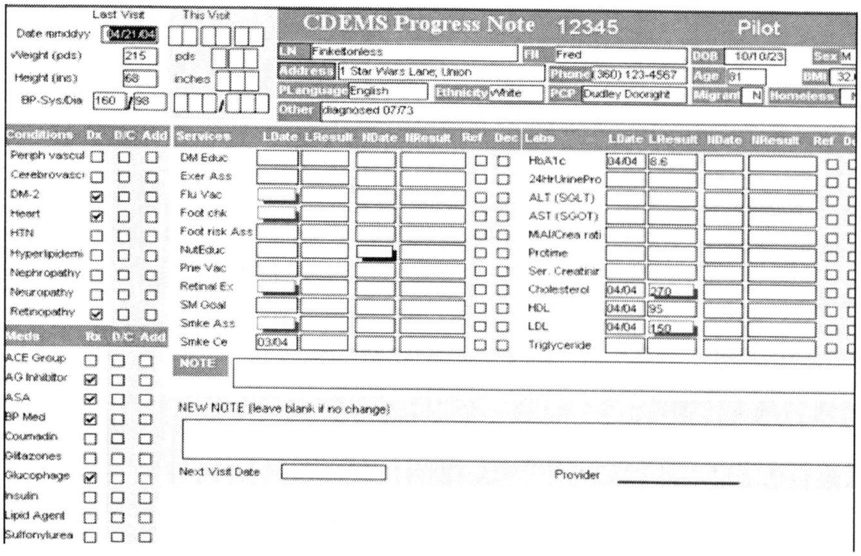

Figure 14.6. CDEMS disease registry (courtesy Washington State Department of Health-Diabetes Prevention & Control Program and Centers for Disease Control - Diabetes Translation Division)

Population Health Navigator (PHN). Population Health Navigator is a program used by the Department of Defense (DOD) to track asthma, beta-blocker use following myocardial infarction, cardiovascular risk factors, breast cancer screening, cervical cancer screening, depression, diabetes, hypertension, COPD, hyperlipidemia, low back pain and high utilizers. Data can be analyzed by physician, clinic or hospital system. Data can be exported to MS Excel for data manipulation. Drawbacks include that it is not integrated into the electronic health record (AHLTA) and data is not available real time (about a 60 day delay). The site is secure and only available to DOD personnel with proper authority.[33]

DocSite Registry. One of the best known web-based commercial registries is Patient Registry by DocSite that will track multiple common diseases. It can be integrated with practice management software, EHRs and e-prescribing systems. Figure 14.7 is a typical clinician report with lab results and due dates. Clinical practice guidelines can be embedded in the registry with the ability to make local modifications. Other features include HL7 links to input lab data, patient education, patient letter generation and the ability to host data locally or on the DocSite server. They integrated the ACP-PIER resource into disease registries in 2008. The charge for this basic registry is $50/clinician/month. They also offer DocSite Enterprise for larger organizations. They continue to evolve and offer integration with pay for performance and medical home models.[34]

DocSite PatientPlanner

Service Activity: Provider Summary

Site: DocSite - Health Care Center
Provider: Dennis Smith

Measure Name	Patient Count	Panel Avg Value	% Met Pt Goal	% Met Pop Goal	% Overdue	Avg Times Checked	Panel Max	Panel Min
HbA1c	68	6.93	56%	56%	0%	5.13	13.00	5.00
HDL	68	45.40	76%	76%	0%	4.49	92.00	24.00
LDL	68	95.49	54%	54%	0%	4.49	251.00	25.00

Provider: Sam Admas

Measure Name	Patient Count	Panel Avg Value	% Met Pt Goal	% Met Pop Goal	% Overdue	Avg Times Checked	Panel Max	Panel Min
HbA1c	83	7.06	49%	49%	0%	5.16	13.00	4.00
HDL	70	41.97	69%	69%	0%	3.80	95.00	16.00
LDL	70	108.63	46%	46%	0%	3.80	207.00	24.00

Provider: Lucy Jones

Measure Name	Patient Count	Panel Avg Value	% Met Pt Goal	% Met Pop Goal	% Overdue	Avg Times Checked	Panel Max	Panel Min
HbA1c	44	7.16	50%	50%	0%	4.57	13.00	4.00
HDL	44	48.84	80%	80%	0%	3.59	104.00	21.00
LDL	44	104.77	45%	45%	0%	3.59	223.00	23.00

Figure 14.7 DocSite Patient Planner registry (Courtesy DocSite)

Remedy MD. This web based site has more than 100 disease registries for clinicians and researchers. Application can capture, aggregate and analyze data from administrative, clinical and genetic information from EHRs as well as imaging applications and portals. Registries are customizable.[35]

MAVIQ. This is the first open source for profit disease management application. Its CarePlus web based module does the following: creates patient lists, manages chronic diseases, recalls patients for preventive health services, tracks patient care plans, has automatic reminders, patient portal for self management and generates quality improvement and compliance reports. Data can be imported from EMR/PM and integrated with HL7 feed. Cost for about 100 patient recalls per day is $179.90 per month.[36]

Patient Electronic Care System (PECYS). This is a disease registry based on Wagner's Chronic Disease Model. It is used frequently by community health centers to manage chronic diseases. Clinical practice guidelines are embedded for decision support.[37]

Key Points

- Chronic diseases are on the rise in the USA

- Chronic diseases are costly so disease management programs are commonplace, but benefits are controversial

- Disease management programs benefit from information technology by creating electronic disease registries

- Most current EHRs are in the process of adding electronic disease management programs

- Electronic disease registries will be helpful in managing patients and reporting results to pay-for-performance programs

Conclusion

For Disease Management programs to succeed there needs to be a mandate to improve the treatment of chronic disease coupled with financial support. Due to the rising costs of chronic diseases, CMS and managed care organizations are interested in new pilot programs. What must be shown is that DM programs improve patient outcomes and save money. It is much easier to show that programs improve processes such as lab tests drawn than improved patient outcomes, like fewer heart attacks or strokes. The Congressional Budget Office in 2004 concluded that there was inadequate evidence that DM programs reduced healthcare spending.[38] Bringing in more patients for preventive care will clearly increase medical costs, at least in the short run. The hope is that the costs will fall long term with preventive care.

The new Medical Home Model must be evaluated extensively to determine if it will impact the quality of care and significantly decrease healthcare costs for chronic diseases. Proof of long term benefit is evolving at this time. Nevertheless, we believe that information technology can aid the study of diseases and populations greatly.

Ultimately, all electronic health records will have comprehensive disease management features that will be customizable for clinicians and administrators. Data will be easier to retrieve and analyze in a real time mode and will be linked to reimbursement. Until that happens, however, we will rely on a variety of disease registries and disease management systems. Even with ARRA reimbursement of EHRs that have disease registries, it will be many years before we understand the true impact of electronic disease management and Meaningful Use reporting.

At this time, models that integrate human (nurse, physician, pharmacist, etc) involvement with technology seem to work better than purely technical solutions for disease management.

References

1. Epstein RS, Sherwood LM. 1996. From outcomes research to disease management: a guide for the perplexed. Ann Intern Med 124: 832-837
2. Crossing the Quality Chasm: A new health system for the 21th century. 2001. National Academies Press http://www.nap.edu/books/0309072808/html (Accessed March 5 2006)

3. Casolino L, Gillies RR, Shortell SM, et al. External incentives, information technology, and organized processes to improve health care quality for patients with chronic diseases. *JAMA.* 2003;289: 434-41.

4. Wu, Shin-Yi and Green, Anthony. *Projection of Chronic Illness Prevalence and Cost Inflation.* RAND Corporation, October 2000

5. Chronic Care Model http://www.improvingchroniccare.org/change/model/components.html (Accessed March 2 2006)

6. Xu S. Advancing Return on Investment Analysis for Electronic Health Investment. JHIM 2007;21:32-39

7. Centers for Medicare and Medicaid Services http://www3.cms.hhs.gov/apps/media/press/release.asp?Counter=1274 (Accessed March 2 2006)

8. Project popHealth http://projectpophealth.org (Accessed March 5 2010)

9. North Dakota Health Plan to Use Health IT in Medical Home Initiative. October 16 2008. www.ihealthbeat.org (Accessed October 16 2008)

10. Medicare Medical Home Demonstration Project http://www.cms.hhs.gov/DemoProjectsEvalRpts/MD/itemdetail.asp?itemID=CMS1199247 (Accessed June 18 2009)

11. The Military Health System Blog. February 24 2009. www.health.mil (Accessed March 2 2009)

12. Rosenthal TC. The Medical Home: Growing Evidence to Support a New Approach to Primary Care. J Am Board Fam Med 2008;21 (5):427-440

13. Meaningful Connections: a resource guide for using health IT to support the patient centered medical home. Patient Centered Primary Care Collabortive. www.pcpcc.net (Accessed April 10 2010)

14. Bates DW, Bitton A. The Future of Health Information Technology in the Patient-Centered Medical Home. Health Affairs 2010 29 (4):614-621

15. Bodenheimer T, Wagner, E H, Grumbach K. Improving Primary Care for Patients With Chronic Illness: The Chronic Care Model, Part 2 JAMA 2002;288:1909-1914

16. Patout CA et al. Effectiveness of a comprehensive diabetes lower extremity amputation prevention program in a predominately low income African-American population Diabetes Care 2000;23:1339-1342

17. HealthPartners. Dr Gail Amundsen (personal communication, August 2006)

18. Gonseth J et al. The effectiveness of disease management programmes in reducing hospital admissions in older patients with heart failure: a systematic review and meta-analysis of published reports Eur Heart Journal 2004;25:150-95

19. Young W et al. A disease management program reduced hospital readmission days after myocardial infarction CMAJ 2003;169:905-10

20. Meyer, M, Kobb R, Ryan P. Virtually Healthy: Chronic Disease Management in the Home. Disease Management 2002;5 (2):87-94

21. Grant RW, Wald JS, Schnipper JL et al. Practice-Linked Online Personal Health Records for Type 2 Diabetes Mellitus. Arch Intern Med 2008;168(16):1776-1782

22. Peikes D, Chen A, Schore J et al. Effects of Care Coordination on Hospitalization, Quality of Care, and Health Care Expenditures Among Medicare Beneficiaries. JAMA 2009;301(6):603-618

23. Lee BJ, Forbes K. The role of specialists in managing the health of populations with chronic illness: the example of chronic kidney disease. BMJ 2009;339:b2395

24. Dell RM, Greene D, Anderson D et al. Osteoporosis Disease Management: What Every Orthopedic Surgeon Should Know. J Bone Joint Surg Am 2009;91Suppl 6:79-86

25. Using Computerized Registries in Chronic Disease http://stage.chcf.org/documents/chronicdisease/ComputerizedRegistriesInChronicDisease.pdf (Accessed March 5 2006)

26. How Registries Can Help Performance Measurement Improve Care. White Paper. June 2010. www.hospitalqualityalliance.org/.../files/Final%20Registries%20paper.pdf (Accessed July 1 2010)

27. Proposed Rules. Federal Register. Vol 75, No. 8. January 13 2010 (Accessed April 11 2010)

28. Chronic Disease Registries: A Product Review May 2004 www.chcf.org (Accessed March 5 2006)

29. Better Chronic Disease Care Through Technology: Health Care Foundations Unveil $4.5 Million Program June 11 2008 www.chcf.org (Accessed June 18 2009)

30. Electronic Health Records versus Chronic Disease Management Systems: A Quick Comparison. California HealthCare Foundation. March 2008. www.chcf.org (Accessed April 11 2010)

31. Khan L, Mincemoyer S, Gabbay RA. Diabetes Registries: Where We Are and Where Are We Headed? Diab Tech & Ther 2009;11 (4): 255-262

32. Chronic Disease Electronic Management Systems www.cdems.com (Accessed June 18 2009)

33. Navy & Marine Corps Public Health Center. http://www-nmcphc.med.navy.mil/Data_Statistics/Clinical_Epidemiology/pophealthnav.aspx (Accessed June 18 2009)

34. DocSite http://www.docsite.com/ (Accessed April 11 2010)

35. RemedyMD www.remedymd.com (Accessed April 11 2010)

36. MAVIQ www.maviq.com (Accessed April 11 2010)

37. PECYS. Aristos Group. www.aristos.com (Accessed April 11 2010)

38. Congressional Budget Office http://www.cbo.gov/showdoc.cfm?index=5909&sequence=0 (Accessed March 5 2006)

15

Quality Improvement Strategies

ROBERT E. HOYT

Learning Objectives

After reading this chapter the reader should be able to:
- Define quality medical care
- State the goal of pay for performance (P4P) programs
- Describe how health information technology could help P4P and quality improvement
- List several Centers for Medicare and Medicaid Services (CMS) pay for performance pilot programs
- List the concerns and limitations of current P4P programs for the average clinician

In the *CMS Quality Improvement Roadmap* they espouse a simple vision *"The right care for every person every time"*. The Roadmap lists six criteria of the right medical care (adopted from the the Institute of Medicine's *Crossing the Quality Chasm*):
- **Safe**: care does not harm patients
- **Effective**: care prevents disease and complications and minimizes suffering, disability and death
- **Efficient**: patients receive care without waste
- **Patient centered**: care is coordinated and continuous; patients are informed and educated and involved in decision making
- **Timely**: patients and staff do not experience unwanted delay
- **Equitable**: care is equal, regardless of race, language, personal resources, diagnosis or condition

The core strategies of the Roadmap can be summarized as follows:
- **Publish quality measurements and information**: use the same performance measures among all healthcare organizations and select those that are the most evidence based
- **Pay for performance**: principles are explained in this chapter and table 15.1
- **Promote health information technology**: the adoption of electronic health records, e-prescribing and health information exchanges
- **Work through partnerships**: select national federal and civilian quality-oriented partners e.g. Agency for Healthcare Research and Quality, National Quality Forum, American Health Quality Association and National Committee on Quality Assurance
- **Improving access to better treatments**: accelerate the availability and effective use of the best treatments [1]

Although this vision derives from the Institute of Medicine, it has been incorporated my most federal, state and civilian healthcare organizations. To accomplish this vision organizations have developed multiple quality improvement strategies e.g. pay for performance, care coordination, patient safety initiatives, e-prescribing, electronic health records, quality performance reporting and clinical practice guidelines. All of these are discussed in detail in other chapters. In this chapter we will focus primarily on one quality improvement strategy, pay for performance.

Pay for Performance Strategy

The Centers for Medicare and Medicaid Services define P4P as a *"quality improvement and reimbursement methodology aimed at changing current payment structure which primarily reimburses based on the number of services provided regardless of outcome. P4P attempts to introduce market forces and competition to promote payment for quality, access, efficiency and successful outcomes".* [1]

There have been numerous studies since the classic *Crossing the Quality Chasm* that confirm we are not getting our money's worth from American Medicine. As an example, a study by the Commonwealth Fund demonstrated that the quality of care delivered to Medicare recipients was not related to the amount of money spent.[2] The Institute of Medicine (IOM) has been consistently critical of the variance in care and serious patient safety issues. As a result, they have repeatedly called for an increase in payments to clinicians who offer higher quality care. These concerns about "value based care" are further aggravated by the fact that the United States has an annual $2.3 trillion dollar health care price tag that continues to rise each year. The IOM released "Rewarding Provider Performance: Aligning Incentives in Medicare" in 2006 that called for a change in reimbursement that would result in higher quality of care delivered. [3]

Statements by organizations such as the IOM have helped support the notion that we need major changes in the field of medicine, to include how we determine reimbursement for care. No incentives for better quality exist under our current system. The more widgets you make, the more you get paid. The widgets don't have to be made well, just well documented. Using information technology data from electronic claims, electronic health records and disease registries we can measure quality parameters faster and without the need to do paper chart reviews. In fact, some authorities feel that P4P should first pay to create an information technology infrastructure and later reimburse for quality.

P4P (also known as value based purchasing) has gained traction in the United States in a surprisingly short period of time. The momentum may in part be due to the 2004 statement made by Mark McClellan, administrator for the Center for Medicare and Medicaid Services in the Wall Street Journal:

> *"In the next five to ten years, pay for performance based compensation could account for 20-30 percent of what the federal programs pay providers"* [4]

As a further example of the rise of P4P programs, Rosenthal et al. in a 2006 article in the New England Journal of Medicine examined the incidence of P4P programs in 252 Health Maintenance Organizations (HMOs). They determined that over half had P4P programs; 90% of programs were for physicians and 38% were for hospitals.[5]

There have been several bills introduced in Congress to address pay for performance but only one has passed thus far. The Tax Relief and Health Care Act of 2006 (HR 6111) implemented a voluntary quality reporting system for Medicare payments tied to claims data. Clinicians who report this information would be eligible for a 1.5% bonus. The new system is called the Physician Quality Reporting Initiative (PQRI) and is discussed later in this chapter.[6]

The concept of P4P is not entirely new as WellPoint (a health plan covering 24 million lives) has had reimbursement based on clinical measurements for about 10 years, but it was not called P4P.[7] Additionally, HealthPartners (Minnesota) has had an Outcomes Recognition Program since 1997.[8] The principles, components and incentive structures of P4P are summarized in table 15.1.

Table 15.1 Principles, components and incentive structure of P4P [1]

Principles	Components	Incentive Structure
Data driven	Evidence based guidelines	Equitable to participants
Patient centered	Consistent measures of access, quality, costs and satisfaction	Timely
Transparent	Coordinated care programs	Sufficient to motivate improvement
Developed through Partnerships	Health information technology	Flexible enough to pay for innovative care processes
Administratively flexible		Structured to avoid unintended consequences

Table 15.2 shows the types of data, clinical scenarios and examples of information technology used in P4P programs.

Table 15.2 Types of data, clinical scenarios and IT support for P4P programs (EHR = Electronic health record, HIE = health information exchange)

Types of data	Clinical Scenarios	IT Support
Utilization data	Emergency room visits	Data repositories, EHRs, HIE
Clinical quality	Women who have had mammograms	Patient lists, disease registries, EHRs, HIE
Patient satisfaction	Percent of patients who would recommend their primary care manager	Online surveys
Patient safety	Percent of patients questioned about allergic reactions	EHRs, e-prescribing module

In spite of the potential of information technology to improve quality numerous issues exist. Most electronic health records (EHRs) are not ready for generating P4P type reports. Ideally, data would be automatically generated from the EHR if the data was inputted into data fields in templates rather than free text. Unfortunately most notes are not written using an electronic template and problem summary lists are not updated often enough to be a data source. Perhaps natural language processing (NLP) will eventually be able to scan a dictated patient encounter and automatically submit a P4P report as well as a coding level. Lab results are easier to report because they are coded by data standards such as LOINC. Similarly, the federal government is not ready to receive voluminous quality reports from EHRs, as part of Meaningful Use requirements.
A February 2007 article by Baker on automated review of quality measures for heart failure using an EHR concluded that the current system lacked the ability to tell why a drug was not started or why it was stopped. Chart reviews were the only way to tell why recommended medications were not used or were discontinued.[9] Furthermore, there is a need to identify acute versus chronic problems and active versus inactive problems in EHRs. Until EHRs are universal, organizations must have a transitional plan like disease registries and disease flow sheets. Healthcare systems may benefit from health information exchange that includes a central data repository (CDR) or data warehouse with a rules engine.[10] Data could be pushed or pulled from the CDR for monthly reports.

Further information about the role of HIT in quality improvement can be found in the chapters on EHRs, disease management and health information exchange.

In order for P4P to be well received there needed to be a set of outpatient clinical performance measures that would be accepted by clinicians. The Ambulatory Care Quality Alliance, the American Academy of Family Physicians, The American College of Physicians, America's Health Insurance Plans and the Agency for Healthcare Research and Quality met in January 2005 and recommended 26 "starter set" measures.[11] The program looked at *processes* and not actual *patient outcomes*. This is similar to the Health Plan Employer Data and Information Set (HEDIS) measurements that are commonly used today as performance measurements.[12] Process measurements check to see if a test was done and not the actual result. This should allow for easier retrieval of data using administrative or insurance claims data. The goal is to eventually have national ambulatory quality measures in place. The following are the categories of the proposed starter set measures: preventive measures (e.g. the percent of women who had mammograms or Pap smears), coronary heart disease, heart failure, diabetes, asthma, depression, prenatal care and measures to address overuse or misuse of drugs. [11]

Pay for Performance Projects

It is estimated that more than 100 organizations have P4P programs in place, in spite of the paucity of studies to prove efficacy or return on investment. Many of the programs are really *pay for reporting* programs, in that, clinicians are being reimbursed for submitting evidence that they checked on an important test, not that the test was optimal or met national recommendations.

Centers for Medicare and Medicaid Services (CMS) Physician Group Practice Demonstration

- First Medicare pay for performance program
- Three year program started in 2005 and involved 10 large physician groups (5,000 physicians and 224,000 patients)
- Used 27 quality measures
- Most purchased EHRs and disease registries to aid reporting
- Practices would keep up to 80% of savings from program
- Overall, most practices improved the quality of care [12-13]

Physician Quality Reporting Initiative (PQRI)

- Medicare program that began in 2007 to reimburse for reporting quality measures
- It is a pay-for-reporting initiative
- You can report individual quality measures, disease/condition-specific measures or reports through disease registries (that could be part of an EHR)
- In 2010 the bonus is 2% (1.5% prior years) of the total Medicare Part B allowable charges during the reporting period. This would amount to about $4,000 per primary care physician
- Data on 30 consecutive patients must be submitted [14]
- Information about how to submit patient data with a disease registry or EHR are available at www.QualityNet.org
- Clinicians are paid the following year and this delay has not been well received according to a physician survey. In addition, clinicians have been slow to receive feedback on their progress from CMS [15]

- A commercial disease registry DocSite is ready to submit reports to Medicare for a flat fee of $350 [16]
- The PQRI experience of one medical group was reported in 2008. In spite of this group having an EHR, they had to create new software in order to create a PQRI report [17]
- For 2010 there are 175 measures and four reporting options: claims-based, registry-based, EHR-based and a new group practice option. The EHR option will require 3 out of 10 measures submission for a bonus incentive payment

Centers for Medicare and Medicaid Services (CMS) Premier Hospital Quality Incentive Demonstration Project

- 270 hospitals began participating in October 2003 as a three year demonstration project
- Hospitals are paid based on compliance with 34 quality indicators in 5 common areas (heart attack, heart failure, pneumonia, bypass surgery and hip/knee disease)
- $7 million given out yearly for 3 years as bonuses
- Program uses the Premier Perspective database, the largest in the nation, currently with 3 billion patient records
- After review of first year data, hospitals scoring in the top 10% received a 2% bonus in Medicare payments. Hospitals scoring in the second 10% received 1% and those below got nothing
- It is possible for hospitals to have a 1-2% decrease in Medicare payments if at year three they have not improved beyond the baseline
- $8.85 million awarded to top 123 performers
- Demonstration project for nursing homes is likely in the future
- Actual improvements in quality measures documented in the first year of the project:
  - Increase from 90 percent to 93 percent for patients with acute myocardial infarction (heart attack)
  - Increase from 86 percent to 90 percent for patients with coronary artery bypass graft
  - Increase from 64 percent to 76 percent for patients with heart failure
  - Increase from 85 percent to 91 percent for patients with hip and knee replacement
  - Increase from 70 percent to 80 percent for patients with pneumonia
- A two year report was published in February 2007 in the New England Journal of Medicine and showed:
  - After adjusting for baseline performance, P4P programs were associated with improvements in the 2.6 - 4.1% range, compared to hospitals that reported but were not part of the program
  - It is unknown what the actual return on investment was for the average participating hospital
  - Patient outcome information is unknown. In other words, did better compliance to guidelines result in fewer deaths and complications?
  - Would the results have been better with a higher incentive?
- A three year report was available on the Premier web site in June 2008 and showed a 15.8% improvement over the previous 3 years. As a result the project was extended to 2009 [18-19]

Centers for Medicare and Medicaid Services (CMS) Medicare Care Management Performance Demonstration

- Passed in late 2006 as part of section 649 of the 2003 MMA

- Three year demonstration project for small to medium sized practices in the states of Arkansas, California, Massachusetts and Utah
- In the first year practices only reported baseline quality data
- Reimbursement was up to $10,000 per physician and $50,000 per practice
- Results implied they need information technology for data retrieval and reporting
- Reimbursement levels may be too low [20]

Surgical Care Improvement Project (SCIP)

- Similar Medicare P4P project for surgical care
- Project looked at post-surgery: site infections, adverse cardiac events, deep vein thromboses (blood clots in the legs) and pneumonia
- Program began in July 2005 [21]
- Many hospitals will rely on labor-intensive retrospective chart reviews, but at least one facility will use a new software application to record live data (concurrent documentation) [22]

Bridges to Excellence

- Is a consortium of employers (General Electric, Procter and Gamble, Verizon Communications, Raytheon Company, UPS, Humana, Ford Motor Company and Cincinnati Children's Hospital Medical Center), health plans (Aetna, Anthem Blue Cross/ Blue Shield of Ohio and Kentucky, Blue Cross /Blue Shield of Illinois, Alabama and Massachusetts, Tufts Health Plan, United Healthcare, Harvard Pilgrim Healthcare and Humana) and physician groups in 10 states
- Their goal is to raise quality and drive down costs
- They target office systems (IT), diabetes, depression, spine care, cardiac care and the medical home model
- Program paid out about $2 million in 2005 to clinician groups who adopted P4P [23]
- A 2009 review article showed a strong correlation between reimbursement and physician participation [24]

Providence Health Systems

- PHS has 17 hospitals in California, Oregon, Washington and Alaska
- Since 2003, PHS has collaborated with Kryptiq (data integrating company) to develop a tool that uses EHR data for disease management
- Program plans to have about 15 disease modules
- Improvement in compliance to national guidelines already seen [25]

CareFirst Blue Cross/Blue Shield

- Will pay up to $20 thousand for installing EHRs
- Part of Bridges to Excellence program
- Plans to spend about $3.6 million over 3 years for P4P [26]

Blue Cross Blue/Shield of Michigan

- Paid $1 million in bonuses in April 2005 to physicians who encouraged patients to use less expensive drugs and follow clinical practice guidelines (CPGs)
- Approximately 2,400 physicians enrolled
- On a trial basis physicians were given 0.5% less in 2006 to create a pool of money [27]

California Pay for Performance Collaboration

- Seven major insurance companies (HMOs) and 225 medical groups have organized towards P4P since 2000
- Serves 6.2 million patients
- P4P bonuses totaled $203 million between 2004 to 2007
- Program used aggregate insurance claims data that was publicly reported and managed by an independent source
- Funded through multiple sources
- Second year data showed gains in all areas of clinical quality
- Groups with more IT did better
- The 2005 measures for P4P were divided into the following categories:
  - Prevention and Disease Management 50%
  - Patient Satisfaction 30%
  - Information Technology 20% [28]
- A follow-on survey reported in 2009
  - Only a handful could provide reports that would drill down to the physician level
  - Reimbursement was 1-5% of salary which was perhaps too low
  - Overall, progress was felt to be modest [29]

HealthPartners Outcomes Recognition Program

- Began in 1997 and later called P4P in 2001
- In 2004 it paid $5.6 million to primary care groups, $1 million to specialty groups and $4 million to hospitals
- Payouts: 25% for patient satisfaction, 75% for measures of diabetes, coronary disease, preventive services, tobacco cessation and use of generic drugs
- Pays for outcome measures and not just process improvement [30]

The British Experience

- Although some P4P programs began in 1990, the current extensive program has existed only since April 2005
- Program involves 10 chronic conditions
- Most data will come from practitioner's computers and the PRIMIS (primary care information services) that provides free of charge:
  - **Training** in information management skills and recording for data quality
  - **Analysis** of data quality, plus a comparative analysis service focused on key clinical topics
  - **Feedback** and interpretation of the results of data quality and comparative analyses
  - **Support** in developing action plans [31]
- Practitioners operate on a points system with a possible total of 1050 points gained [32]
- Study published in 2009 showed early improvement in quality measures followed by a plateau. An unintended consequence of this program was a reduction in continuity of care [33]

Clear Choice Health Plans

- Opted to use three measures of quality for P4P awards program:
  - Use of evidence based medicine by accessing the medical resource UpToDate

- Appropriate ordering of images by accessing the American College of Radiology web site that lists appropriateness criteria
- Self improvement by accessing a reporting web site www.managedcare.com [34]

State Medicaid Programs

- A majority of states plan for P4P programs in the next 5 years, according to a survey by the Commonwealth Fund
- Alabama, Alaska, Arizona, Massachusetts, Minnesota, New York, Pennsylvania and Utah will offer incentives to clinicians who adopt EHRs and/or e-prescribing
- Most Medicaid directors stated that their emphasis was on quality and not on saving costs [35]

P4P Concerns and Limitations

The following are some of the concerns about P4P programs expressed primarily by physicians and their organizations:

- Does it discriminate against practices without EHRs?
- Are EHRs sophisticated enough to provide accurate measures of quality?
- Should data be public?
- Should reporting be voluntary?
- Will it cause clinicians to "dump" non-compliant or sicker patients?
- Will it result in higher quality care or long term return on investment?
- Will it adjust for sicker, poorer and more elderly patients?
- Much of the practice of medicine does not have identified quality measures, so P4P will not apply
- Will the motive be money and not quality?
- Is a bonus of 5-10% of yearly compensation adequate for P4P programs?
- Is the extra work to report actually worth the small payment?
- Do P4P programs favor large medical groups?
- Should data come from a centralized data repository?
- Will P4P work outside health maintenance organizations (HMOs)?
- Should bonuses be paid for improvement even if results do not meet national goals?
- At this time, the majority of P4P reimbursements go to primary care physicians and not specialists or hospitals
- Waiting on "report cards" occasionally takes a long time and impedes next year's improvement [36-42]
- Physicians are still skeptical

Current status of pay for performance programs

On a positive note, P4P has the potential to blend evidence based medicine, disease management, clinical practice guidelines, electronic health records and information technology. Nevertheless, P4P programs have their share of obstacles. There have been multiple P4P articles written that are both pro and con in the lay press, but little written in the medical literature. In a 2005 article by Rosenthal et al. in the Journal of The American Medical Association, they reported a study comparing P4P in a large California physician group compared to a control group. The measurements studied were Pap smears, mammograms and Hgb A1C (diabetic) testing. There was very little improvement noted except for modest improvement in Pap smear testing. In spite of the $3.4 million payout by the health plan, the conclusion of the study was:

> *"Paying clinicians to reach a common, fixed performance target may produce little gain in quality for the money spent and will largely reward those with higher performance at baseline".*[43]

A 2006 systematic review by Petersen et al. was unable to prove that P4P incentives have been shown to improve the quality of care.[44]

Until better evidence is available we should proceed cautiously and be sure performance measures are fair and equitable as outlined by the Medical Group Management Association (MGMA) 2005 principles:

- The primary goal must be improving health quality and safety
- Participation by practices should be voluntary
- Physicians and professional organizations should be involved in P4P design
- P4P measures must be evidence based, broadly accepted, clinically relevant and continually updated
- Physicians should have the ability to review and correct performance data
- P4P must reimburse physicians for any administrative burden
- P4P must reward physicians' use of electronic health records and decision support tools [45]

Key Points

- US Medicine is the most expensive in the world, yet many important quality outcomes such as infant mortality demonstrate worse results than other countries who spend less money

- Civilian and federal insurers are looking at reimbursing for quality, instead of quantity

- Measuring quality will be difficult and controversial but will likely benefit from information technology, particularly the electronic health record

- Multiple pay for performance demonstration projects are underway

- It is unknown whether current pay for performance programs will have any significant impact on improving the quality of medical care

Conclusion

The jury is out regarding pay for performance programs. No one disputes the need to change reimbursement to better reflect performance and not just volume. Current fledgling P4P programs measure process and not actual patient outcomes which dilutes the significance of any results. Information technology is mandatory in order to make reporting rapid, accurate and paperless. Electronic health records with disease registries and pre-formatted pay for performance reports seem to have the most potential. Importantly, further research will need to determine how much pay will be required for how much performance.

References

1. CMS Quality Improvement Roadmap. Executive summary https://www.cms.gov/CouncilonTechInnov/downloads/qualityroadmap.pdf (Accessed July 16 2010)
2. Leatherman S, McCarthy D. Quality of Health Care for Medicare Beneficiaries: A Chartbook 2005. The Commonwealth Fund http://www.cmwf.org/publications/publications_show.htm?doc_id=275195 (Accessed March 2 2006)
3. Rewarding Provider Performance: Aligning Incentives in Medicare IOM September 2006 www.iom.edu (Accessed October 22 2006)
4. Wall Street Journal September 17 2004 (Accessed October 15 2005)
5. Rosenthal MB et al. Pay for performance in commercial HMOs. NEJM 2006;355:1895-902
6. CMS Quality/Pay for Performance Initiatives. https://www.do-online.org/index.cfm?PageID=gov_regqualityperform (Accessed February 19 2008)
7. WellPoint www.wellpoint.com (Accessed November 2 2007)
8. HealthPartners www.healthpartners.com (Accessed November 2 2007)
9. Baker, DW, Persell SD, Thompson JA et al. Automated Review of Electronic Health Records to Assess Quality of Care for Outpatients with Heart Failure. Ann Intern Med 2007;146:270-7
10. White paper: Pay for performance Information Technology Implications for Providers. First Consulting Group Feb 2005 www.fcg.com (Accessed November 20 2007)
11. Agency for Healthcare Research and Quality. Recommended Starter Set. http://www.ahrq.gov/qual/aqastart.htm (Accessed November 7 2007)
12. Medicare Physician Group Practice Demonstration http://www.cms.hhs.gov/DemoProjectsEvalRpts/downloads/PGP_Fact_Sheet.pdf (Accessed June 19 2009)
13. Physician Groups Earn Performance Payments for Improving Quality of Care for Chronic Illnesses. September 23 3008 www.psqh.com (Accessed September 24 2008)
14. Physician Quality Reporting Initiative. www.cms.hhs.gov/pqri (Accessed April 15 2010)
15. Survey: Medicare PQRI Data Not Useful. September 8 2008) www.healthdatamanagement.com (Accessed June 19 2009)
16. PQRI. DocSite www.docsite.com (Accessed April 15 2010)
17. Wintz R, Rosenthal B, Zadem SZ. The Physician Quality Reporting Initiative: A Practical Approach to Implementing Quality Reporting. Advances in Chronic Kidney Disease 2008;15(1):56-63
18. CMS/Premier Hospital Quality Incentive Demonstration Project http://www.premierinc.com/quality-safety/tools-services/p4p/hqi/index.jsp (Accessed June 20 2008)
19. Lindenauer PK et al. Public Reporting and Pay for Performance in Hospital Quality Improvement. NEJM 2007;356:486-496
20. CMS Demonstration Site http://www.cms.hhs.gov (Accessed June 19 2009)
21. Martin CB. Medicare's Pay for Performance Legislation: A newsmaker interview with Thomas Russell MD www.Medscape.com   September 15 2005 (Accessed February 20 2006)
22. Brennan KC, Spitz G. SCIP Compliance and the role of concurrent documentation. www.psqh.com January/February 2008 (Accessed February 28 2008)
23. Bridges to Excellence www.bridgestoexcellence.org/ (Accessed June 19 2009)

24. de Brantes PS, D' Andrea BG. Physician's Respond to Pay for Performance Incentives: Larger Incentives Yield Greater Participation. Am J of Man Care 2009;15(5):305-310

25. Endrado P. Pay for performance tools evolve as market shifts www.healthcareitnews.com September 26 2005 (Accessed February 9 2006)

26. Pay for performance www.acponline.org/weekly/2005/4/5/index.html (Accessed February 22 2006)

27. Merx K. Win-win program for docs, patients. Detroit free press. April 25 2005 www.freep.com (Accessed February 1 2006)

28. California Pay for Performance Collaboration http://www.pbgh.org/programs/documents/PBGH_ProjSummary_P4P_03_2005.pdf (Accessed February 11 2006)

29. Damberg CL, Raube K, Teleki SS et al. Taking Stock of Pay for Performance: A Candid Assessment From the Front Lines. Health Affairs 2009;28(2):517-525

30. Apland BA, Amundson GM. Financial Incentives, an indispensable element for quality improvement. Patient Safety & Quality Healthcare. Sept/Oct 2005. www.psqh.com (Accessed February 2 2006)

31. Primus + University of Nottingham http://www.primis.nhs.uk/pages/default.asp (Accessed November 20 2007)

32. Roland M. Linking Physicians' Pay to the Quality of Care — A Major Experiment in the United Kingdom NEJM 2004;351:1448-1454 (Accessed February 7 2006)

33. Campbell SM, Reeves D, Kontopantelis E et al. Effects of Pay for Performance on the Quality of Primary Care in England. NEJM 2009;361:368-78

34. Patmas MA. A Novel Pay for Performance Program www.uptodate.com/p4p.ppt (Accessed February 20 2007)

35. Majority of State Medicaid Programs Plan for Pay for Performance Standards. HealthDailyNews. April 12 2007 www.healthfinder.gov/news (Accessed February 21 2008)

36. Rohack JJ. The Role of Confounding Factors in Physician Pay for Performance Programs. Johns Hopkins Advanced Studies in Medicine 2005;5:174-75 (Accessed January 20 2006)

37. Audet AM et al. Measure, Learn and Improve: Physicians' Involvement in Quality Improvement Health Affairs 2005;24:843-53

38. Raths D. Pay for Performance. Healthcare Informatics Feb. 2006; 48-50

39. Colwell J. Market forces push pay for performance. ACP Observer May 2005

40. Shaw G. What can go wrong with pay for performance incentives ACP Observer March 2006

41. Colwell J. Pay for performance takes off in California ACP Observer Jan/Feb 2005

42. Bodenheimer T et al. Can money buy Quality? Physician response to pay for performance. http://hschange.org/CONTENT/807/ (Accessed February 23 2006)

43. Rosenthal MB et al. Early Experience with Pay-for-Performance: From Concept to Practice JAMA. 2005;294:1788-1793

44. Petersen LA et al. Does Pay for Performance Improve the Quality of Health Care? Ann of Int Med 2006;145:265-272

45. Medical Group Management Association. http://mgma.com (Accessed January 4 2008)

Patient Safety and Health Information Technology

ROBERT E. HOYT
TODD R. JOHNSON

Learning Objectives

After reading this chapter the reader should be able to:

- Identify why patient safety is a national concern
- Define medical error, adverse event and preventable adverse event
- Compare and contrast how information technology can potentially improve or worsen patient safety
- Compare and contrast the private and governmental patient safety programs
- List the various technologies that are likely to improve medication error rates
- Identify the obstacles to widespread implementation of patient safety initiatives

The 1999 Institute of Medicine report *To Error is Human* defined patient safety as *"freedom from accidental injury"*. They defined medical error as the *"failure of a planned action to be completed as intended or the use of a wrong plan to achieve an aim"*. Errors can involve different aspects of medical care such as diagnosis, treatment and preventive care. Fortunately, not all errors result in an injury. Errors that result in injury may be termed *"preventable adverse events"* and adverse events are defined as *"an injury resulting from a medical intervention"*. Furthermore, medical errors can be errors of commission or omission. Importantly, not all medical errors are preventable so we will limit our discussion to those errors felt to be preventable.

This same IOM report estimated that at least 98,000 inpatients die every year and 1,000,000 are injured due to preventable errors.[1] The mortality and morbidity rate may have been actually higher as outpatient adverse events were not reported.[10] While McDonald and others argue that the methodology used to report these statistics was flawed, most agree that American medicine is not as safe as it should be.[2] The 2001 Institute of Medicine report *Crossing the Quality Chasm* stated that there has been little progress since the first report. No congressional action has taken place since the first IOM report. The current medical system was described as an *"era of Brownian motion in health care"*. Furthermore, the IOM commented on three categories of inappropriate use of medical services that decrease medical quality and hence patient safety:

Overuse: the delivery of care of little or no value (e.g. widespread use of antibiotics for viral infections; also 32% of carotid artery surgeries are inappropriate and 32% are equivocal)
Underuse: the failure to deliver appropriate care (e.g. vaccines, cancer screening, beta blockers post heart attack, etc)
Misuse: the use of certain services in situations where they are not clinically indicated (e.g. magnetic resonance imaging for routine low back pain)

The IOM has long been an advocate of using information technology to improve healthcare quality and patient safety. They clearly state that safety is the first domain of medical quality. The 2001 IOM report recommended that we *"improve access to clinical information and support clinical decision making"* and *"create a national information infrastructure to improve health care delivery and research*". Also, one of their goals was to eliminate handwritten notes in the following decade.[3]

Until recently, there has been a paucity of articles written about patient safety and most articles have dealt specifically with medication errors and not errors occurring in other areas of medical practice. A 2003 article ranked the most common types of medical errors made by American family physicians: prescribing medications, getting the correct laboratory test for the correct patient at the correct time, filing system errors, dispensing medications and responding to abnormal test results. [4]

According to Dr. Leape, an early patient safety advocate, the only specialty that has experienced dramatic advances in patient safety is anesthesiology, with less than one death in 200,000 patients undergoing anesthesia. [5]

Most authorities believe that errors occur more often due to inadequate systems and not inadequate individuals. Most of these errors arise because our system of medical care including training, staffing, financial incentives, as well as local and federal policies, was not designed to prevent errors or mitigate their effects. Some authorities believe that about 50% of medical errors are preventable with better systems. [6] Also, our fee-for-service system does not reimburse based on quality or patient safety.

Other industries such as the airlines have dramatically reduced mishaps thru initiatives such as *"crew resource management"* (CRM).[7] CRM training focuses on interpersonal communication, situational awareness, leadership and decision making. This technique has been so successful hospitals often incorporate CRM as part of management training. In particular, operating rooms employ a CRM-based check list prior to initiating surgery.

In addition to the obvious increased mortality and morbidity that results from medical errors there is a resulting increase in litigation. It was estimated in 2003 that US malpractice costs totaled $27 billion.[8]

This chapter will discuss how health information technology (HIT) may improve patient safety, largely through improving the quality of care delivered. It is important to stress, however, that HIT can also create new types of medical errors. HIT is a facilitator of many medical processes but it cannot correct existing faulty processes.

In the next section we will list several well known reports on the quality of medical care in the United States, with comparisons to other countries.

The United States Medical Report Card

World Health Organization (WHO). Table 16.2 shows that the US spends a greater percent of their gross national product (GNP) on healthcare compared to other countries, but life expectancy is not substantially better. Based on available outcome measures, the United States is generally in the bottom half and its relative ranking has been declining since 1960.[9]

Table 16.2 Percent of Gross National Product (GNP) spent on healthcare and life expectancy in years

Country	GNP % Spent on Healthcare						Life expectancy 2007 (years)
	1997	1998	1999	2000	2001	2006	
Australia	8.5	8.6	8.7	8.9	9.2	8.7	82
France	9.4	9.3	9.3	9.4	9.6	11	81
Japan	6.8	7.1	7.5	7.7	8	8.1	83
United Arab Emirates	3.6	4	3.7	3.5	3.5	2.5	78
United Kingdom	6.8	6.9	7.2	7.3	7.6	8.2	80
Tanzania	4.1	4.4	4.3	4.4	4.4	6.4	52
USA	13	13	13	13.1	13.9	15.3	78
Uruguay	10	10.2	10.7	10.9	10.9	8.2	75

Rand Study 2003

Some have argued that one of the causes of decreased patient safety is the failure of US physicians to follow national clinical practice guidelines. A telephone survey of 13,275 adults living in 12 urban areas in the US was conducted that looked at quality indicators for 30 acute and chronic conditions including preventive care. The conclusion was that participants received only 55% of recommended care. It is unknown, however, how often patient non-compliance or lack of finances played a role in patients not receiving the recommended care. It cannot be assumed that the primary reason for low compliance to guidelines was an insufficient effort on the part of clinicians.[10]

Institute of Medicine (IOM) Reports

Earlier in this chapter we commented on the important IOM patient safety reports *To Err is Human* and *Crossing the Quality Chasm*. Their recommendations are listed in the executive summary:
- Congress should create a Center for Patient Safety within the Agency for Healthcare Research and Quality
- A nationwide reporting system for medical errors should be established
- Volunteer reporting should be encouraged
- Congress should create legislation to protect internal peer review of medical errors
- Performance standards and expectations by healthcare organizations should include patient safety
- FDA should focus more attention on drug safety
- Healthcare organizations and providers should make patient safety a priority goal
- Healthcare organizations should implement known medication safety policies [3]

The IOM's 2004 *Patient Safety: Achieving a New Standard for Care* repeated the same recommendations.[11]

HealthGrades 2010 Patient Safety Excellence Awards

This organization reviews Medicare data each year and rates hospitals, in terms of patient safety. The most recent report covered data from 2006-2008. They estimate that the top ranking hospitals represent, on average, a 43% lower risk of a patient safety adverse event compared to the lowest ranking hospitals. They awarded 238 hospitals with the excellence award and this represented about 5% of the nation's hospitals. The awardees are listed on their web site. Samantha Collier MD of HealthGrades believes that the hospitals that traditionally have excellent safety scores have a *"culture of safety"* and they are the ones that have all of the mechanisms including technology in place to prevent and track patient safety issues.[12]

Commonwealth Fund Study 2009

This was a survey of primary care physicians (>10,000 from the United States, Australia, Canada, Germany, New Zealand, United Kingdom, France, Italy, Netherlands, Sweden and Norway). The following is a synopsis of this international study. Compared with other countries surveyed, US physicians:
- Reported the highest percent (58%) of patients claiming difficulty affording medication
- Provided the lowest after hours support (29%) for patients
- Were last in use of EHRs
- Had one of the lowest rates of using teams to treat chronic diseases

- Scored well for patient waiting time to see specialists
- Were among the lowest in incentives for primary care [13]

Why is the USA Healthcare report card unfavorable?

In *Crossing the Quality Chasm* the IOM states the problem of poor medical care is based on the:

- Growing complexity of science and technology
- Increase in chronic conditions, e.g. obesity, diabetes and heart failure
- Poorly organized delivery systems that are not organized around patient safety
- Constraints on exploiting the revolution in information technology [3]

According to the Agency for Healthcare Research and Quality, there is too much variation in medical care within the United States as demonstrated by variation in coronary angiography rates between states (Fig. 16.1).[14] The implication is that some areas may perform too many surgeries or procedures compared to other areas that may perform far fewer. As a rule, urban and affluent areas have more specialists and better insurance coverage influencing this disparity.

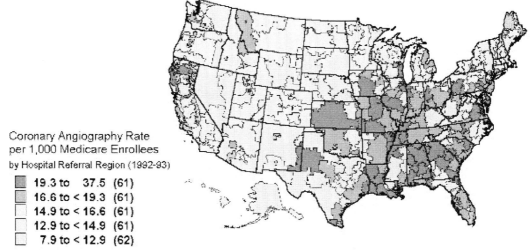

Figure 16.1 US Coronary Angiography Rate (adapted from AHRQ web site)

In addition, there are significant differences between early landmark trials resulting in national recommendations and widespread implementation, as shown in table 16.3.[15]

Table 16.3 Current Rate of Use of Landmark Trial Results

Clinical Procedure	Landmark Trial	Current rate of use
Flu vaccine	1968	64% (2000)
Pneumococcal Vaccine	1977	53% (2000)
Diabetic eye exam	1981	48.1% (2000)
Mammography	1982	75.5% (2001)
Cholesterol Screening	1984	69.1% (1999)

Barriers to Improving Patient Safety through Technology

- **Organizational.** US Medicine is primarily a decentralized system with no unifying philosophy. Many small physician groups have no loyalty to hospitals or other healthcare organizations. They do not interact with other physician offices or healthcare organizations and do not share data. In summary, the US system was not designed for quality and hence information technology may not solve the problems
- **Financial.** Who will pay for what? It is estimated that it will cost $500-700 billion dollars over the next 10 years to have a full fledged interoperable electronic health record nationwide. This is 3-4% of the total health care budget which is a lower percentage than what other industries spend on technology. In 1996 the healthcare industry spent about $543 per worker as compared to $12,666 per worker spent by security brokers and other industries for information technology [16]
- **Trained HIT Workforce.** How can we train and implement technologies such as electronic health records for everyone without expanding the number of health IT workers? It has been estimated that we need between 50-75,000 trained HIT workers over the next 5 years. The lack of trained clinical informaticians is also covered in the first chapter
- **Privacy.** HIPAA concerns continue to be an obstacle to many developments in technology
- **Security.** Several major episodes of identity theft have alarmed patients and staff and have contributed to the slow adoption of electronic health records and other technologies
- **Behavioral.** Some would argue that the medical profession has *"mural dyslexia"* or the failure to see the handwriting on the wall. This is due to a fear of change and an underlying skepticism by the medical profession. It is also unclear if physicians and the public view patient safety as a major issue. An article published in 2002 showed that neither group ranked medical errors at the top of problems facing medicine. Furthermore, 70% of the public and 25% of physicians thought errors should be reported to a state agency and only 19% of physicians and 46% of the public thought that computerized physician order entry (CPOE) was an effective way to reduce errors [17]
- **Error reporting.** Reporting continues to be voluntary and inadequate at best. In a recent study of over 90,000 voluntary electronic error reports from 26 hospitals, most were from nurses and only 2% were reported by physicians.[18] A survey of over 1000 physicians revealed that 45% did not know if their institution had an error reporting system. Seventy percent thought the current reporting systems were inadequate. Physicians believed reporting would improve if information was kept confidential, nondiscoverable, it was quick to input and it was nonpunitive. [19] Clearly there are legal and licensure issues associated with error reporting. Currently, there is no universal method to standardize error reporting in the US. A new FDA portal will be discussed in another section. [20] Previously, alerts to physicians about defective devices and drug alerts were mailed. To improve the situation the Health Care Notification Network was created that will e-mail alerts as well as public health emergencies and bioterrorism events [21-22]

What is the Federal Government doing about patient safety?

Agency for Healthcare Research and Quality (AHRQ)

The AHRQ is an agency under the Department of Health and Human Services. One of its goals is to provide grant money for research to reduce medical errors. This includes research using information technology. Importantly, the agency developed the AHRQ Patient Safety Network that includes an e-mail newsletter with medical literature reviews of patient safety articles.[14]

Patient Safety and Quality Improvement Act S.544

This patient safety initiative became law in 2005 and called for voluntary reporting of medical errors without recrimination. The Department of Health and Human Services maintains databases to collect data. This act is limited by the fact that there is no definition of "error" and data can only be used in a criminal case and with the physician's permission. It is unclear what the incentive is for the average physician or hospital to report errors.[23]

Centers for Medicare and Medicaid Services (CMS)

Effective October 1, 2008 Medicare stopped reimbursing hospitals for complications they deemed preventable. At this time, this new policy does not affect physicians. The list of non-reimbursable complications included:
- Objects left in a patient during surgery and blood incompatibility
- Catheter-associated urinary tract infections
- Pressure ulcers (bed sores)
- Vascular catheter-associated infections
- Surgical site infections
- Serious trauma while hospitalized
- Extreme blood sugar derangement
- Blood clots in legs or lungs [24]

The Food and Drug Administration

The FDA is extending its MedWatch program to include safety reports on all FDA regulated products (drugs, devices, biologics, dietary supplements, human food, animal feed and veterinary products). The new program known as MedWatchPlus will partner with the National Institute of Health (NIH) will include a public web site and data collection tools. This will offer a single reporting portal and repository for data analytics. The program will be released in phases and slated to be complete by 2011. [20] The Center for Devices and Radiological Health (CDRH) is part of the FDA and is responsible for the pre-market approval of all medical devices.

What are the States doing about patient safety?

Florida became the first state to openly report a range of cost and quality measures for both hospitals and outpatient facilities. Two web sites were created. The first was www.FloridaCompareCare.gov that provides broad coverage of data such as infection and mortality rates, in addition to costs for common operations. The second site www.MyFloridaRx.com lists pricing information for the 50 most common drugs prescribed in Florida. At this point there is inadequate evidence that simply reporting quality and cost to patients affects their choices or outcomes.

What are private agencies doing about patient safety?

The Joint Commission

In 2008 the Joint Commission recommended that there be one national infrastructure to measure and track quality improvement data.[25] In addition they warned healthcare organizations about new technologies that potentially cause new patient safety concerns if there is inadequate training and testing.[26] Four of the twelve applicable ambulatory 2010 National Patient Safety Goals have health

IT implications: improve the accuracy of patient identification, improve the effectiveness of communication among caregivers, improve the safety of using medications and accurately and completely reconcile medications across the continuum of care.

Institute for Healthcare Improvement (IHI)

The IHI instituted a plan in December 2004 to save 100,000 lives from medical errors by getting hospitals to incorporate at least one of six safety measures. A report on June 14th 2006 estimated that 122,300 deaths have been prevented through the adoption of new safety measures by more than 3,000 participating hospitals over an 18th month period.[27] It should be noted, however, that none of these methods directly involved information technology.

LeapFrog Group

LeapFrog is a consortium of healthcare purchasers that demand better quality. One of the four areas they promote is the adoption of inpatient computerized physician order entry (CPOE). They maintain survey safety data from over 1,000 hospitals who volunteered to submit data, as well as a calculator to determine return on investment (ROI) for hospital pay for performance programs. A consumer can search hospital overall patient safety and safety related to specific procedures via a search engine on their site. [28]

HealthGrades

HealthGrades is an organization that rates different aspects of medical care. Some of the reports are associated with charges ($7.95). Hospital reports compare 28 surgical procedures or diagnoses by state. Physician reports compare disciplinary action, board certification and patient opinions. Nursing homes are also rated. Medical cost reports compare cost information for 56 common procedures to include doctor, hospital, lab and drug costs and include out of pocket costs and average health plan payments.[12]

Health Information Technology and Patient Safety

Medication error reduction remains the most important patient safety area impacted by healthcare IT. Safety organizations mandate that patients receive the "five rights" before medication administration: right drug, right patient, right dosage, right route and right time. It has been shown that injury from medications or ADEs accounts for up to 3.3 % of hospital admissions.[34] To compound the issue, serious ADEs reported to the FDA increased about 2.6 fold from 1998 to 2005, as did fatalities due to medications.[29-30] While the IOM cited a study that 7,000 deaths occurred in 1993 due to medication errors, Rooney maintains that 31% of deaths cited were actually due to drug overdoses.[31] Fortunately, 99% of medication errors do not result in injury. About 30% of ADEs are felt to be preventable and of those about 50% are preventable at the ordering stage.[32] It is worth noting that CPOE does not prevent errors of administration or timing.[33]

In spite of the fact that more drugs are prescribed for outpatients, inpatient drug use is more dangerous. Intravenous (IV) medications are associated with 54% of ADEs and 61% of serious or life threatening errors.[34] The 2007 monograph by the Institute of Medicine entitled Preventing Medication Errors made several salient points:

- On average, a hospital patient is subject to one medication error per day
- About 1.5 million preventable ADEs occur yearly with about 400,000 preventable ADEs occurring in inpatients

- Estimated cost of $5,857 per inpatient error resulting in about $3.5 billion in 2006 dollars due to longer length of stay and additional services, but excluding litigation
- Estimates are probably low, based on how statistics were collected [35]

Technology has great potential in reducing medication errors, but it is not a panacea. As an example, an article by Oren on technology and medication errors concluded that well controlled studies are lacking, tend to be reported only at a select number of universities and patient outcomes are lacking.[36] Moreover, in a systematic review on the impact of HIT on quality, efficiency and cost reduction the point is made that four institutions are responsible for the majority of what is written on the subject; Brigham and Women's Hospital, Regenstrief Institute, Veterans Administration Hospital system and LDS Hospital/Intermountain Healthcare.[37]

An article in Health Affairs in mid-2008 reported on adoption of medication safety related HIT by 4,561 non-federal hospitals in 2006. The IT applications studied were: electronic medical records, clinical decision support, CPOE, bar coding medication dispensing (BarD), medication dispensing robot, automated dispensing machine, electronic medication administration records (eMAR) and bar coding at medication administration (BarA). They concluded the following:

- Larger and urban hospitals had much higher adoption rates
- On average, only 2.24 of eight applications were adopted per hospital
- One fourth of hospitals had not adopted any of the eight technologies
- Teaching hospitals had higher rates of adoption
- The most widely adopted application was the automated dispensing machine and least adopted was bar coding for medication administration (BarA) [38]

A 2007 survey by the American Society of Health-System Pharmacists evaluated the adoption of pharmacy IT in the United States. They reported the following conclusions:

- 50% of respondents had at least one component of an EHR; 6% were paperless; 12% had CPOE with clinical decision support; 40% had digital documentation; it was a challenge for EHRs to connect to pharmacy IT systems and be available on all hospital units
- 24% had barcode medication administration systems
- 44% used smart pumps
- 83% used automated dispensing cabinets for drugs
- 10% used pharmacy robots
- 21% used e-prescribing
- 10% had a completely electronic medication reconciliation application
- 8.5% had electronic medication administration records (eMar)
- 54% use a paper-based system
- 20% had an eMar bundled with barcoding and electronic nursing documentation
- 36% had pharmacy IT personnel [39]

For the sake of completeness we will mention that diagnostic errors are also a concern in regards to patient safety. A Harvard Medical Practice Study suggested that 17% of preventable errors were due to a misdiagnosis.[40] Older studies utilizing autopsies for inpatients that died revealed about 35-40% of patients died due to the wrong diagnosis.[41] Unfortunately, we currently request autopsies infrequently so we can no longer correlate pre and post-mortem diagnoses. Although we don't have any evidence at this point that technology improves diagnostic accuracy, it seems likely that better imaging and better and faster online educational resources result in improved diagnostic accuracy.

Technologies That Have the Potential to Decrease Medication Errors

Computerized Physician Order Entry

- **Inpatient CPOE**. This functionality was recommended by the IOM in 1991. Most studies so far have looked primarily at inpatient CPOE and not ambulatory CPOE. A 1998 study by David Bates showed that CPOE can decrease serious inpatient medication errors by 55% (relative risk reduction).[42] Many of the studies showing reductions in medication errors by the use of technology were reported out of the same institution. Other hospital systems are unlikely to enjoy the same optimistic results. Inpatient CPOE is covered in more detail in the chapter on EHRs

- **Outpatient CPOE**. There is a greater chance for a medication error written for outpatients because of the number of outpatient prescriptions written. Inpatient prescriptions, however, are more dangerous, particularly intravenous blood thinners, opiates and chemotherapy. Kuo et al. reported medication errors from primary care settings. Seventy percent of medication errors were related to prescribing, 10% were administration errors, 10% were documentation errors, 7% dispensing errors and 3% were monitoring errors. ADEs resulted from 16% of medication errors with 3% hospitalizations and no deaths. In their judgment, 57% of errors might have been prevented by electronic prescribing.[43] CPOE may have other safety implications, besides medication safety. Casolino et al. reported on how often patients fail to hear about lab results such as mammograms, Pap smears and stool specimens for blood. They concluded that about 1 in 14 abnormal tests are not adequately reported to patients and/or not documented in the chart. Of the 23 practices studied the failure rate varied from 0%-25% and the most significant problem was that some practices failed to have an "*iron clad*" plan to notify patients. The worst case scenario was when a practice had a mixed system of either a paper record and electronic lab results or vice versa. This study reinforces the concept that safety processes and work flow must be worked out ahead of time and apparent to all clinicians or problems will occur, regardless as to whether you use a paper-based or electronic system [44]

- **Clinical Decision Support**. Computerized drug alerts have obvious potential in decreasing medication errors but they continue to be refined. Kuperman has divided drug alerts into basic and advanced as demonstrated in table 16.4. [45]

  Table 16.4 Basic and Advanced Drug alerts

Basic	Advanced
Drug allergy	Dose adjustment for renal disease
Dosage guidance	Geriatric dosing
Formulary decision support	Medication-related laboratory testing
Duplicate drug orders	Drug-disease contraindications
Drug-drug interactions	Drug-pregnancy checking

According to a systematic review by Kawamoto et al. successful alerts need to be automatic, integrated with CPOE, require a physician response and make a recommendation.[46] In an interesting study of all alert overrides for 3 months at Brigham and Women's Hospital in Boston they noted that 80% of alerts were over-ridden because: 55% of respondents stated they were "*aware of the issue*"; 33% stated "*patient doesn't have this allergy*" and 10% stated "*patient already taking the medicine*". Only six percent of patients experienced ADEs from alert overrides but half were serious. Their conclusion was that alert overrides are common but don't usually result in serious errors.[47] In a newer study at the same institution of outpatient alerts, they found better acceptance when alerts were interruptive only for critical situations. Sixty seven percent of interruptive alerts were accepted which represents an improvement. Many alerts were still incorrect so further improvements are needed.[48] In a third study from Brigham and Women's Hospital, critical lab alerts were automated to call a physician's cell phone. This strategy led to 11% quicker treatment and reduced the duration of a dangerous condition by 29%.[49]

A 2009 systematic review of computerized drug alerts demonstrated benefit (improved prescribing behavior or error reduction) with 23 out of 27 types of drug alerts. They point out that there is a need for standardized outcome measurements for measuring medication errors so studies can be compared and contrasted accurately.[50]

Non-drug related alerts for inpatients have a variable success record. A computer-alert program at the Brigham and Women's Hospital alerted physicians about the risk of blood clots in legs and was able to show a substantial improvement in the preventive measures used, as well as a decrease in the actual number of blood clots reported in the legs and lungs compared to a group who did not receive alerts.[51] WakeMed Health and Hospitals created a robust library on their Intranet as decision support for IV drugs. They reasoned that making clinical content more accessible at the point of care would decrease pharmacy questions and promote evidence based nursing. They posted local hospital guidelines and standard drug information that used infobuttons to quickly link to content. A survey showed that 87% of nurses used the information and 95% believed the content improved productivity and efficiency.[52] More information about clinical decision support and alerts can be found in the chapters on EHRs and e-prescribing

- **Accurate Drug Histories**. The significance of having prior prescribing information available at the time a prescription is written should not be underestimated. Researchers at Henry Ford Health System reported a study in which clinicians were given six months of prescription claims data compared to a control group with no such information. Those with the additional information were more likely to change dosages (21% vs. 7%); add drugs (42% vs. 14%) and discontinue drugs (15% vs. 4%). Also, physicians with prior drug histories detected non-compliance in about 1/3 of patients versus none in the control group.[53] Another important issue concerning medication error reduction is the ability to reconcile all outpatient medications when a patient is admitted to a hospital. In many instances the information given by the patient is not correct. Lau reported that 61% of patients had at least one drug missing and 33% had two or more drugs missing on initial admission interview.[54] EHRs, HIOs and pharmacy claims data all offer the opportunity to provide additional patient drug history. Medication reconciliation will be discussed later in this chapter.

There is a web site www.ICERx.org that provides physicians caring for evacuees from a disaster with necessary prescription histories through a secure web portal. This service is offered through the SureScripts network

Automated Inpatient Medication Dispensing Devices
- Devices are kept on nursing units
- Like ATM machines, these devices communicate with pharmacy computers and dispense medications stocked by the pharmacy
- User must have password to access
- Device keeps medication records

- Unfortunately, there is no evidence that these systems reduce errors or affect outcomes, in spite of their high price tags [55]

Home Electronic Medication Management Systems

- At least one company is in the process of developing an ATM-like machine to administer medications to the elderly at home
- Medications are loaded into the machine as a 6x9 inch blister pack with storage for up to 10 medications for one month
- The device is connected to the pharmacy via the Internet so they can monitor compliance and adjust doses
- The device (EMMA) gives a visual and audible alert when it is time to take a medication [56]

Pharmacy Dispensing Robots

- Studies suggest that robotic systems save space, decrease manpower, increase the speed to fill a prescription and decrease errors
- Robots are very helpful when there is a shortage of pharmacists or staff. Technology allows pharmacists to have more of a supervisory role
- Ideally, systems would receive electronic prescriptions from outpatient and inpatient areas, then be checked by both the EHR and the pharmacist, then labels are printed and the prescription filled [57]
- Robots are available in different models that handle a variety of drugs (50 to 200), giving pharmacies financial flexibility [58]

Electronic Medication Administration Record (eMAR)

- Eliminates legibility issues
- There is no need to rewrite the MAR when medications are changed or discontinued
- Ready access to the patient's chart to see what medications the patient is on
- EMAR can provide allergy and timing alerts
- Application is available to nurses and physicians who usually make separate rounds
- Program can be web based and can be wireless [59]

"Smart" Intravenous (IV) Infusion Pumps

- Intravenous sedatives, insulin, anticoagulants and narcotics pose the highest risk of harm from medication errors [60]
- Early pump versions allowed for constant infusion rates without programmable alerts
- Smart pumps can be programmed to deliver the correct amount of IV drugs and are associated with drug libraries and alerts that the dose differs from hospital guidelines. This feature is known as a "dose error reduction system" (DERS). This would be particularly important if there were a decimal point error or the units of administration such as mg/hour were incorrect. The end result is that the infusion will not begin until the discrepancy is corrected
- It is estimated that 37% of hospitals are now using smart infusion pumps
- As an added benefit some pumps also wirelessly transmit data so that specific events can be captured and studied
- Smart pumps will eventually link to eMars, CPOE and pharmacy IT systems

- Evidence so far indicates that smart infusion pumps avert serious IV medication errors. It is important to realize that even a small reduction in errors that involve dangerous IV drugs is an important advance [61]
- In 2005 Rothschild et al reported on a controlled trial of smart infusion pumps in 744 cardiac surgery patients. They found that serious medication errors were unchanged compared to a control group. This was thought to be due to the fact that the default data entry interface bypassed the error reduction system, leading many nurses to not consult the drug library. Also, alert overrides were common and there were many undocumented verbal orders. It would be important for hospitals to set the drug library as the default for the program. An unanticipated bonus of this program was the fact that the memory system of the infusion pump was a treasure trove of information, pointing out future areas of training and changes in nursing protocols [62]
- Recently a smart pump with built in bar coding was introduced by Alaris [63]

Calculators

- Johns Hopkins University created a web based pediatric total parenteral (IV) nutrition (TPN) calculator and as a result reduced medication errors in half with an annual projected saving of $60-80,000
- The infusion calculator was associated with 83% fewer errors [64]
- Other web based and handheld medical calculators are available but little is known regarding their impact on patient safety

Bar coding and RFID

Bar Coded Medication Administration (BCMA)

BCMA involves a variety of elements: bar code printers, scanners, a network (wired or wireless) to connect to a server, server with bar coding software and integration with the pharmacy information system and any CPOE system. A typical linear bar code is most common but newer two dimensional bar codes exist that encode more information in a smaller space and can be read from different angles (figure 16.2).

Figure 16.2 Bar coded label with 2-D code (Courtesy LaserBand)

Figure 16.3 demonstrates a wrist band with linear bar coding and pictures.

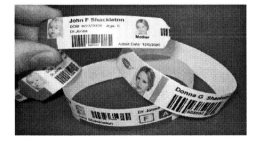

Figure 16.3 Bar coded ID bracelets (courtesy ENDUR ID)

How does a BCMA system work?

- A standard scenario would be for a nurse to scan his/her ID bar code, the patient's bar code and the medication's bar code. This information could be sent wirelessly to the program server at which time the software determines that the correct medication is going to the correct patient at the correct time. In general, the system will generate a warning or an approval

- Studies have shown that about 35% of medication errors occur at the administration stage. Further breakdown of errors that might be prevented by BCMA include: dose omission (21%), wrong patient (4%), wrong time (4%), wrong route (1%) [65]

- Most healthcare organizations use three linear barcodes: codes 128, 39 and Reduced Space Symbology (RSS). Two dimensional barcodes are available that can store 3,000 characters of patient information

- Bar codes can be placed on patient ID bands, medications, vials of blood and transfusion bags

- FDA mandated that drug companies apply bar codes on unit dose medications and blood components by April 2006. Barcodes must contain the national drug code (NDC) that can be used to indentify medications

- The price tag is likely to be $300K- $1 million for hospitals to adopt barcode technology

- There are very few studies looking at patient outcomes with this technology. Poon et al studied dispensing errors before and after implementation of a BCMA at the Brigham and Women's Hospital Pharmacy. They demonstrated that the target dispensing error rate dropped by 0.25% to 0.018% (93 % relative risk reduction). [66] A May 2010 follow on study by the same author from the same hospital concluded that bar coding coupled with an eMAR reduced medication errors at the transcription and administration stages. In addition there was a reduction in potential adverse drug events, with true adverse drug events (documented harm to patients) was not reported [67]

- A study showed an error rate decrease from 1.0% to .2% and improvement in pharmacy stock ordering time [68]

- BCMA in an adult medical intensive care unit was reported in 2009. It had the potential to improve these standard drug errors: wrong drug, wrong administrative time, wrong route, wrong dose, omissions, administration of a drug with no order and documentation error. This study showed an improvement in only administration time errors after implementation of a BCMA system [69]

- Veteran's Affair hospitals have had bar coding since 1999 in their 161 hospitals. Once scanned, the software confirms that the correct medication in the correct dose and frequency has been given to the correct patient. It also updates the electronic medication record. As a result of this technology one VA hospital was able to decrease medication errors by 66% over 5 years [70]

- Bar coding is also used for laboratory specimen labeling. As an example, an inpatient's ID bracelet is scanned and it confirms that this patient requires a certain blood test. A mobile printer prints labels that are attached to the tube of blood at the bedside. [71] A study published in February 2008 from a Pediatric Oncology hospital demonstrated a decrease from 0.03% to 0.005% in mislabeling errors after one year of implementation. The incidence of unlabeled specimens continued to be the same, after implementation. There were a few misreads due to the curvature of the wrist band, that will likely be prevented with a 2 dimensional bar code band. They estimated that the cost of the system added $1.75 to each specimen processed [72]

- Sutter Health implemented barcode medication administration at its 25 hospitals at a cost of $25 million. After one year there were: 28,000 medication mix-ups averted and 9% could have resulted in moderate to serious harm and 1% were life threatening [73]
- Southwestern Vermont Health Care organization provided some valuable lessons learned in the May/June 2008 Patient Safety & Quality Healthcare magazine [74]
- AHRQ has funded pilot programs in 11 states for bar coding. In spite of some successes they concluded that implementation is not easy [75]
- Problems with BCMA include
  o High cost
  o Nurse work flow issues
  o Some meds need to be re-packaged in order to be read
  o Scanners are not interoperable so institutions may have to buy different scanners [76]

  o Koppel reported on the 15 types of "workarounds" hospitals develop to make BCMA work [77]
  o In June 2010 Young et al reported a systematic review of BCMA and concluded this technology inconsistently decreased ADEs. They were able to indentify several new types of medication errors that were not part of the "5 rights" approach [78]

Radio Frequency Identification (RFID)

- Unlike bar coding, RFID can be read-only or read-write capable
- RFID tags can be read if wet or thru clothing. Better for blood and IV bags
- Drug companies and Walmart want RFID to record and track all inventory. They are also seeking to decrease the counterfeit drug market which may total $2 billion yearly [79]
- A scanner must interface with an established database to identify the object with the RFID tag
- Tags are cheap but transceivers (scanners) are expensive
- Scanners can be part of a PDA or laptop computer. A PDA with Socket CF reader-scan card can read RFID and bar codes [80]
- Tags can be active (needs battery, larger, more memory, longer range and more expensive) or passive (smaller, cheaper, short range and no battery) (Figure 16.4)

Figure 16.4 Passive RFID tag on back of a drug label (courtesy CPTTM)

- RFID tags can be low, medium or high frequency

- RFID systems can track patients within a hospital with an active tag that works like a transmitter and gives location and time
- RFID tracking will allow for better business and time analysis
- RFID systems may replace or complement bar coding
- One company offers RF-scanning of the surgical patient before the wound is closed to be sure no surgical sponges are left in [81]
- In 2007 the Mayo clinic began to use passive RFID tags, attached to specimen bottles, used to hold GI endoscopy biopsies. The RFID system was provided by 3M and over 30,000 specimens have been processed. The RFID holds a unique patient number stored in a database that must match. The error rate prior to RFID was 9.2%/100 bottles and .55%/100 bottles after transition [82]
- A June 2008 article in JAMA raised concerns that when an active or passive RFID tag is read by the scanner it emits electromagnetic inference (EMI). They reported frequent potentially hazardous incidents in a non-clinical scenario when devices like pacemakers and ventilators were exposed to EMI, even at distances greater than 12 inches. Although they tested only RFID tags produced by two vendors, there should be a note of caution with all RFID devices around critical equipment [83]

Figure 16.5 demonstrates an implantable FDA approved RFID chip has been used in both humans and animals. The chip contains a 16 digit number that is read by

Figure 16.5 Implantable RFID chip (courtesy VeriChip)

a scanner. The number is used to locate the web based medical record. The market may be primarily for infants, the unconscious and the demented [84]

Medication Reconciliation.

It is well known that when patients move from hospital-to-hospital, from physician-to-physician or from floor-to-floor, medication errors are more likely to occur. Home medications are occasionally forgotten or incompletely recorded. The Joint Commission mandated in January 2006 that hospitals must reconcile a list of patient medications on admission, transfer and discharge. Medication reconciliation is now part stage I Meaningful Use criteria. A report of "errors of transition" concluded the following: 66% occurred at transition to another level of care e.g. ICU, 22% occurred on admission and 12% occurred on discharge. [85] If all medical offices, pharmacies and hospitals had the same EHR or were connected to a shared health information organization, then the answer would be simpler and electronic. Instead, we find completely disparate systems that are not interoperable. Patients can compound the issue by using multiple pharmacies, taking alternative drugs and not keeping records. Multiple IT solutions are available but none are comprehensive because of the disparate process. Several IT initiatives are worth mentioning:

- RelayHealth will offer IntegrateRx-Medication History as part of its patient portal and will be based on payer or pharmacy claims data. The system can be integrated with an EHR or e-prescribing program [86]
- Standalone software programs such as MedsTracker have appeared on the scene and have the capability of communicating with an EHR. Once medications have been

reconciled, at discharge a list can be given to patients and physicians and faxed to a pharmacy [87]

- E-prescribing companies such as DrFirst offered medication reconciliation in 2008, based on pharmacy claims data [88]

- Healthcare systems such as Partners Healthcare will develop their own systems to be part of their EHRs. Only a minority of organizations have the ability to build an interoperable system [89]

While pharmacy claims data derived from pharmacy benefits managers makes sense, it will not help the uninsured who do not have records. Also, many patients take herbal medications they fail to report and are not retrievable electronically.

Electronic Prescribing: to be covered in the next chapter

Key Points

- Patient safety is a major issue facing US Medicine today. Far too many people die from medical errors each year

- There is great hope that information technology, particularly clinical decision support as part of the electronic health record, will improve patient care and safety

- There is some evidence that clinical decision support and alerts may reduce medication errors

- Bar code medical administration also appears to reduce some medication related errors but is expensive and complicated

- A dedicated and focused patient safety strategy and culture must accompany any deployment of information technology

Conclusion

Better studies are needed before we can expect widespread purchase and implementation of technology to improve patient safety. Until then, we will have to rely on anecdotal and limited studies. Surprisingly, there is not a national database or method to store and analyze medical errors.[90] Moreover, CEOs and CIOs will be looking for a reasonable return on investment. However, if improved patient safety means a larger market share, fewer law suits or a better hospital ranking by the state or federal government, then adoption will likely occur. According to HealthGrades, there is evidence that the highest ranked hospitals for quality have lower mortality rates.[91] Additionally, it appears that the most wired hospitals also have lower mortality rates but it is too early to establish clear-cut cause and effect.[92] One could also draw on the experience of the Veterans Affairs hospitals to show how their electronic health record has markedly improved the quality of care and efficiency.[93] Is their dramatic systemic improvement solely due to their EHR or is it due to the visionary Dr Kiser who saw the need for modernization and the establishment of a culture of quality and safety? A study by Menachemi et al. published in December 2007 looked at 98 Florida hospitals' IT adoption and patient outcome measures and concluded there was a definite correlation. They felt that IT systems for clinicians provided up-to-date guidelines at the point of care. The relationship between HIT and patient outcomes is likely to be more complicated and involves more than just technology, such as the effects of better leadership, training, etc.[94]

A top to bottom management approach is needed to improve patient safety. That is, healthcare leadership, administrative policies, physicians, patients and payers need to be united and aligned to improve patient safety. HIT can also help with collecting and analyzing patient safety data, but without the full support of the many players involved, it will not improve patient safety.

References

1. To Err is Human: building a safer health system. 2000. IOM http://www.nap.edu/books/0309068371/html/ (Accessed October 10 2009)
2. McDonald CJ, Weiner M, Hui, SL. Deaths Due to Medical Errors are Exaggerated in Institute of Medicine Report JAMA 2000;284(1): 93-95
3. Crossing the Quality Chasm: A New Health System for the 21st Century (2001) Institute of Medicine (IOM) http://lab.nap.edu/books/0309072808/html/3.html (Accessed October 4 2009)
4. Types of medical errors commonly reported by family physicians. American Family Physician 2003 67(4):697
5. Leape LL. Preventing Medical Injury. Quality Review Bull. 1993;19(5):144-149
6. Vilamovska AM, Conklin A. Improving Patient Safety: Addressing Patient Harm Arising form Medical Errors. RAND Policy Insight. April 2009 Vol 3. Issue 2
7. Airline industry since 1970 http://www.airsafe.com/airline.htm (Accessed October 4 2005)
8. Langreth R .Fixing Hospitals. Forbes June 20 2005
9. World Health Organization http://www.who.int/whosis/whostat/EN_WHS09_Full.pdf (Accessed October 4 2009)
10. McGlynn EA et al. The Quality of Health Care Delivered to Adults in the United States NEJM 2003:2635-2645
11. Patient Safety: Achieving a new standard of care http://www.nap.edu/catalog/10863.html (Accessed April 19 2006)
12. HealthGrades Excellence Awards 2010. http://www.healthgrades.com/ (Accessed April 19 2010)
13. Davis K et al. Mirror, Mirror on the Wall: An International Update on the Comparative Performance of American Health Care. May 2007 http://www.commonweathfund.org (Accessed June 4 2007)
14. Agency for Healthcare Research and Quality www.ahrq.gov (Accessed October 16 2009)
15. Balas EA, Boren SA. Managing Clinical Knowledge for Health Care Improvement. Yearbook of Medical Informatics 2000
16. Health Professions Education: A Bridge to Quality (2003) Board on Health Care Services (HCS) Institute of Medicine (IOM) http://darwin.nap.edu/books/0309087236/html/29.html (Accessed February 11 2006)
17. Blendon RJ et al. Views of Practicing Physicians and the Public on Medical Errors. NEJM 2002;347:1933-40
18. Milch CE et al. Voluntary Electronic Reporting of Medical Errors and Adverse Events. J of Gen Int Med 2006
19. Garbutt J, Waterman AD, Kapp JM et al. Lost Opportunities: How Physicians Communicate About Medical Errors. Health Affairs 2008;27(1):246-255
20. FDA Developing Web Portal to Ease Adverse Event Reporting. October 24 2008. www.ihealthbeat.org (Accessed October 28 2008)
21. Health Care Notification Network www.hcnn.net (Accessed June 20 2009)

22. Pizzi R. Cardiologists and OB-GYNs join online national drug alerts network. Healthcare IT News July 22 2008. www.healthcareitnews.com (Accessed June 20 2009)

23. S. 544 Patient Safety and Quality Improvement Act of 2005 http://www.whitehouse.gov/omb/legislative/sap/109-1/s544sap-h.pdf (Accessed June 20 2007)

24. Baker B. Hospitals works with admitting doctors on documentation. ACP Internist. October 2008 p. 9

25. Joint Commission on Accreditation of Healthcare Organizations www.jointcommission.org (Accessed March 5 2008)

26. Joint Commission Alert: Prevent Technology-Related Health Care Errors. Patient Safety & Quality Healthcare December 17 2008 www.psqh.com (Accessed December 17 2008)

27. Institute for Healthcare Improvement www.ihi.org (Accessed April 19 2010)

28. Leapfrog www.leapfroggroup.org (Accessed April 15 2010)

29. Fattinger K et al. Epidemiology of drug exposure and adverse drug reactions in two Swiss departments of internal medicine. Br J Clin Pharm 2000;49(2):158-167

30. Moore TJ et al. Serious Adverse Drug Events Reported to the Food and Drug Administration, 1998-2005. Arch Intern Med. 2007;167:1752-1759

31. Rooney Cl. Increase in US medication-error deaths. Letter to the Editor Lancet 1998;351:1656-1657.

32. Bates DW et al. Incidence of Adverse Drug Events and Potential Adverse Drug events: Implications for Prevention. JAMA 1995;274:29-34

33. Fitzhenry F et al. Medication Administration Discrepancies Persist Despite Electronic Ordering. JAMIA 2007;14:756-764

34. Averting Highest Risk Errors Is First Priority. Patient Safety & Quality Healthcare May/June 2005

35. Preventing Medication Errors. Committee on Identifying and Preventing Medication Errors, Aspden P, Wolcott A, Bootman KL, Cronenwett, LR (eds). Institute of Medicine. The National Academies Press. Washington, DC 2007

36. Oren E, Shaffer ER and Guglielmo JB. Impact of emerging technologies on medication errors and adverse drug events. Am J Health Syst. Pharm 2003;60:1447-1458

37. Chaudhry B et al. Systematic Review: Impact of Health Information Technology on Quality, Efficiency and Costs of Medical Care. Ann of Int Med 2006;144:E-12-E-22

38. Furukawa MF, Raghu TS, Spaulding TJ, Vinze A. Adoption of Health Information Technology for Medication Safety in US Hospitals, 2006. Health Affairs 2008;27:865-875

39. Pedersen CA, Gumpper KF. ASHP national survey on informatics: Assessment of the adoption and use of pharmacy informatics in US Hospitals—2007. Am J Health-Syst Pharm 2008;65:2244-2264

40. Patient Safety Primer. Diagnostic Errors. AHRQ. http://psnet.ahrq.gov (Accessed March 23 2009)

41. Leape LL. Error in Medicine. JAMA 1994 272(23):1851-1857

42. Bates DW et al. Effect of computerized physician order entry and a team intervention on prevention of serious medication errors. JAMA 1998;280:1311-1316

43. Kuo GM, Phillips RL, Graham D. et al. Medication errors reported by US family physicians and their office staff. Quality and Safety In Health Care 2008;17(4):286-290

44. Casolino LP, Dunham D, Chin MH et al. Frequency of Failure to Inform Patients of Clinically Significant Outpatient Test Results. Arch Intern Med 2009;169(12):1123-1129

45. Kuperman GJ, Bobb A, Payne TH et al. Medication-related clinical decision support in computerized provider order entry systems: A review. J Am Med Inform Assoc 2007;14 (1):29-40

46. Kawamoto K et al. Improving clinical practice using clinical decision support systems: a systematic review of trials to identify features critical to success. BMJ 2005 330: 765-772

47. Hsiegh TC et al. Characteristics and Consequences of Drug Allergy Alert Overrides in a Computerized Physician Order Entry System. JAIMA 2004;11:482-491

48. Shah NR et al. Improving Acceptance of Computerized Prescribing Alerts in Ambulatory Care. JAMIA 2006;13:5-11

49. Bates DW, Gawande AA. Patient Safety: Improving Safety with Information Technology NEJM 2003;348:2526-2534

50. Schedlbauer A, Prasad V, Mulvaney C et al. What Evidence Supports the Use of Computerized Alerts and Prompts to Improve Clinician's Prescribing Behavior? J Am Med Inform Assoc 2009;16:531-538

51. Kucher N et al. Electronic Alerts to Prevent Venous Thromboembolism among Hospitalized Patients. NEJM. 2005;352-969-977

52. Smith A. Online support for IV drug administration. Patient Safety & Quality Healthcare March/April 2007:50-54

53. Bieszk N et al. Detection of medication non-adherence through review of pharmacy claims data. Am J Health Syst Pharm 2003; 60:360-366

54. Lau HS et a.l The completeness of medication histories in hospital medical records of patients admitted to general internal medicine wards. Br J Clin Pharm 2000;49:597-603

55. Murray M. Automated Medication Dispensing Devices http://www.ahrq.gov/clinic/ptsafety/chap11.htm (Accessed April 23 2006)

56. In Range Systems www.inrangesystems.com) (Accessed November 24 2007)

57. Hospital Pharmacist http://www.pjonline.com/pdf/papers/pj_20050618_automateddispensing.pdf (Accessed April 10 2006)

58. ScriptPro www.scriptpro.com (Accessed November 15 2007)

59. Ascend eMAR www.hosinc.com (Accessed April 23 2006)

60. Winterstein AG, Hatton RC, Gonzalez-Rothi R. Identifying clinically significant preventable adverse drug events through a hospital's database of adverse drug reaction reports. Am J of Health Sys Pharm 2002;59:1742-1749

61. Vanderveen T. Smart Pumps: Advanced and continuous capability. Patient Safety & Quality Healthcare. Jan/Feb 2007. p40-48

62. Rothschild JM, Keohane CA, Cook EF. A Controlled Trial of Smart Infusion Pumps to Improve Medication Safety in Critically Ill Patients. Critical Care Medicine 2005;33 (3):533-540

63. Vanderveen T. IVs First, a New Barcode Implementation Strategy. Patient Safety & Quality Healthcare May/June 2006

64. Ball MJ, Merryman T, Lehmann CU. Patient Safety: A tale of two institutions. J of Health Info Man 2006;20:26-34

65. Cummings J, Bush P, Smith D, Matuszewski K. Bar-coding medication administration overview and consensus recommendations. Am J Health-Syst Pharm 2005;62:2626-2629

66. Poon EG et al. Medication Dispensing Errors and Potential Adverse Drug Events before and after Implementing Bar Code Technology in the Pharmacy. Ann of Int Med 2006;145:426-434

67. Poon EG, Keohane CA, Yoon CS et al. Effect of Bar Code Technology on the Safety of Medication Administration. NEJM 2010;362 (18):1698-1707

68. Chester M, Zilz D. Effects of bar coding on a pharmacy stock replenishment system. Am J Hosp Pharm 1989;46:1380-5

69. DeYoung JL, Vanderkooi ME, Barlfletta JF. Effect of bar code assisted medication administration on medication error rates in an adult medical intensive care unit. Am J Health Syst Pharm 2009;66(12):1110-5

70. Coyle GA, Heinen M. Evolution of BCMA within the Department of Veterans Affairs. Nurs Admin Q 2005;29:32-38

71. Murphy D. Barcode basics. Patient Safety & Quality Healthcare. July/August 2007:40-44

72. Hayden RT et al. Computer-Assisted Bar-Coding System Significantly Reduces Clinical Laboratory Specimen Identification Errors in a Pediatric Oncology Hospital. J Pediatr 2008;152:219-24

73. Bar codes help Sutter avoid medication errors November 24 2004 www.ihealthbeat.org (Accessed October 7 2005)

74. Lanoue E, Still CJ. Patient Identification: Producing a Better Bar coded Wristband. Patient Safety &Quality Healthcare. May/June 2008. pp 12-16

75. Decisionmaker Brief: Bar-Coded Medication Administration (BCMA) http://healthit.ahrq.gov (Accessed September 24 2008)

76. Gee T. Bar coding: Implementation Challenges. Patient Safety & Quality Healthcare. March/April 2009 pp 22-26

77. Koppel R, Wetterneck T, Telles JL et al. Workarounds to Barcode Medication Administration Systems: Their Occurrences, Causes and Threats to Patient Safety. JAMIA 2008;15:408-423

78. Young J, Slebodnik M, Sands L. Bar Code Technology and Medication Administration Error. J Patient Saf 2010;6:115-120

79. ABC News. Counterfeit Drugs, Real Problems. September 14 2008. http://tiny.cc/zdP8b. (Accessed July 15 2009)

80. Iglesby T. Ready for Prime Time? Patient Safety & Quality Healthcare May/June 2006 p.50-5

81. RF Surgical Systems. http://www.rfsurg.com/benefits.htm (Accessed January 4 2008)

82. Study:RFID Technology Can Reduce Errors During Biopsy Analysis. October 8 2008 www.ihealthbeat.org (Accessed October 9 2008)

83. van der Togt R et al. Electromagnetic Interference From Radio Frequency Identification Inducing Potentially Hazardous Incidents in Critical Care Medical Equipment. JAMA 2008;299(24):2884-2890

84. Verichip http://www.4verichip.com/nws_10132004FDA.htm (Accessed April 5 2006)

85. Clancy, C. Medication Reconciliation: Progress Realized, Challenge Ahead. Patient Safety & Quality Healthcare. July/August 2006. www.psqh.com (Accessed July 20 2007)

86. RelayHealth. www.relayhealth.com (Accessed July 20 2007)

87. MedsTracker. www.designclinicals.com/meds-tracker.html (Accessed July 20 2007)

88. DrFirst and MEDITECH to deliver sophisticated mediation reconciliation, a strategic alliance to integrate. EHR Consultant. January 7 2008 www.emrconsultant.com (Accessed January 12 2008)

89. Hamann C et al. Designing a medication reconciliation system. AMIA 2005 Proceedings p 976 (Accessed January 12 2008)

90. Garbutt J et al. Lost Opportunities: How Physicians Communicate About Medical Errors. Health Affairs January/February 2008:246-255

91. Study: Hospitals rated in top 5% have mortality rates 27% lower. Patient Safety & Quality Healthcare March/April 2006: 57

92. Annual list of most-wired hospitals released. 7/12/2005. www.ihealthbeat.org (Accessed April 5 2006)

93. Stires D. Technology has transformed the VA. CNN Money http://money.cnn.com 5/25/2006 (Accessed May 5 2006)

94. Menachemi N et al. Hospital Adoption of Information Technologies: A Study of 98 Hospitals in Florida. J of Healthcare Man Nov/Dec 2007;52(6)

17

Electronic Prescribing

ROBERT E. HOYT

Learning Objectives

After reading this chapter the reader should be able to:

- List the confirmed and potential benefits of electronic prescribing
- Identify the problems and limitations of handwritten prescriptions
- Describe the SureScripts-RxHub network
- Enumerate the Medicare reimbursement regulations for e-prescribing
- List the obstacles to widespread e-prescribing

Introduction

E-prescribing is simply the generation of a digital prescription that can be transmitted to a pharmacy over a secure network. The reason we are moving in this direction can be demonstrated by the prescription shown in figure 17.1 that resulted in a patient death as the pharmacist interpreted the drug prescribed as *Plendil* and not *Isordil*. This was the first medical malpractice case successfully prosecuted due to illegible handwriting. The jury awarded $450,000 in damages with 50% of responsibility on the cardiologist and 50% on the pharmacist.[1]

Cases such as this led the Institute for Safe Medication Practices (ISMP) to push to "*eliminate handwritten prescriptions within three years*". The ISMP points out that up to 7,000 Americans die each year due to medication errors resulting in a cost of about $77 billion annually.[2] A majority of these errors are due to illegible handwriting, wrong dosing and missed drug-drug or drug-allergy reactions. E-prescribing has been promoted as a solution to improve patient safety, decrease costs and streamline the prescription process.

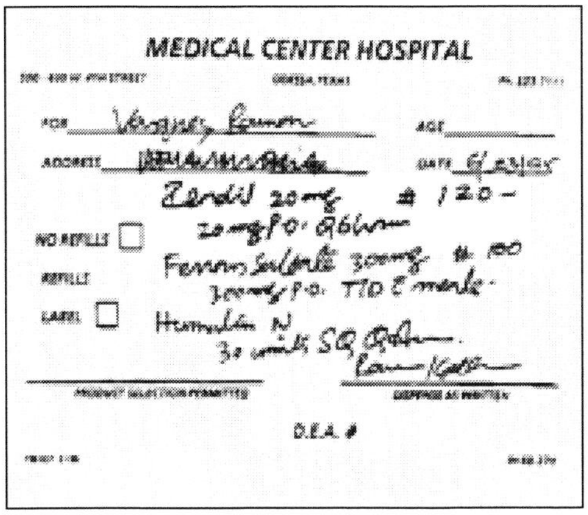

Figure 17.1 Illegible prescription

In 2006, the Institute of Medicine stated that all prescriptions should be electronic by the year 2010.[3] Since 2000, Delaware, Florida, Idaho, Washington, Montana, Tennessee, and Maryland have enacted laws requiring legible prescriptions. Montana fines up to $500 for each illegible script. Washington State took a further step by outlawing prescriptions written in cursive.[4]

Approximately 3 billion prescriptions are written annually in the United States but the majority are paper-based. This trend is changing, partly due to the multiple advantages of e-prescribing:

- Legible and complete prescriptions that help eliminate handwriting errors and decrease pharmacy "callbacks" (150 million yearly) and rejected scripts (30%) [5]
- Abbreviations and unclear decimal points are avoided
- The wait to pick up prescriptions would be reduced
- Fewer duplicated prescriptions
- Better compliance with fewer drugs not filled or picked up
- Should reduce workload for pharmacists
- Timely notification of drug alerts and updates
- Better use of generic or preferred drugs
- The ability to check plan-level and patient-level formulary status and patient copays
- E-prescribing can interface with practice and drug management software
- The process is secure and HIPAA compliant
- It is the HIT platform for future clinical decision support, alerts and reminders. It could integrate decision support related to both disease states and medications
- Digital records improve data analysis of prescribing habits
- Programs offer the ability to look up drug history, drug-drug interactions, allergies and compliance
- While entering an e-script is slower than writing a paper script, clinicians have options to speed up the process like batch refills and choosing from lists of drugs most commonly prescribed in a practice
- Provides a single view of prescriptions from multiple clinicians
- Applications have the ability to check eligibility, co-pays and it can file drug insurance claims
- Reduced cost. A 2005 study suggested that e-prescribing reduced labor costs $0.97 for a new prescription and $0.37 for a renewed prescription [6]

The Medical Group Management Association published an important report in 2004 documenting the time and money spent by physicians and staff to refill medications, verify proper formulary choices, etc. It demonstrated that non-electronic prescribing can be time intensive and expensive if you factor in the calls back and forth to a pharmacy. For an average 10 man medical group there were an average of 7 phone calls per day, 63% for refills at an annual price tag of $157,000 for the time spent by the office staff and physicians to handle pharmacy related calls. [7]

The Center for Information Technology Leadership's 2003 *Report on the Value of Computerized Physician Order Entry in Ambulatory Settings* estimated that e-prescribing would save $29 billion annually from fewer medication errors; reduced overuse, misuse and adverse drug event related hospitalizations and more cost effective selection of generic or less expensive medications.[8]

The concept of e-prescribing is gathering momentum in the United States and all states now approve eRx. Figure 17.2 shows the growth of e-prescribing in the United States through the year 2009. There was an increase of 181% between 2008 and 2009 and this involved both new prescriptions and refills. There was also an increase in the number of connected pharmacies.

Figure 17.3 shows the breakdown for e-prescribing software as to whether it was part of an EHR or a standalone program and whether it was used for routing prescriptions, providing prescription benefits/eligibility, providing prescription histories, or all three. It demonstrates that EHR-related e-prescribing was used more often for all three software purposes. It also shows the percent of active prescribers who used an EHR to generate a prescription versus a standalone e-prescription. [9] Clearly, e-prescribing as part of an EHR is the dominant mode of electronic prescribing. Vermont and Rhode Island are two states with statewide e-prescribing initiatives.

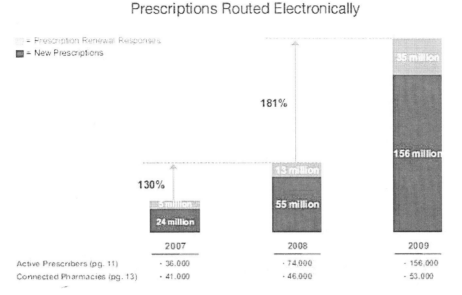

Figure 17.2 Number of e-prescribers by year (courtesy SureScripts)

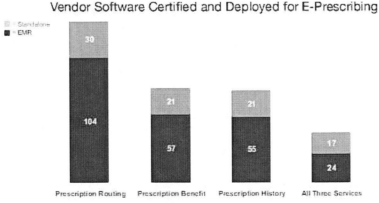

Figure 17.3 E-prescribing software and platform used (courtesy SureScripts

E-prescribing is available in two modes: 1) a standalone software program installed on a PDA/smartphone, computer or available as a web-based application 2) eRx integrated into an electronic health record. The standalone choice is simpler, less expensive and easier to learn but the EHR mode offers greater patient information at the time of prescribing. With the web-based application, patient data resides on a remote server and not on your computer. Another point worth mentioning is, although an increase in e-prescribing is occurring, many clinicians choose to print out an electronic script and then fax it to a pharmacy, thus negating several of the benefits of electronic prescribing. To date, most EHR vendors offer e-prescribing functionality.

In order for standalone e-prescribing to function well, patient lists need to be uploaded into the system to eliminate manually inputting of information for new patients at the time of e-prescribing. Patient lists can derive from practice management systems or a practice can pay for an interface to be built to upload practice patient demographics.

One of the strongest driving forces behind e-prescribing was the Medicare Prescription Drug Improvement and Modernization Act (MMA) of 2003 that allocated about $50 million in 2007 to support e-prescribing systems and allowed health plans to offer incentives to adopt information technology. Congress also approved the exceptions that allowed for donations of e-prescribing technologies to physicians from hospitals and other entities. Initial standards for e-prescribing were in place by January 2006 when the Medicare prescription benefit began. Standards were finalized in 2009.[10] Under this Act, prescription drug plans were required to offer e-prescribing, although the option was voluntary for physicians. This program has tremendous clout as 40% of all scripts written are covered by Medicare. The database created for this program would be the largest ever related to prescribing. The Act also offered drug plans higher reimbursement for physicians who e-prescribe.[11-12]

In 2008 the Medicare Improvements for Patients and Providers Act (MIPPA) was passed that codified the following:

- In 2009, clinicians who e-prescribed for Medicare patients could receive a 2% bonus for years 2009 and 2010; a 1% bonus in 2011 and 2012 and a 0.5% bonus in 2013
- For those clinicians who do not e-prescribe, Medicare would enact a penalty of 1% in 2012, 1.5% in 2013 and 2% thereafter
- It is estimated that this bonus would amount to roughly $1000-$4000 per year for the average clinician
- In order to be a "qualified" e-prescribing program the software must:
    o Generate a complete active medication list electronically
    o Transmit prescriptions electronically and generate alerts for drug-drug interactions, etc
    o Provide information on lower cost alternatives
    o Provide information on tiered medications, patient eligibility and authorizations
    o Program must use the NCPDP Script 8.1 standard by April 2009
- Computer-generated faxes are approved until January 1, 2012. After that date, they must be truly electronic or carried by the patient or manually faxed
- For calendar year 2010 there were some changes. An e-prescribing code (G8553) only had to be used when a visit resulted in an e-prescription. Clinicians needed to report this code at least 25 times during the reporting period to be eligible. E-prescribing from nursing homes or home care also count and a qualified group practice can qualify for reimbursement rather than at the individual level [13]

All EHR systems certified in 2008-2011 by CCHIT meet these requirements and they began to certify standalone systems in 2009.[14] SureScripts® lists the standalone and EHR vendors who have certified products and the exact functionality that was certified on their web site.[9]

On March 31st 2010 the DEA published a final rule that allowed for controlled drugs (narcotics, etc) to be prescribed electronically. As the rule stands, users will have to provide two-factor authentication (i.e. two means of identification): something you know (e.g. password), something you have (e.g. hard token or smart card) or something you are (e.g. finger print).[15] It will be up to EHR and e-prescribing vendors to possibly modify their software to comply with all DEA regulations. In addition, clinicians will need to apply to a private entity that verifies identities.

As pointed out in chapters 1 and 2, the HITECH Act, as part of the American Recovery and Reinvestment Act, will include reimbursement by Medicare and Medicaid for EHRs, starting in 2011. To achive Meaningful Use eligible physicians will have to utilize e-prescribing.

Surescripts Network

SureScripts was founded in 2001 by the National Association of Chain Drug Stores (NACDS) and the National Community Pharmacists Association (NCPA) to improve the quality, safety and efficiency of the overall prescribing process. One of the strongest motivating forces behind this collaboration was the need to reduce the number of physician call backs. Figure 17.4 demonstrates how the network ties together the multiple parties related to e-prescribing.

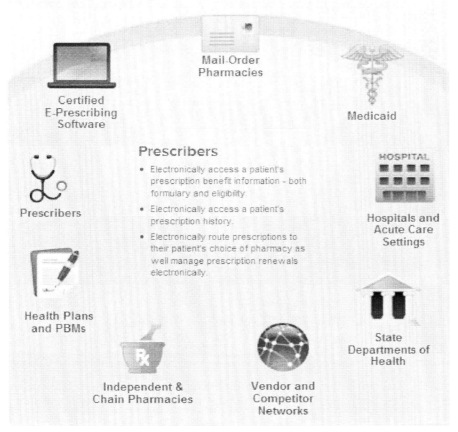

Figure 17.4 Surescripts Network connections (Courtesy Surescripts)

The Surescripts Network is the largest network to link electronic communications between pharmacies and physicians, allowing the electronic exchange of prescription information. In 2009 about 90% of major pharmacy chains and about 50% of independent pharmacies were certified to connect to the network. By 2007 all 50 states and Washington DC were connected to their network. Surescripts works with software companies that supply electronic health records (EHRs) and electronic prescribing applications to physician practices and pharmacy technology vendors to connect their solutions to the network. Although the network is free to physicians, pharmacies pay a small amount (21.5 cents per transaction) to the software vendor like DrFirst and they in turn pay Surescripts. Vendors must complete a certification process that establishes rules that safeguard the prescribing process, to include patient choice of pharmacy and physician choice of therapy.

SureScripts created the Center for Improving Medication Management to promote research into improved prescribing through technology.[9,16] In 2008 Surescripts merged with RxHub, a pharmacy benefits managing operation to create one organization.

How Does E-prescribing Work?

Typically, a patient arrives at a physician's office with a new medical problem. As part of the check-in process or during the visit with the physician the patient's prescription benefits and potential co-pays can be reviewed. If a medication is prescribed, the prior prescription history is available, known drug allergies are posted and the formulary choices offered by the insurance company can be reviewed. The electronic prescription is forwarded electronically to the patient's pharmacy of choice where it is queued until it is filled by the pharmacist.

In the case of the renewal or refill of a chronic medication the patient usually contacts their pharmacy and the pharmacist sends a secure message to the appropriate physician. The clinician can generally renew or refill a single or multiple medications with a mouse click. The electronic approval is then sent back to the pharmacy for processing.

E-Prescribing Initiatives

E-Script Pilot Study studied 100 physicians in the Washington, DC area using DrFirst software in 2005. 0.3% scripts generated serious drug interactions or allergy alerts. The study estimated an annual avoidance cost of $100,000 and an average savings to health plans of $29 per prescription filled due to improved use of generics and drugs of choice. [17]

Maryland Safety through Electronic Prescribing Initiative created a consortium of 27 health organizations. The goal was to expedite the adoption of electronic prescribing in order to reduce medication errors. They will use multiple training modalities such as conferences, workshops, etc. [18]

Wellpoint is a large US health plan covering 34 million members. They offered physicians either a free office automation package or a free e-prescribing system using Dell Axim PDAs. Only 2,700 physicians out of the 25,000 contacted signed up for the e-prescribing alternative. This outcome was a reminder that it is not enough to provide a service free. You must provide training and physicians have to believe in the long run it will save time or money for them or their patients. [19]

South East Michigan E-Prescribing Initiative. Three major auto makers have partnered with the three largest health insurers to promote e-prescribing. The project uses e-prescribing software including RxHub, hosted on PCs or PDAs. The goal is to eventually sign on 17,000 physicians. Hardware was not provided and it is unclear who will pay for the software.[20] Clinician satisfaction has been high and the program was extended. An analysis of 3.3 million prescriptions in 2007 demonstrated:

- Alerts of potential drug reactions were sent for about 1/3 of prescriptions, with physicians changing 41% of prescriptions
- The program warned physicians on about 100,000 potential drug allergies
- When an alert stated that a drug was not on formulary, a physician changed the prescription almost 40% of the time
- 40% of clinicians only used e-prescribing
- With every 1 per cent shift to generic drugs General Motors saved almost $20 million [21]

Nevada Project. Sierra Health Services and Clark County Medical Society collaborated to provide Allscripts e-prescribing software for its 5,000 physicians, for free, for 10 years. Maintenance fees were waived for two years. According to one report, utilization of generic drugs increased from 53% to 64% (estimated cost saving of $5 million). Although they stated that medical errors were down, specifics were not provided.[22]

Tuft's Health Plan Experience (2001-2002). The study involved 226 network physicians and allied health providers. As a result of the initiative, 8.9 fewer safety errors per physician per year were reported. Cost saving of $.30-.40 per member per month (PMPM) due to the use of generics and preferred drugs were noted. Clinicians and office staff reported that e-prescribing saved as much as 2 hours per day for the doctor's office and up to one hour per day for the pharmacist.[23]

Massachusetts's e-Rx Collaborative. Program was formed by Blue Cross/Blue Shield, Tufts Health Plan and the Neighborhood Health Plan. Two e-prescribing vendors (Zix and DrFirst) provided a PDA or PC based program to about 3000 physicians, paid for by the Collaborative. Software worked with a Pocket PC PDA using PocketScript or with a web based program on the desktop and can operate wired or wirelessly.[24]

Florida Medicaid awarded a contract to drug reference company Gold Standard to provide PDAs with drug information. The program began in 2001 with 500 users with the goal of expansion to 3000 physicians. The program and PDAs were free to clinicians and featured wireless technology. A major goal was to promote use of more cost effective drugs using a preferred formulary. It is important to note that the average cost for a one month supply of a generic drug is $22.79; whereas the average for a brand name drug is $77.29.[25] As a result, 39 states require a generic drug for Medicaid patients when available. Florida is the first state to provide PDAs at no cost to try to reduce Medicaid drug costs. It is anticipated that the program will pay for itself in approximately five years.[26-27] Mississippi has a similar program for Medicaid patients that equipped 225 physicians with handheld e-prescribing devices. The state believes they are saving about $1.2 million in costs monthly as a result of the program.[28]

Florida Blue/Cross Blue Shield partnered with Prematics to offer free e-prescribing in their networks, as part of Availity (see chapter on Consumer Health Informatics). Prematics is providing CarePrescribe® web-based or loaded on PDAs, as well as a printer, Internet connectivity and training at no cost to physicians.[29]

Horizon Blue/Cross Blue Shield of New Jersey partnered with three e-prescribing vendors to deliver eRx software to 665 physicians. They have written over 400,000 prescriptions at a cost of over $5 million dollars to Horizon Blue Cross/Blue Shield. Results so far show a savings of 30-60 minutes daily for prescribers.[30] More state initiatives are noted in a monograph entitled *Advancing Healthcare in America,* 2009-2010 located on the SureScripts web site. [9]

National E-prescribing Patient Safety Initiative (NEPSI) was announced in early 2007 and represents one of the most significant e-prescribing initiatives thus far. In 2008 they had 11,000 licensed prescribers. They offer free web based e-prescribing software (eRx Now™) to every physician in the United States. The two main sponsors are Allscripts and Dell Computers. Other coalition partners include Google, Aetna, Cisco, Fujitsu, Intel, Microsoft, Sprint, WellPoint, Wolters Kluwer Health and others. The program includes important features like drug-drug interactions, benefit and formulary status information and only requires 15 minutes of training. Quest Diagnostics has joined the network such that lab work can be ordered and results returned. Prescription histories can be stored on Google Health or Microsoft HealthVault. In order to upload patient demographics for the entire practice, they offer the ability to upload patient data from a .csv file free or a fee-based service.[31-32]

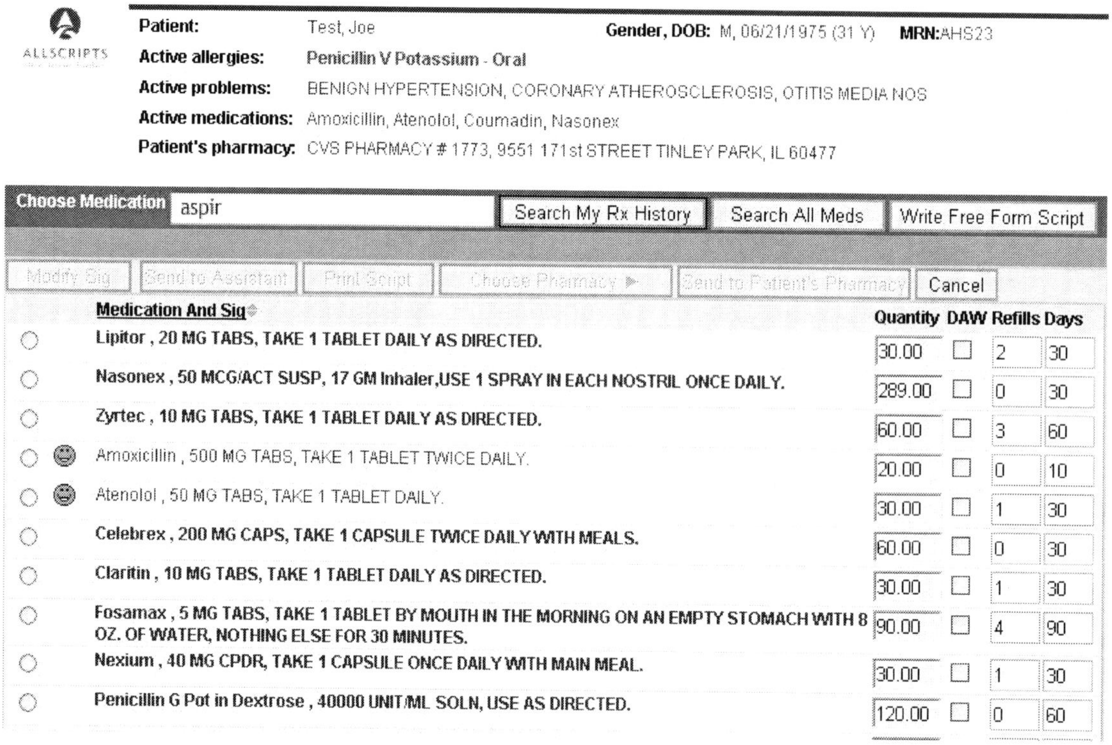

Figure 17.5 NEPSI e-prescribing program screen shot (Courtesy NEPSI)

Get RxConnected Initiative was launched in 2008 by five physician groups and the Center for Improving Medication Management in an effort to promote e-prescribing. The site gives technology guidance, physician testimonies and a list of connected pharmacies. The initiative provides free e-prescribing and support efforts to secure an electronic connection to local pharmacies. [33]

eHealth Initiative is a non-profit multi-stakeholder organization that tackles healthcare issues. They have a robust section on their web site on electronic prescribing. In October 2008 they published *A Clinician's Guide To Electronic Prescribing* that covers everything a clinician would need to get started. They also have monographs for payers and consumers that are very current and comprehensive.[34]

ePrescribe Florida was created to promote eRx in Florida in an effort to improve patient safety and increase use of generic medications. Its steering committee consists of Blue Cross/Blue Shield of

Florida (BCBSF), Humana, AvMed, SureScripts-RxHub, Florida Academy of Family Physicians (FAFP), Florida Medical Association (FMA) and Walgreens.[35]

Electronic Prescribing Studies

This section will discuss several interesting articles about e-prescribing. The chapters on EHR and patient safety also discuss aspects of computerized physician order entry (inpatient and ambulatory) and clinical decision support. More detail about drug alerts and patient safety can be found there.

A medication error study reported in 2008 evaluated the effect of electronic prescribing on medication errors and adverse drug events (ADEs) in a systematic review with the following conclusions:

- 23 of the 25 studies analyzed showed a relative risk reduction in medication errors of 13%-99%
- 6 of the 9 studies that analyzed the effects on *potential* adverse drug events (injury) showed a relative risk reduction of 35%-98%
- 4 of 7 studies that analyzed the effects on adverse drug events showed a relative risk reduction of 30%-84%
- Although they concluded that e-prescribing can reduce medication errors and ADEs, better studies are needed [36]

A **handwritten versus computerized prescription study** was conducted in an emergency room setting. Computer written scripts were three times less likely to include errors and five times less likely to require a pharmacist's clarification. This resulted in a decrease in wait time for patients and call back time by pharmacists to physicians.[37]

A **commercial e-prescribing study** was published in 2007 in *Health Affairs* that reported on physicians' experiences using commercial e-prescribing systems in the time period from November 2005-March 2006. The following are some of its conclusions:

- Two thirds of the practices used the e-prescribing module of their EHR, so this article is less relevant to standalone e-prescribing programs
- It was a major effort to upload and maintain a complete medication list
- Most could not view medication records from other physicians
- There was little decision support available for prescribing
- Many had no access to formulary information
- Only the practices with standalone e-prescribing had pharmacy connections
- Many pharmacies were not ready for electronic transmission of prescriptions, to include faxes
- EHR e-prescribing modules required continued IT support to make it work
- Greatest time saving was for the renewal or refill of multiple medications
- It is assumed that many of their observations have since been corrected [38]

The effect of e-prescribing with formulary decision support medication use and cost was reported in December 2008. They concluded that e-prescribing increased the use of generic medications (tier 1) by 3.3%. The estimated that this would result in savings of about $845,000 per 100,000 patients treated yearly.[39]

Overrides of medical alerts in ambulatory care study was reported in 2009 in the *Archives of Internal Medicine*. They analyzed the safety alerts generated by over 2800 clinicians in three states who used a common e-prescribing system. Almost 7 percent of e-prescriptions generated alerts and clinicians accepted only 23% of allergy alerts and 9.2% of drug interaction alerts. Most clinicians viewed the alerts as a nuisance rather than an educational tool. In fact, clinicians accepted high severity alerts at about the same frequency as low severity alerts. They concluded that much work needs to be done with ambulatory e-prescribing alerts before there is widespread acceptance.[40]

Reducing the prescribing of heavily marketed medications randomly allocated 14 internal medicine practices to receive usual care, computerized prescribing alerts or alerts plus group educational sessions. They were able to show that alerts kept the prescribing of heavily marketed medications (sleeping pills) stable compared to an increase in the usual care group. The educational session did not cause additional improvement. [41]

Evaluation of e-prescribing by chain pharmacy staff study. Most studies have reported the benefits and limitations of e-prescribing from the clinician's perspective. This study evaluated 422 chain pharmacies located in six states by survey. Pharmacists and their staff reported that the safety, effectiveness and efficiency of patient care were somewhat improved but problems were also reported. Of the written comments, prescribing errors were the most commonly reported negative features. Selecting the wrong drug was the most common error (29%), followed by the wrong directions (28%). Delays in receiving the e-prescription was the second most commonly reported observation. They concluded with 11 recommendations; the most significant are as follows:

- Clinicians, not office staff should submit e-prescriptions
- Clinicians should use decision support when available
- Prescriptions should be bundled together for each patient (e.g. prescription 1 of 4)
- Pharmacists need an alerting mechanism to let them know an e-script has been received
- Pharmacy workflow should change so that e-scripts are not printed then again entered electronically
- Pharmacists should have the ability to electronically message the clinician for clarification as needed
- E-prescribing applications should have standard formats and procedures
- Patients need to be better educated about e-prescribing by office staff [42]

How to prepare for E-prescribing

The following are steps a clinician should take prior to adopting e-prescribing:
- Decide on standalone (Web, PC or PDA) or part of EHR. This might also depend on whether you see a large number of Medicare or Medicaid patients, as both e-prescribing and EHRs (chapter 2) are potentially reimbursable
- Be sure you check state and national e-prescribing initiatives that are either discounted or free, found on the SureScripts web site [9]
- Be certain the e-prescribing and/or EHR product is qualified or certified by a recognized organization
- Review *Clinician's Guide to Electronic Prescribing*, mentioned previously and/or SureScripts that will address:
  o Assessing your practice readiness
  o Defining your practice needs
  o Costs and financing options
  o Selecting a system
  o Deployment

      o   SureScripts has a vendor matrix as well as a search engine for pharmacies ready for eRx [9]

E-Prescribing Obstacles and Issues

In spite of steady progress towards e-prescribing, adoption and implementation challenges remain:

- Adoption rate is still low but on the rise. More work needs to be done to improve adoption by independent pharmacies, state Medicaid programs and small practices
- Will government support of eRx and EHRs be adequate in the short and long term? At this time physicians are reluctant to pay because they believe others such as pharmacists and PBMs benefit more from eRx. Implementation will also require training and upgrades of office technology to accommodate wireless transmission and other issues such as uploading patient demographics. Physicians don't want to lose time if eRx doesn't integrate with other office IT systems. Workflow and responsibilities will have to change to accommodate e-prescribing. Set up costs to upload patient demographics and IT support add to the cost. Cost for a standalone e-prescribing program is in the $800-$1000/clinician/year range [45]
- Stage 1 Meaningful Use objectives should increase the use of EHR-related e-prescribing
- E-prescribing with an EHR is superior to standalone applications because standalone applications tend to lack decision support. Regardless, many EHR e-prescribing programs still only generate a paper script or one that can be faxed to a pharmacy
- E-prescribing is slower than paper scripts, but not when you factor in time spent calling back pharmacists or playing "phone tag" [44]
- New errors have appeared due to the newness of electronic prescribing, in spite of improved legibility [45]
- Will voice recognition, as part of e-prescribing, make the process faster? In 2008 MedVoice partnered with Nuance to create software for multiple mobile platforms. A sample dictation might be "*write a script for Jim Labowski; Zocor 20mg. Take each evening. Dispense 90, give three refills*"[46]
- Most authorities at this time have an attitude of cautious optimism [47]
- In their monograph *Advancing Healthcare in America* SureScripts makes 8 recommendations to support continued growth of e-prescribing [9]

Key Points

- Electronic prescribing has many advantages over paper prescriptions, yet the adoption of this technology has been slow for multiple reasons

- Electronic prescribing should decrease medication errors and be cost-effective overall for the average clinician

- Seventy percent of electronic prescriptions in 2009 were from EHRs so we anticipate that Medicare/Medicaid reimbursement for EHRs will help e-prescribing adoption

- Obstacles such as who will pay for the software, integration with other office systems and incomplete pharmacy networks are a few of the known issues

Conclusion

E-prescribing as well as electronic networks between pharmacies and clinicians' offices are becoming a reality. Although the networks and prescribing software are readily available, not all of the pieces of the puzzle connect. Evidence so far indicates that e-prescribing should save time and money and hopefully result in improved patient safety. To date it appears that e-prescribing clearly decreases *medication errors* but may have a small effect in reducing *adverse drug events*. In March 2008 the normally skeptical Congressional Budget Office determined that if Medicare physicians were required to use e-prescribing then there would be a savings of $0.2 billion for the time period 2008-2013.[48] One can assume that the large increase in e-prescribers is due to federal government reimbursement, but it is likely also due to the steady increase in adoption of technology in general. Will we see a continued increase in eRx as physicians make the decision to adopt EHRs for reimbursement?

More research is needed to evaluate barriers to adoption. A research obstacle is the delay between when e-prescribing is studied and the eventual publication. Studies reported in 2010 usually reflect studies performed in 2006-2007 and therefore don't take into account newer improved technologies and incentives.[49]

References

1. Poison in Prescription. Allscripts 3/19/2001 http://www.allscripts.com/ahsArticle.aspx?id=297&type=News%20Article (Accessed November 2 2005)
2. A Call to Action: Eliminate Handwritten Prescriptions within 3 years! White Paper. Institute for Safe Medication Practices www.ismp.org (Accessed November 3 2005)
3. IOM Report Calls for E-prescribing www.ihealthbeat.org July 21 2006 (Accessed July 24 2006)
4. State Initiatives to Avoid Prescription Drug Errors December 2006 www.ncsl.org/programs/health/rxerrors.htm (Accessed February 22 2007)
5. RxHub www.rxhub.net (Accessed February 22 2007)
6. Rupp, MT. E-Prescribing: The Value Proposition. America's Pharmacist. April 2005. www.surescripts.com (Accessed November 5 2007)
7. The Medical Group Management Association www.mgma.com/about/default.aspx?id=280 (Accessed November 5 2007)
8. The Value of Computerized Provider Order Entry in Ambulatory Settings 2003 www.citl.org/research/ACPOE.htm (Accessed February 22 2007)
9. SureScripts www.surescripts.com (Accessed June 22 2010)
10. Medicare Modernization Act www.cms.hhs.gov/medicarereform/ (Accessed November 8 2005)
11. Sarasohn-Kahn. Medicare's broadening shoulders www.ihealthbeat.org July 29 2005 (Accessed July 30 2005)
12. Healthcare IT News www.healthcareitnews.com Nov 8 2005 (Accessed November 8 2005)
13. Centers for Medicare and Medicaid http://www.cms.gov/ERXincentive/01_overview.asp (April 22 2010)
14. Commission for Certification of Health Information Technology http://www.cchit.org/ (Accessed April 22 2010)
15. E-prescriptions for controlled substances, final rule. Federal register. March 31 2010. http://edocket.access.gpo.gov/2010/pdf/2010-6687.pdf (Accessed April 22 2010)

16. The Center for Improving Medication Management. www.thecimm.org (Accessed November 23 2007)
17. E-Script Pilot Study www.healthdatamanagement Feb 7 2005 (Accessed February 10 2005)
18. Maryland Consortium to advance e-prescribing effort www.ihealthbeat.org 2/09/2006 (Accessed February 10 2006)
19. Wellpoint e-prescribing www.healthcareitnews.com April 25 2005 (Accessed April 28 2005)
20. Automakers' E-prescribing program reduces errors, costs www.ihealthbeat.org 2/23/2006 (Accessed February 25 2006)
21. Kosmetatos, S. Online rx program helping cut errors. The Detroit News. Oct 29 2007. (Accessed October 29 2007)
22. Ackerman K. Nevada Physicians Use E-Prescribing to Reduce Medical Errors. www.ihealthbeat.org 10/14/05 (November 20 2005)
23. Getting Physicians Connected http://www.tuftshealthplan.com/pdf/epresribing_results_summary.pdf (Accessed July 10 2006)
24. Kowalczyk L. Take one and stop scribbling on pads of paper www.Boston.com Jan 10 2005 (Accessed January 12 2005)
25. Generic Pharmaceutical Association www.gphaonline.com (Accessed April 3 2006)
26. Florida Medicaid Expands PDA, E-prescribing Software Program www.ihealthbeat.org 8/12/04. (Accessed August 13 2004)
27. More Florida Docs Could Get Handheld Prescribing Devices www.ihealthbeat.org 5/06/05 (Accessed May 7 2005)
28. Mississippi Medicaid Considers Expanding E-Prescribing Program. September 14 2007. www.ihealthbeat.org (Accessed September 14 2007)
29. Availity and Prematics Launch First Comprehensive E-prescribing Service for Florida Physicians. www.psqh.com/enews/0908m.html (Accessed September 24 2008)
30. Blue-Cross/Blue-Shield http://www.horizon-bcbsnj.com/newsroom/news_releases.asp?article_id=614&urlsection (Accessed November 8 2005)
31. NEPSI www.nationalerx.com (Accessed April 22 2010)
32. eRxNow. Allscripts. www.allscripts.com/products/physicians-practice/eprescribing-med-servicesk/erxnow/default.asp (Accessed January 18 2007)
33. Get Connected. www.getrxconnected.com (Accessed March 5 2008)
34. Ehealthinitiative. www.ehealthinitiative.org (Accessed July 1 2009)
35. EPrescribe Florida www.eprescribeflorida.com (Accessed July 10 2009)
36. Ammenwerth E et al. The Effect of Electronic Prescribing on Medication Errors and Adverse Drug Events: A Systematic Review. JAMIA PrePrint June 25 2008 doi:10.1197/jamia.M2667 (Accessed July 2 2008)
37. Bizovi KE et al. The Effect of Computer-Assisted Prescription Writing on Emergency Department Prescription Errors. Acad Emerg Med 2002;9:1168-1175
38. Grossman JM, Gerland A, Reed MC and Fahlman C. Physicians Experiences Using Commercial E-Prescribing Systems. Health Affairs 2007;26 (3):w393-w404
39. Fischer MA, Vogeli C, Stedman M et al. Effect of Electronic Prescribing With Formulary Decision Support on Medication Use and Cost. Arch Intern Med 2008;168(22):2433-2439
40. Issac T, Weissman JS, Davis RB et al. Overrides of Medication Alerts in Ambulatory Care. Arch Intern Med 2009;169(3):305-311
41. Fortuna RJ, Zhang F, Ross-Degnan DR et al. Reducing the Prescribing of Heavily Marketed Medications: A Randomized Controlled Trial. J Gen Intern Med 24(8):897-903

42. Rupp MT, Warholak TL. Evaluation of e-prescribing in chain community pharmacy: Best-practice recommendations. J Am Pharm Assoc 2008;48:364-370

43. Brooks P, Sonnenschein C. E-Prescribing. JHIM 2010;24(2):53-59

44. Hollingsworth W et al. The Impact of e-Prescribing on Prescriber and Staff Time in Ambulatory Care Clinics: A Time-Motion Study. JAMIA 2007;14:722-730

45. Pharmacists Newsletter www.pharmacistsletter.com December 8 2009 (Accessed March 1 2009)

46. MediVoice LLC Selects Nuance Voice Control to Power New E-Prescription Application. www.nuance.com (Accessed March 2 2008)

47. Miller R et al. Clinical Decision support and electronic prescribing systems: at time for responsible thought and action. JAMIA 2005;12:403-9

48. Miller M. Financial Returns from E-prescribing, saving Medicare $2.1 billion. Health Policy and Communications Blog. July 23 2008 http://tiny.cc/TLly1 (Accessed July 15 2009)

49. Wang JC, Mihir HP, Schueth AJ et al. Perceptions of Standards-based Electronic Prescribing Systems as Implemented in Outpatient Primary Care: A Physician Survey. J Am Med Inform Assoc 2009;16:493-5

18

Telehealth and Telemedicine

ROBERT E. HOYT

Learning Objectives

After reading this chapter the reader should be able to:

- State the difference between telehealth and telemedicine
- List the various types of telemedicine such as teleradiology and teleneurology
- List the potential benefits of telemedicine to patients and clinicians
- Identify the different means of transferring information with telemedicine such as store and forward
- Enumerate the most significant ongoing telemedicine projects

According to the Office for the Advancement of Telehealth (OAT), Telehealth is defined as:

> *"the use of electronic information and telecommunications technologies to support long-distance clinical health care, patient and professional health-related education, public health and health administration"* [1]

Similar to the term e-health, telehealth is an extremely broad term. A review by Oh et al. found 51 definitions for e-health, suggesting that the term is too general to be useful and the same is probably true regarding telehealth. [2] One could argue that Health Information Organizations (HIOs), Picture Archiving and Communication Systems (PACS) and e-prescribing are also examples of telehealth if they exchange healthcare information between distant sites. Clearly, telehealth is the broader term that incorporates clinical and administrative transfer of information, whereas telemedicine relates to remote transfer of only clinical information. In this chapter we will use the term telemedicine instead of telehealth and define it as follows:

> *"the use of medical information exchanged from one site to another via electronic communications to improve patients' health status"* [3]

Telemedicine was postulated in the 1920s when an author from Radio News magazine demonstrated how a doctor might examine a patient remotely using radio and television. Ironically this was proposed before television was even available. [4] (Figure 18.1) The first instance of remote monitoring has been attributed to monitoring the health of astronauts in space in the 1960's. [5] Very rudimentary telemedicine has been conducted using telephone communication for the past fifty years or more. With the advent of the Internet and video conferencing many new modes of communication are now available. Traditionally we have seen telemedicine arise to primarily communicate with remote rural or disadvantaged populations. This is largely due to shortages of specialists in rural areas. The goal of telemedicine ultimately is to provide timely and high quality medical care remotely. In addition, telemedicine can result in high patient satisfaction due to better access to specialty care and less time lost from work.

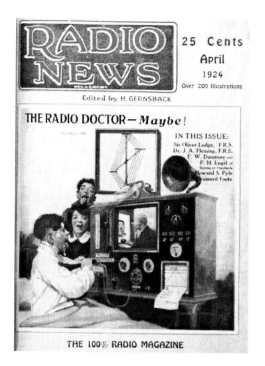

Figure 18.1 Early Telemedicine (Courtesy Radio News)

Telemedicine Communication Modes

In this chapter we will mention multiple ways patients can receive remote care, starting from simple e-mail to complex video teleconferencing. In the past several years we have seen new telemedicine technologies and business models appear with more on the way. Table 18.1 shows several of the communication modes used in telemedicine, along with pros and cons.

Table 18.1 Telemedicine Communication Modes

Communication Mode	Pros	Cons
Patient-Portal secure-messaging	Asynchronous. Able to attach photos. Response can be formatted with template. Could use VoIP. Audit trail is available	Not as personal as live visit. Usually not connected to EHR or other information
Telephone	Widely available, simple and inexpensive. Real-time	Not asynchronous. Unstructured. No audit trail. Only real-time
Audio-Video	Maximal input to clinician. Can include review of x-rays, etc. Perhaps more personal than just messaging	Currently, most expensive in terms of networks and hardware but that is changing

Telemedicine Transmission Modes

There are three major types of telemedicine transmission:

- **Store-and-forward.** Images or videos are saved and sent later. As an example, a primary care physician takes a picture of a rash with a digital camera and forwards it to a dermatologist to view when time permits. This method is commonly used for specialties like dermatology and radiology. This could also be referred to as asynchronous communication

- **Real time**. A specialist at a medical center views video images transmitted from a remote site and discusses the case with a physician. This requires more sophisticated equipment to send images real time and often involves two way interactive televisions. Telemedicine also enables the sharing of images from peripheral devices such as stethoscopes, otoscopes, etc. This would be an example of synchronous communication

- **Remote monitoring**. A technique to monitor patients at home, in a nursing home or in a hospital for personal health information or disease management

Telemedicine Categories

We are going to arbitrarily divide Telemedicine into the categories noted below. It should be pointed out that virtual patient visits (e-visits) and patient portals could have been included in this chapter but we elected to keep them in the chapter on Consumer Health Informatics.

Traditional Telemedicine: Teleradiology, Teledermatology, Telecardiology, Telepathology, Telesurgery, Telepsychiatry, Teleneurology, Teleophthalmology, Telepharmacy, etc
Telerounding of inpatients
Telehomecare and Telemanagement

Traditional Telemedicine

Currently, in the United States there are over 200 telemedicine programs that are operational in 48 states.[6] Most programs consist of a central medical hub and several rural spokes. Programs attempt to improve access to services in rural and underserved areas, to include prisons. This reduces travel time and lowers the cost for specialists and patients alike. Programs have the potential to raise the quality of care delivered and help educate remote rural patients and physicians. The most commonly delivered services are mental health, dermatology, cardiology and orthopedics.

Non-traditional Telemedicine can also be found in the international boating world where sailors can access a remote medical resource site. After registration they can call, fax or e-mail the site for advice on medical treatment while at sea.[7] Similarly, Virgin Atlantic Airlines will equip all of its aircraft with telemedicine devices for emergencies. Satellite technology will transmit the patient's vital signs to MedAire Centre in Arizona for interpretation by medical experts.[8]

- **Teleradiology**. The military has taken the lead in this area partly due to the high attrition rate of radiologists and to enhance radiology support for military deployments. By 2007 most Army x-rays became digital, which helped the storage, transmission and interpretation of images. With this newer technology a computerized tomography (CT) scan performed in Afghanistan can be read at the Army medical center in Landstuhl, Germany. Another example of military teleradiology can be found on the Navy hospital ships Mercy and Comfort where digital images can be transmitted to shore based medical centers.

  o In the civilian sector, NightHawk Radiology Services help smaller hospitals by supplying Radiology services located in the United States, Switzerland and Australia. All are board certified; most trained in the United States and carry multiple state licenses. Currently, they cover 1350 hospitals in the USA. They list a staff of about 125 radiologists on their web site. The turnaround time for an image to be read is less than 30 minutes, using a large VPN service, with an average cost of $55. They offer conventional radiology as well as CT, MRI, Ultrasound and Nuclear Medicine interpretation.[9]

o Another more common but important example of teleradiology is the practice of radiologists reading films after-hours at home. They must have high resolution monitors and high speed connections to the Internet but with this set up and voice recognition software; they can be highly productive at home. This is becoming the standard practice for radiologists. Instead of driving in or staying at the hospital at night to interpret images, they can deliver interpretations while at home

- **Telesurgery**. The initial approach was to "telementor" surgeons performing operations in remote sites. In 2001 surgeons in New York were able to successfully perform laparoscopic cholecystectomies (gallbladder removal) on six pigs located in Strasbourg, France. This was followed one year later by the uneventful remote removal of the gallbladder in a 68 year old woman; the first case of telesurgery in a human [10]

- **Teleneurology**. Many regions lack neurologists to see patients with stroke-like symptoms to determine if they need clot-busting drugs (thrombolytics) or need to be transferred to a higher level of care. This is, in part, due to the increased malpractice risk and decreased reimbursement situation of treating emergency patients. With the advent of telemedicine, the case can be discussed real time and the patient and their x-rays can be viewed remotely by a stroke specialist. One company, REACH Call Inc. developed a web based solution that includes a complete audio-visual package so neurologists can view the patient and their head CT (CAT scan). REACH Call Inc. was developed by neurologists at the Medical College of Georgia. Because the program is web based, the physician can access the images from home or from the office. Likewise, the referring hospital only has to have an off-the-shelf web camera, a computer and broadband Internet connection.[11] Specialists-on-Call is a Massachusetts based organization that has 40 part time or full time neurologists on board to handle emergency consults via telemedicine for about 60 private community hospitals. Their capabilities include the ability to transfer head CT images and bidirectional audio and video conferencing with remote physicians and families. To accomplish this they have an infrastructure that consists of a PACS, a call center, an electronic health record and videoconferencing equipment. The cost for this service is not inexpensive, for a 200 bed hospital, it would cost $400 per day and $40,000 for initial installation fees. It is unknown if third party payers will eventually reimburse for this service.[12-13] A Teleneurology study is reported later in this chapter.

- **Telepharmacy**. Like teleradiology, this field arose because of the shortage of pharmacists to review prescriptions. Vendors now sell systems with video cameras to allow pharmacists to approve prescriptions from a remote location. This is very important at small medical facilities or after-hours when there is not a pharmacist on location.[14] The North Dakota Telepharmacy Project operates 36 remote sites where pharmacy technicians receive approval for a drug by distant pharmacists via teleconferencing. In this manner a full drug inventory is possible even in small rural communities and the pharmacists still perform utilization reviews and other services remotely [15]

- **Telepsychiatry**. Several studies have indicated that telepsychiatry is equivalent to face-to-face psychiatry for most patients.[16] The American Psychiatric Association promotes telepsychiatry, primarily for remote or underserved areas, using live video teleconferencing. During a telesession, there can be individual or group therapy, second opinions and medication reconciliation. In general, virtual visits help team medicine and patient satisfaction has been good. On the American Psychiatric Association web site, there is a listing of telepsychiatry programs and web resources covering reimbursement and other helpful topics.[17] Another telepsychiatry trend that is appearing is the use of free commercial-off-the-shelf (COTS) audiovisual programs such as Skype. Voyager Telepsychiatry uses this popular program to hold virtual telepsychiatry sessions.[18] One of the most important areas for telepsychiatry will involve military members who return from war with Posttraumatic Stress Disorder (PTSD) and Traumatic Brain Injury (TBI). About 40% of veterans live in rural areas, where transportation may be an issue. The VA has opened three Veterans Rural Health Resource Centers in Iowa, Utah and Vermont to help develop and evaluate telemedicine programs.[19] More on telepsychiatry will be presented later in this chapter

- **Teledermatology.** With the advent of good quality digital cameras and cell phones with medium quality cameras, the concept of teledermatology was born. The Teledermatology Project (http://telederm.org), created in 2002, has the goal of providing free worldwide dermatology expertise, particularly for third world countries and the underserved. Physicians can easily obtain a teleconsultation and diagnostic and therapeutic advice using the store and forward mode. A 2003 survey indicated that there were 62 teledermatology programs in the United States.[20-21] For more details on teledermatology, we recommend a very current e-Medicine article [22] and a December 2009 review by the California HealthCare Foundation [23]

- **E-Mail Teleconsultation**. Audio and video teleconferencing is not the only way to communicate between remote clinicians. The Army has established a Teleconsultation service for deployed military clinicians, based on e-mail communication. The service is available 24/7 for all branches of the military with most responses completed in less than six hours. Almost every specialty is available to the military physician while on ships, the battlefield or part of humanitarian or disaster relief operations. The most common specialty consult requested is dermatology and the most common location requesting from has been Iraq. The program is administered as part of the Office of the Surgeon General Teleconsultation Program [24]

Telerounding

This is a new concept developed to help address the shortage of physicians and nurses. Telerounding is being rolled out in facilities with reasonably good reviews in spite of obvious criticisms that it further compromises the already strained doctor-patient relationship.

- **Robot Rounds.** A study in 2005 in the Journal of the American Medical Association showed that surgeons could make a second set of rounds using a video camera at the patient's bedside (InTouch Robots). A physician assistant makes the actual rounds, backed up by the attending physician remotely via the robot. Robot units are 5 ½ feet tall, weigh 220 lbs and have a computer monitor as a head. The cost is more than $100,000 each or they can be leased for $5000 monthly plus $5000 per viewing station. At this time they are being used in 20 plus hospital systems in the United States. They can move around and can project x-ray results to the patient. Ellison et al reported on Urological patients who either received face-to-face rounds post-operatively or robotic telerounding. The concluded that robotic rounding was safe and well received by patients. Two-thirds of patients stated they would rather see their own physician remotely than a stranger making rounds in person [25-27]

- ***E-ICU Rounding.*** In the United States it is predicted that we need approximately 35,000 intensivists (physicians who specialize in ICU care), but we only have 6,000. Moreover, in spite of the fact that hospital beds are not increasing, ICU beds are. Therefore, remote monitoring makes sense particularly during nighttime hours when physicians might not be present. The Leapfrog Group has advocated care delivered by intensivists for all ICUs as one of its four patient safety recommendations; but this goal remains elusive.[28] Hospitals that use e-ICUs believe there are patient safety and financial benefits but both need to be proven. An e-ICU service may be less expensive than recruiting full time intensivists. Also, because ICU care can cost $2500 daily, any cost saving modality that positively affects length of stay or mortality will gain market attention. Avoiding law suits in the ICU also means a cost saving. In a study in 2004, monitors were placed in two large ICUs serving 2000+ patients over 2 years. ICU and hospital mortality and length of stay were compared before and after intervention. Patients were monitored remotely from 12 pm-7 am. The results showed that mortality was 9.4% compared to 12.9% for conventional care. The length of stay in the ICU was 3.63 days compared to 4.35 days for conventional care. It is estimated that over 100 hospitals now have e-ICU programs, even though there is no reimbursement by insurers.[29]

The leading vendor in this area is VISICU with about 150 customers. It was founded by two intensivists from Johns Hopkins in 1998. In December 2007 VISICU was purchased by Phillips Electronics Healthcare division.

Sentara Healthcare System in Norfolk, Virginia reported a 27% decrease in ICU mortality, a 17% decrease in length of stay and a savings of $2,150 per patient using e-ICUs.[30] VISICU extended support of care outside the ICU in December 2007. Their plan is to use the eCareMobile™ unit (figure 18.2) to monitor sick patients on medical surgical floors, emergency departments, step-down units and post anesthesia units. At Parkview Health in Fort Wayne Indiana, e-ICU physicians will assist rapid response teams that attend to deteriorating patients. Thus far, there has been a 32% reduction in floor-based cardiac arrests and improved nursing satisfaction.[31]

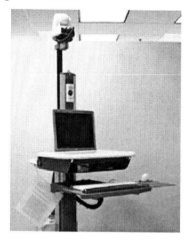

Figure 18.2 eCareMobile™ unit (courtesy VISICU)

The cost for e-ICUs is considerable in spite of the potential benefits. The University of Massachusetts Memorial Health Care network spent $8 million to create a virtual ICU network to connect eight intensive care units. Specialists can now remotely view electronic health records, nursing notes, test results and video images of patients as well as access the latest clinical practice guidelines. They will soon add two-way video feeds so patients' families can communicate with specialists.[32] Sutter Health paid more than $25 million to establish its VISICU e-ICU system. Based on their analysis they have saved about $2.6 million in treatment costs by preventing deaths due to sepsis. In addition, they estimate that if sepsis is treated early, the ICU stay is shortened by four days.[33] Banner Health plans to expand the e-ICU concept to all of its 390 ICU beds and the e-ICU control center staff consists of 12 full time intensivists and one nurse for every 35 patients. Benefits of the program include improved nursing retention and a reduction in insurance costs of $1 million in the past year.[34] It is unfortunate that many understaffed rural hospitals will not be able to afford these services unless they are part of a larger network or there is reimbursement by insurers.

More information can be obtained from a 2007 consensus statement from the University Health System Consortium Intensive Care Unit Telemedicine Task Force.[35] eICU studies are reported later in this chapter. The bottom line is that further research is needed to provide the kind of detail necessary to determine the benefit of this type of telemedicine. For example, is the benefit greater for a small hospital with limited ICU expertise compared to a large integrated ICU system with an abundance of intensivists?

Telehomecare and Telemanagement

Telehomecare means that vital signs, weights, blood sugars, etc can be sent via a wired or wireless mode from homes to physicians' offices, health information exchanges, etc. At least 55 companies offer technology to monitor patients at home and the list continues to grow and include large companies such as Intel and General Electric. There are multiple reasons telemonitoring is burgeoning:

- Chronic illnesses are on the rise and will likely increase hospitalizations, readmissions and unnecessary emergency room visits. Measures like home monitoring might decrease this trend. The goal is to intervene immediately rather than wait till the next appointment
- Medicare changed reimbursement to home health agencies from the number of visits to a diagnoses based system, leading to decreased reimbursement for visiting nurses
- Telemonitoring programs allow for audio and visual contact with patients at home and therefore can reduce home visits by a nurse or physician. Nurses can make visits only if there is a problem, such as a change in symptoms or vital signs
- One consulting organization predicts a nursing shortage of 800,000 and a physician shortage of 85,000 to 200,000 by the year 2020 [36]
- When "baby boomers" turn 65 they will be tech savvy and more likely to demand services like telemonitoring
- Monitoring may be possible using the ubiquitous cell phone and new microsensors
- Linking home monitoring devices to EHRs with decision support and health information exchanges will increase the functionality of this new technology. The potential to save costs is attractive but elusive and will require high quality confirmatory studies
- CMS will administer a Medicare Medical Home Demonstration project to test the "medical home" and "hospital at home" concepts. Medical groups will be paid for coordination of care, health information technology, secure e-mail and telephone consultation and remote monitoring. Details are preliminary and available on the CMS site.[37] For additional information on the patient-centered medical home model see the chapter on disease management

Examples of home telemonitoring systems

Health Buddy is a FDA approved device that is certified by the National Committee for Quality Assurance and used by the Veterans home telemedicine programs. Health Buddy is used by over 12,000 patients and has been shown in one study (of limited design) to increase medication compliance and reduce outpatient visits. [38-39] The Centers for Medicare and Medicaid Services will test the system with about 2,000 patients with chronic diseases in

Oregon and Washington State with the goal of reducing medical costs by 5%.[40] Features include:

- Data is sent via phone lines
- Device comes with desktop decision support software
- Program covers 45 disease protocols
- Device connects to a glucometer, BP machine, weight scales and peak flow meter for asthmatics
- Program is interactive with patients; it asks questions daily

HoneyWell HomMed has over 15,000 monitors currently in use and more than 300,000 patients have been monitored. Features include: [41]

- Voice messages to patients in multiple languages
- Digital weight scale, blood pressure, oximetry, glucometer, peak flow meter, blood tests (PT/INR), temperature and EKG
- Data is transmitted via phone lines
- LifeStream Connect™ is a new option that interfaces with EHRs

More Telemonitoring Systems

- *MyCareTeam*: a fee for service diabetic portal that stores data on Google Health
  - www.mycareteam.com
- *MediCompass*: patient portal for diabetes, asthma and cardiovascular disease. Provides a dashboard view, a registry, an executive view and messaging www.medicompass.com
- *Healthanywhere*: allows for uploading of data by BlackBerry, home monitor and two way videoconferencing http://www.igeacare.com/HealthAnywhere/index.htm
- *Intel Health Guide*: Intel along with General Electric has entered the telehomecare market with a comprehensive program. At this time they are conducting pilot projects with Aetna, Erickson retirement communities and Scan Health Plan. The proposed system will include interactive software, a multimedia library, two-way video conferencing, home device uploads and patient alerts/reminders http://www.intel.com/healthcare/ps/healthguide/index.htm

Telemanagement

More and more companies are developing home monitors and sensors that will transmit information to a physician's office or other healthcare organization. They perceive an increased need based on our graying population, more chronic diseases and expensive home care. Programs will be interactive and include patient education for issues such as drug compliance. This data may interface with an electronic health record, health information organization (HIO) or web site for others to evaluate. Some predict that houses will be wired with multiple small sensors known as "motes" that will monitor daily activities such as taking medications and leaving the house. The information would be transmitted to a central organization that would notify the patient and/or family if there was non-compliance or a worrisome trend. At this point the patient or families will have to pay for the systems. Some have already complained about the perceived lack of privacy and the potential for too many alerts or alarms.[42-44] There is some evidence that telemanagement can result in improved outcomes and cost savings, particularly in patients with diabetes, heart failure and chronic obstructive lung disease. In a monograph by the Medical College of Georgia, prepared for the Advanced Medical Technology Association, they reviewed the medical evidence for telemedicine and telemanagement. Studies suggest that these technologies can reduce hospitalizations, emergency room visits and pharmacy utilization, thus paying for the technology. For telehomecare to be successful there must be adequate broadband access, reimbursement for physicians and nurses and financial resources to acquire and maintain the necessary medical technologies.[45]

Telemedicine Projects

The following provides a sampling of some interesting telemedicine projects:

- **Informatics for Diabetes Education and Telemedicine (IDEATel)**: The largest government sponsored telemedicine program in the US. The project evaluated approximately 1650 computer illiterate patients living in urban and rural New York State. Patients received a home telemedicine unit that consisted of a computer with video conferencing capability, access to a web portal for secure messaging and education and the ability to upload glucose and blood pressure data. These same subjects were assigned a case manager who was under the supervision of a diabetic specialist. They used the Veterans Affairs clinical practice guidelines on diabetes. They were compared to a control group that didn't receive the home monitoring system. The results of this project are reported in the next section [46]

- **Georgia Telemedicine Network**: The first state-wide effort to link 36 rural hospitals and clinics with specialists at eleven large urban hospitals. Project created partnerships among Wellpoint (Blue Cross/Blue Shield) and the state government. Importantly, telemedicine consults were reimbursed as office visits due to a new Georgia law and 20 specialties were felt to be appropriate for telemedicine [47-48]

- **University of Texas Medical Branch at Galveston:** Program is the largest telemedicine system in the world with 300 locations and 60,000 annual telemedicine sessions. Sixty per cent of visits deal with a prison population. They also offer specialty services in neurology, addiction medicine and psychiatry [49]

- **VA Rocky Mountain Healthcare Network**: In Colorado veterans with heart failure, diabetes and emphysema were enrolled in a telemedicine program. The VA reported a 53% reduction in hospital stays resulting in a $508,000 savings overall. Outpatient visits dropped 52% and overall estimated savings of the program was $1.2 million dollars [50]

- **Intel Telemedicine**: Project will provide the real time video technology to service 105 rural clinics and hospitals for the municipal government of Zhanjiang, China [51]

- **Teleburn Project**: University of Utah Burn Center used telemedicine to treat burn patients in three states. Specialists can view videos or digital photos of burn patients for initial determination or follow up. The demonstration project was funded by the Department of Commerce [52]

- **TeleKidcare**: Urban project operated by the Kansas University Medical Center to deliver care at school so that children do not have to leave school to receive medical care. Reasons for televisits include: 47% ear, nose and throat problems, 31% for behavioral problems and 10% for eye related problems. In spite of private and government support they do not have a long term business plan for sustainability [53]

- **TelePediatrics and TeleDentistry**: The University of Rochester created the Health-e-Access program in 2001. The program was initially set up to connect pediatricians to inner city child care centers and elementary schools using telemedicine and two-way video conferencing via the Internet. The program has allowed the children and their parents to not leave the centers or their jobs. The program was started by grants but insurers have been willing to cover this initiative, presumably because it cuts down on emergency room visits. The director of the project has stated that he believes about 28% of pediatric visits to the emergency room in upstate New York could have been treated with telemedicine [54]

- **Veteran's Teleretinal Program**: Since 2000 the Veterans Health Affairs has operated a VHA Teleretinal Imaging project at 104 sites. Because 20% of veterans have diabetes they felt they had to conduct retinal screening for diabetic damage even though they might not have retinal experts (ophthalmologists) at each clinic. They took high resolution retinal images and "stored and forwarded" them to eye specialists who would later made a determination [55]

- **California Central Valley Teleretinal Program**: Using a non-proprietary, open source web based program (EyePACS) images can be forwarded to an ophthalmologist for interpretation. Images are stored on a SQL Server and images are viewed with a web browser. A simple software program on the PC allows for uploading images to the server. There is e-mail notification to the consultant and back to the individual who sent the images. California will expand the program from the Central Valley to the entire state to serve 100 clinics and approximately 100,000 patients [56-57]

- **Colorado Telehealth Network**: is funded by the FCC to create the largest statewide fiber optic broadband network in the US. As of 2008, 72 hospitals, 118 clinics and 184 mental health clinics plan to participate [58]

- **Joint Telemedicine Network**: The US Army has used telemedicine for many years for remote consultation and more recently for transmission of images. Up to about 2009 there were substantial bandwidth limits such that patients might arrive in Landstuhl Germany before their images due to long transmission times. This has improved substantially in the 2009-2010 time frame. As a result of this recent improvement and the problem of mental health issues in deployed troops, the Army is evaluating the use of videoconferencing to extend the reach of mental health [59]

- **Connected Care**: In 2009 UnitedHealth Group and Cisco announced that they would partner to build the first national telehealth network based on open standards so they could connect with electronic health records and other IT platforms. There are currently six Connected Care programs being piloted across the United States [60]

- **Northwest Telehealth**: This initiative has 65 sites using about 100 telemedicine devices. They offer the following services: clinical care (15 specialties), teleER, telepharmacy, distance education, administrative and operational planning/coordination [61]

- **Federal Communications Commission (FCC)**: In 2006 they announced a $400 million budget for pilot projects to promote broadband networks in rural areas. The goal is to create networks for public healthcare organizations and non-profit clinicians that will eventually connect to a national backbone. The network could be used for telemedicine or other medical functions in rural areas. In 2007 the FCC created a $417 million fund that would support pilot projects to connect more than 6,000 hospitals, research centers, universities and clinics. The FCC paid up to 85% of the cost to design, engineer and construct the networks. Internet2 or the LambdaRail Network will be used. Many of the projects will involve multi-state areas and most will enhance telemedicine. Much of the funding will come from the Universal Service Fund that derives from a fee added to consumers and telecommunication companies.[62] The New England Telehealth Consortium announced in January 2008 that it will use the $24.7 million in FCC grant money to link 555 clinics, physician offices, hospitals, public health offices and universities in Maine, Vermont and New Hampshire. The network will act like a second Internet to allow the transmission of records and x-rays and the creation of videoconferences.[63]

In early 2009 Congress directed the FCC to develop a National Broadband Plan with goal of providing broadband access to every American and funded it as part of the American Recover and Reinvestment Act (ARRA). In 2010 the FCC posted the Plan on a new web site. [64]

Telemedicine Studies

- A 2001 systematic review of telemedicine by Roine et al looked at reported patient outcomes, administrative changes or economic assessments. Of 1124 potential articles, 50 were felt to fit criteria for review. Most of the studies reviewed pilot projects and were of low quality. They felt that teleradiology, teleneurosurgery (looking at head CT scans before transfer), telepsychiatry, the transmission of echocardiograms, the use of electronic referrals to enable e-mail consultations and video teleconferencing between primary and secondary clinicians had merit. They also felt that it was impossible to state the economic value of telemedicine based on current evidence [65]

- The US Department of Veterans Affairs operates perhaps the largest telehomecare networks in the world (37,500 patients as of 2009). This is partly due to the fact that the VA has transitioned from inpatient to outpatient and home care. Also, with so many active duty members returning injured from the war zone they will eventually need telehomecare. Their Care Coordination/Home Telehealth program is also a disease management program. The VA currently runs three programs: telehomecare, teleretinal and a video teleconferencing services that link 110 hospitals and 380 clinics. Data from home devices inputs into the VA's EHR.[66] A study of 17,000 VA home telehealth patients was reported in late 2008. Although the cost per patient averaged $1600, it was considerably less expensive than in-home care. They utilized individual care coordinators who each managed a panel of 100-150 general medical patients or 90 patients with mental health related issues. They promoted self management, aided by secure messaging systems and a major goal was early detection of a problem to prevent an unnecessary visit to the clinic or emergency room. 48% were monitored for diabetes, 40% for hypertension, 25% for heart failure, 12% for emphysema and 1% for PTSD. Patient satisfaction was very high. This study showed a 19% reduction in hospitalizations and a 25% reduction in the average number of days hospitalized [67]

- The Electronic Communications and Home Blood Pressure Monitoring study compared home blood pressure (BP) monitoring along with a BP Web portal, with and without the assistance of a pharmacist. The web portal was integrated with an enterprise EHR. In the group that received assistance from the online pharmacist, they showed significantly more patients achieving control than those who were monitored and had web portal access but no interaction with a pharmacist. Results might not pertain to other diseases and requires patients to have Internet access and pharmacists to be able to have EHR access [68]

- Web-based care for diabetes was evaluated by the same group (Group Health) who evaluated hypertension control in the above paragraph. They compared a group of Type 2 diabetes who received "usual care" with another group who had access to a web portal linked to an EHR. The web-based program included secure e-mail messaging with clinicians, feedback on blood sugar results, educational web resources and an interactive online diary to record diet, etc. After one year the control of diabetes, based on a glycated hemoglobin was marginally better (decrease of .7%) but there was no difference in blood pressure or cholesterol control between the two groups. There was no correlation between improvement and the number of times the web portal was accessed. They only used one care manager so it is unknown if their results would have been different with multiple care managers [69]

- Group Health conducted another study of 1,500+ diabetics aged >18 years old to determine if those who used secure messaging with their clinician had better blood sugar, blood pressure and cholesterol control. Only 19% of patients chose to message their physician. Those that did had better blood sugar control, but not better control of blood pressure or cholesterol but had a <u>higher</u> rate of outpatient visits. Patients were not randomized for this study and the study was not prospective, so results are more difficult to interpret [70]

- The one year results of the IDEA TEL were published in late 2007 and showed mild improvement in blood sugars, cholesterol and blood pressure compared to the control project. Patient and physician satisfaction were positive but detailed cost data was lacking. Ironically, Medicare claims were higher in the study patients than in the control group, for unclear reasons. [71] The five year results were published in 2009 and although they showed some statistically significant improvement in blood sugar, cholesterol and blood pressure control, they were of doubtful clinical significance. Importantly, users of this technology had a dropout rate greater than 50%. [72] In 2010 a final report from this group concluded that *"telemedicine case management was not associated with a reduction in Medicare claims".* [73]

- Teleneurology or telestroke care was evaluated by a study by Meyer in 2008. They compared the outcomes of patients with a possible impending stroke and consultation by telephone, versus full video teleconferencing. Correct treatment decisions were made more frequently (98% versus 82%) for the teleconferencing sessions, but patient outcomes were the same. There was no difference in death rates or hemorrhaging after the clot busting drugs (thrombolytics) were administered. [74] An excellent review article on stroke telemedicine was published by Demaerschalle et al. in the January 2009 issue of the *Mayo Clinic Proceedings.* [75] The jury is out whether stroke telemedicine is cost effective or a reasonable choice, compared to telephonic consultation [76]

- An study was reported at the 2009 Society of Critical Care Medicine by Avera Health who reported the following results from a 30 month implementation of VISICU:
  - Rural hospitals estimated a 37.5 percent reduction in the number of patients requiring transfer, representing a cost savings of more than $1.2 million
  - Reduced length of stay in Intensive Care Units saved an estimated $8 million
  - ICU and hospital mortality rates were 65-80 percent lower than predicted outcomes after implementation of the e-ICU, compared to 50 percent lower than predictions before implementation
  - 90 percent of rural hospital clinical leaders surveyed reported being more comfortable caring for critically ill patients with e-ICU
  - 90 percent of rural hospital leaders surveyed agreed that patients and families are comfortable staying in the hospital with the added e-ICU care
  - 100 percent of rural physicians surveyed agreed that better, safer care can be supported by a remote critical care team [31]

In spite of the many potential virtues of the e-ICU, a 2009 article by Berenson et al. expressed the opinion that the actual value of e-ICUs was far from proven and there was a major interoperability issue between the e-ICU software and critical ICU systems like IV fluids and mechanical ventilation.[77]Another article in 2009 by Thomas et al. evaluated the medicare care in six ICUs before and after the implementation of an e-ICU system. They concluded that there was not an overall improvement in mortality or length of stay [78]

- A study reported in JAMA in 2009 looked at whether telephone delivered care for post cardiac bypass depression by nurses would be equivalent to usual care. In this randomized controlled trial telephonic collaborative care was superior in terms of mental health-related quality of life, physical functioning and mood symptoms at 8 month follow up [79]

- A study out of the UK evaluated almost 300 patients with depression by randomly assigning them to either usual care or online "cognitive behavioral therapy" (CBT) plus usual care. At 4 months follow up 38% of those who received CBT over the Internet recovered from depression compared to 24% in the usual care group. The gains were also confirmed at the 8 month point [80]

- An international meta-analysis of 10 randomized controlled trials looked at remote patient monitoring (RPM) of heart failure patients. They concluded RPM reduced the risk for all cause mortality and hospitalization for heart failure. The number needed to treat (discussed in evidence based medicine chapter) was 50 for all cause mortality and 14 for heart failure hospitalization [81]

In summary, the studies on telemedicine are mixed and are of low quality. It does seem like the addition of a skilled healthcare worker such as a nurse or pharmacist is necessary to experience benefits from a telemedicine program. In this early stage of telemedicine, the technology by itself does not seem to produce significant benefit.

Telehealth Organizations and Research

- **Office for the Advancement of Telehealth (OAT):** falls under Health Resources and Services Administration (HRSA) that is an agency of the Department of Health and Human Services. Its goal is to promote telemedicine in rural/underserved populations, provide grants, technical assistance and "best practices" [1]

- **American Telemedicine Association (ATA):** a non-profit international organization with paid membership that began in 1993. Individual state telemedicine policies are included on their web site. Goals of the ATA are as follows:
    - "Educating government about telemedicine as an essential component in the delivery of modern medical care
    - Serving as a clearinghouse for telemedicine information and services
    - Fostering networking and collaboration among interests in medicine and technology
    - Promoting research and education including the sponsorship of scientific educational meetings and the Telemedicine and e-Health Journal
    - Spearheading the development of appropriate clinical and industry policies and standards"[3]

- **USDA Rural Development Telecommunications Program**. The USDA has a program to finance the rural telecommunications infrastructure. In 2007 there were grants and loans totaling $128 million to achieve the goals of broadband access for distant learning and remote medical care. The USDA Rural Development agency has funded several e-ICU programs in the US, including the study by Avera Health noted above.

- **The Agency for Healthcare Research and Quality** has funded a number of telemedicine projects looking at virtual ICUs, telewound projects, cancer management, medication management, heart failure management and others [83]

Barriers to Telemedicine

- Limited reimbursement. Most telemedicine networks are created with federal grants. Medicare will reimburse if there is a formal consultation linked by live 2-way video teleconferencing and the patient resides in a professional shortage area. Medicaid at the federal level does not reimburse for telemedicine. Medicare will reimburse physicians, nurse practitioners, physician assistants, nurse midwives, clinical nurse specialists, clinical psychologists and clinical social workers. The originating sites can be offices, hospitals, skilled nursing homes, rural health clinics and community mental health centers. Medicare reimburse for telemedicine services for initial inpatient care, outpatient care, pharmacologic management, end stage renal disease-related visit and psychiatric diagnostic interviews. Clinicians at the remote site submit claims using the correct CPT or HCPCS codes as well as the telemedicine modifier GT. [84] In 2010 Medicare added three new outpatient and inpatient HCPCS codes for Telemedicine.[85] Many private insurers don't cover telemedicine, but a few provide the same coverage as face-to-face visits. In 2010 Virginia became the 12th state to require insurers to cover telemedicine for interactive audio or video visits [86]
- Limited research showing reasonable benefit and return on investment
- High cost or the limited availability of high speed telecommunications
- Bandwidth issues, particularly in rural areas where telemedicine is most needed. VPN connections slow the process further
- High resolution images or video require significant bandwidth, particularly if x-rays or images or pills have to be read by remote clinician. Telepsychiatry may require lower resolution
- State licensure laws when telemedicine crosses state borders. Some states require participating physicians to have the same state license
- Lack of standards
- Lack of evaluation by a certifying organization
- Fear of malpractice as a result of telemedicine. Who is going to evaluate telemonitoring data 24/7?
- Sustainability is a concern due to an inadequate long term business plan
- Lack of sophistication on the part of the patient, particularly in the elderly and under-educated [87]

Key Points

- Telehealth is a neologism that relates to long distance clinical care, education and administration

- Telemedicine refers to the remote practice of medicine using technology

- Almost all specialties now have telemedicine initiatives

- In spite of the lack of reimbursement, virtual ICUs have gained in popularity because they have perceived benefits

- Telehomecare is a new telehealth initiative that has appeared due to the graying of the US population and the increase in chronic diseases

- Lack of uniform reimbursement, lack of standards and lack of high quality outcome studies have impacted the adoption of telemedicine

Conclusion

Telemedicine is still in its infancy in most areas of the country. The barriers are largely financial due to the high cost to set up the system and the lack of reimbursement in many cases. With the price of telemedicine systems dropping, telemedicine for patients in rural areas may eventually be more cost-effective than referral to distant urban specialists. As is common with other informatics areas, there continues to be overly optimistic hype. In a late 2007 monograph The Center for Information Technology Leadership maintained that implementation of nationwide telemedicine will reach a breakeven point in 5 years with a total annual net benefit of $4.28 billion. Their predictions are based on models and not actual studies.[88] If the FCC and ARRA initiatives are successful and/or HIOs flourish, we may then have the infrastructure required for telemedicine throughout the United States. Transmission of large images and the ability to compare old and new imaging studies will be greatly aided by Internet2 or the LambdaRail. If future studies prove there is substantial return on investment then it is a matter of time before more payers support telemedicine. At this point, it seems premature for a healthcare system to provide patients with a home computer and Internet access without clear cut medical or financial benefit.

As with other areas of Medical Informatics, studies tend to show improvement in patient care only if it is combined with human supervision. Further down the road with more telemedicine experience less human intervention may be required.

References

1. Office for the Advancement of Telehealth http://www.hrsa.gov/telehealth/ (Accessed July 1 2009)
2. Oh H, Rizo C et al. What is eHealth?: a systematic review of published definitions. J Med Inter Res 2005; 7 (1) e1
3. American Telemedicine Association http://www.atmeda.org/about/aboutata.htm (Accessed July 1 2009)
4. Telemedicine: A guide to assessing telecommunications in health care. Marilyn Field ed. National Academies Press 1996. http://www.nap.edu/catalog/5296.html (Accessed September 25 2006)

5. Delivering Care Anytime, Anywhere: Telehealth Alters the Medical Ecosystem. California HealthCare Foundation Monograph. November 2008 (Accessed December 4 2008)

6. Puskin DS HHS Perspective on US Telehealth www.ieeeusa.org/volunteers/committees/mtpc/Saint2001puskin.ppt (Accessed December 6 2006)

7. Jacobs M. Telemedicine. Sail Magazine June 2006. pp. 60-61

8. Virgin Atlantic to Equip Airplanes with Telemedicine devices. www.ihealthbeat.org May 24 2006 (Accessed May 25 2006)

9. Nighthawk Radiology www.nighthawkrad.net (Accessed June 20 2009)

10. Marescaux J et al. Transatlantic robot-assisted telesurgery. Nature 2001;413:379-380

11. REACH Call www.reachcall.com (Accessed December 7 2007)

12. Teleneurology Helps Combat Specialist Shortage, Wait Times. July 17 2007 www.ihealthbeat.org. (Accessed July 18 2007)

13. Specialists oncall www.specialistsoncall.com (Accessed June 21 2009)

14. Envision Telepharmacy www.envision-rx.com (Accessed March 3 2007)

15. North Dakota Telepharmacy Project. http://telepharmacy.ndsu.nodak.edu (Accessed November 24 2007)

16. O'Reilly R, Bishop J, Maddox K et al. Is telepsychiatry equivalent to face to face psychiatry? Results from a randomized equivalence trial. Psychiatr Serv 2007;58(6):836-843

17. American Psychiatry Association http://www.psych.org/Departments/HSF/UnderservedClearinghouse/Linkeddocuments/telepsychiatry.asp

18. Telepsychiatry www.telepsychiatry.com (Accessed June 21 2009)

19. Joch A. Tele-therapy. Government Health IT November 2008 p.30-31

20. Telederm Project www.telederm.org (Accessed June 22 2009)

21. Soyer HPk Hofmann-Wellenhof R, Massone C et al. Telederm.org:Freely Available Online Consultations in Dermatology www.Plosmedicine.org 2005 2(4) (Accessed June 10 2009)

22. Teledermatology. E-Medicine http://emedicine.medscape.com/article/1130654-overview (Accessed June 27 2008)

23. Armstrong AW, Lin SW, Liu et al. Store-and-Forward Teledermatology Applications. December 2009. California HealthCare Foundation www.chcf.org (Accessed February 10 2010)

24. Army Telecommunications Program. www.cs.amedd.army.mil/teleconsultation.aspx. (Accessed May 24 2009)

25. Ellison LM, Nguyen M, Fabrizio MD, Soha A. Postoperative Robotic Telerounding Arch Surg 2007;142(12):1177-1181

26. Roberts R. Robots on Rounds. Kansas Business Journal Sept 5 2005 (Accessed September 10 2005)

27. Robotic Doctor Makes Rounds in Baltimore. www.Ihealthbeat.org Feb 27 2006 (Accessed March 10 2006)

28. Leapfrog. www.leapfroggroup.org (Accessed September 10 2005)

29. Breslow MJ. Effect of a multiple-site intensive care unit telemedicine program on clinical and economic outcomes: An alternative paradigm for intensivist staffing. Crit Care Med 2004;32:31-38

30. E-ICU Solution http://www.visicu.com/index_flash.asp (Accessed September 15 2005)

31. VISICU Introduces Critical Care Without Walls. Patient Safety and Quality Healthcare. December 7 2006 www.psqh.com/enews/1206r.shtml (Accessed December 8 2006)

32. Massachusetts Hospital System Taps eICU To Offset Staff Shortages. www.ihealthbeat.org November 19 2007. (Accessed November 24, 2007)
33. Sutter Health Taps eICU System to Combat Sepsis-Related Deaths. June 26 2007. www.ihealthbeat.org. (Accessed June 26 2007)
34. Monegain B. All eyes on critically ill. Healthcare IT News. January 2008.
35. Cummings J et al. Intensive Care Unit Telemedicine: Review and Consensus Recommendations. American Journal of Medical Quality. 2007;22:239-250.
36. Healthcare Staffing Growth Assessment. Staffing Industry Strategic Research. June 2005 http://media.monster.com/a/i/intelligence/pdf (Accessed September 25 2006)
37. Medicare Medical Home Demonstration http://www.cms.hhs.gov/DemoProjectsEvalRpts/downloads/MedHome_TaxRelief_HealthCareAct.pdf (Accessed June 20 2008)
38. Health Buddy® www.healthbuddy.com (Accessed June 13 2009)
39. Cherry JC, Moffatt TP, Rogriquez C and Dryden K. Diabetes Disease Management Program for an Indigent Population Empowered by Telemedicine Technology. Diab Tech Ther 2002;4:783-791
40. McGee, MK.Medicare Center Tests Telemedicine in Treating Chronic Illnesses. InformationWeek July 6 2005 (Accessed June 20 2009)
41. Honeywell HomMed www.HomMed.com (June 20 2009)
42. Ross P. Managing Care thru the Air. IEEE Spectrum December 2004 pp. 26-31
43. Advances in home monitoring technology Wall Street Journal Dec 12 2005 (Accessed December 15 2005)
44. Telemanagement Center for aging services technologies www.agingtech.org (Accessed March 10 2006)
45. Stachura ME, Khasanshina EV. Telehomecare and remote monitoring: An outcomes overview. www.advamed.org. (Accessed November 24 2007)
46. Informatics for Diabetes Education and Telemedicine http://www.ideatel.org/ (Accessed September 10 2007)
47. Rural Georgia Hospitals To Receive Telemedicine Funds www.ihealthbeat.org December 2 2004 (Accessed September 10 2006)
48. Georgia Taps Telemedicine for Rural Health care. www.ihealthbeat.org. March 26 2007. (Accessed March 27 2007)
49. Texas Telemedicine Used To Treat Rural Patients, Prisoners. November 30 2006. www.ihealthbeat.org. (Accessed December 1 2006)
50. Austin M. Telehealth a virtual success. Denver Post October 24 2005 (Accessed November 10 2006)
51. Intel Announces Chinese Telemedicine Program. November 1 2006 www.ihealthbeat.org (Accessed November 2 2006)
52. Telemedicine Network Facilitates Burn Care in Three States www.ihealthbeat.org February 17 2006 (Accessed February 18 2006)
53. Egan, C. School-Based Telemedicine Program Holds Promise for Student Health. September 14 2007 www.ihealthbeat.org. (Accessed September 14 2007)
54. University of Rochester Medical Center www.urmc.edu (Accessed May 12 2008)
55. VHA Teleretinal Imaging Program www.va.gov/occ/Teleret.asp (Accessed February 7 2007)
56. CHCF Expands Project to Prevent Diabetes-Related Blindness through Telemedicine. www.chcf.org (Accessed December 18 2007)
57. EyePACS. www.eyepacs.org. (Accessed December 19 2007)

58. Merill M. Colorado telehealth network set to be largest in the US. August 20 2008 www.healthcareitnews.com (Accessed September 1 2008)
59. JTMN Moves Information Faster. Federal Telemedicine News. December 9 2009. http://telemedicinenews.blogspot.com/2009/12/jtmn-moves-information-faster.html (Accessed April 24 2010)
60. Connected Care. Cisco. www.cisco.com (Accessed May 2 2010)
61. Northwest Telehealth www.nwtelehealth.org (Accessed June 26 2010)
62. Federal Communications Commission www.fcc.gov (Accessed December 20 2007)
63. Network Aims to Link Health Care Sites Across New England. January 08 2008. www.ihealthbeat.org. (Accessed January 8 2008)
64. National Broadband Plan. www.broadband.gov (Accessed March 31 2010)
65. Roine R, Ohinmaa, Hailey D. Assessing telemedicine: a systematic review of the literature. CMAJ 2001;165(6):765-71.
66. Buxbaum P. True Believer. Government Health IT. March/April 2009 p.13
67. Darkins A, Ryan P, Kobb R et al. Care Coordination/Home Telehealth: the Systematic Implementation of Health Informatics, Home Telehealth and Disease Management to Support the Care of Veteran Patients with Chronic Conditions. Telemedicine and e-Health. 2008;14(10):1118-1126
68. Green BB, Cook AJ, Ralston JD et al. Effectiveness of Home Blood Pressure Monitoring, Web Communication, and Pharmacist Care on Hypertension Control: A Randomized Controlled Trial. JAMA 2008;299(24):2857-2867
69. Ralston JD, Hirsch IB, Hoath J et al. Web-Based Collaborative Care for Type 2 Diabetes. Diabetes Care 2009;32(2):234-239
70. Harris LT, Haneuse SJ, Martin DP et al. Diabetes Quality of Care and Outpatient Utilization Associated with Electronic Patient-Provider Messaging: A Cross-Sectional Analysis. Diabetes Care 2009;32:1182-1187
71. Shea S. The Informatics for Diabetes and Education Telemedicine (IDEATEL) Project. Trans Amer Clin Clim Assoc 2007;118:289-300
72. Shea S, Weinstock RS, Teresi JA et al. A Randomized Trial Comparing Telemedicine Case Management with Usual Care in Older, Ethnically Diverse, Medically Underserved Patients with Diabetes Mellitus: 5 Year Results of the IDEATel Study. JAMIA 2009;16:446-456
73. Palmas W, Shea S, Starren J et al. Medicare payments, healthcare service use and telemedicine implementation costs in a randomized trial comparing telemedicine case management with usual care in medical underserved participants with diabetes. J Am Med Inform Assoc 2010;17:196-202
74. Meyer BC, Raman R, Hemmen T et al. Efficacy of site-independent telemedicine in the STRokEDOC trial: a randomized, blinded, prospective study. August 3 2008 www.thelancet.com/neurology e-publication (Accessed August 10 2008)
75. Demaerschalk BM, Miley ML, Kiernan TJ et al. Stroke Telemedicine. Mayo Clin Proc 2009;84(1):53-64
76. Berthoid J. Help from afar: telemedicine vs. telephone advice for stroke. ACP Internist April 2009 p. 19
77. Berenson RA, Grossman JM, November EA. Does Telemonitoring of Patients—The eICU—Improve Intensive Care? Health Affairs 2009;28(5):w937-947
78. Thomas EJ, Lucke JF, Wueste L et al. Association of Telemedicine for Remote Monitoring of Intensive Care Patients With Mortality, Complications and Length of Stay. JAMA 2009;302(24):2671-2678
79. Rollman BL, Belnap BH, Hum DB et al. Telephone-Delivered Collaborative Care for Treating Post-CABG Depression. JAMA 2009;302(19):2095-2103

80. Kessler D, Lewis G, Kaur S, et al. Therapist-delivered internet psychotherapy for depression in primary care: a randomized controlled trial. Lancet 2009;374:628-634

81. Klersy C, De Silvestri A, Gabutti G et al. A meta-analysis of remote monitoring of heart failure patients. J Am Coll Cardiol 2009;54:1683-94

82. USDA Rural Development http://www.rurdev.usda.gov/ (Accessed June 20 2009)

83. Using Telehealth to Improve Quality and Safety Findings from the AHRQ Health IT Portfolio. www.healthit.ahrq.gov (Accessed January 5 2009)

84. Majerowicz A, Tracy S. Telemedicine. Briding Gaps in Healthcare Delivery. Journal of AHIMA. May 10 2010. 52-56

85. Centers for Medicare and Medicaid Services www.cms.hhs.gov/telmedicine (Accessed June 20 2009)

86. Virginia Gov. Signs Bill Requiring Insurers To Cover Telemedicine. April 6 2010. www.ihealthbeat.org (Accessed April 12 2010)

87. Butterfield S. Remote monitoring: Out of sight, right in line. ACP Observer March 2010 p. 1,10

88. The Value of Provider-to-Provider Telehealth Technologies. November 2007.www.citl.org. (Accessed November 20 2007)

Picture Archiving and Communication Systems (PACS)

ROBERT E. HOYT

Learning Objectives

After reading this chapter the reader should be able to:
- Describe the history behind digital radiology and the creation of picture archiving and communication systems
- Enumerate the benefits of digital radiology to clinicians, patients and hospitals
- List the challenges facing the adoption of picture archiving and communication systems
- Describe the difference between computed and digital radiology

The following is a detailed definition of PACS:

"Systems that facilitate image viewing at diagnostic, reporting, consultation, and remote computer workstations, as well as archiving of pictures on magnetic or optical media using short or long-term storage devices. PACS allow communication using local or wide-area networks, public communications services, systems that include modality interfaces, and gateways to healthcare facility and departmental information systems." [1]

PACS History

Digital imaging appeared in the early 1970's by pioneers such as Dr. Sol Nudelman and Dr. Paul Capp. The first reference to PACS occurred in 1979 when Dr. Lemke in Berlin published an article describing the functional concept. In 1983 a team lead by Dr Steven Horii at the University of Pennsylvania began working on the data standard Digital Imaging and Communications in Medicine (DICOM) (see chapter on data standards) that would facilitate image sharing. The US Army Medical Research and Materiel Command installed the first large scale PACS in the US in 1992. [2] The University of Maryland hospital system was the first to go "filmless" in 1999. While PACS had many early contributors the father of PACS in the United States is felt to be Andre Duerinckx MD PhD. [3]

Many hospitals and radiology groups have made the transition from analog to digital radiography. To their credit, radiologists have pushed for this change for years but have had to wait for better technology and financial support from their healthcare organizations. Early pioneers understood that a digital system would mean no more bulky film jackets, frequently lost films and slow retrieval. We are now at a point where the technology is mature and widely accepted but cost is still an issue at smaller healthcare organizations. Initially, hospitals purchased film digitizers so routine x-rays could be converted to the digital format and this was followed by scanning the digital image directly into the PACS.

Importantly, with the increasing use of electronic health records (EHRs), there is a need to integrate PACS with EHRs, hospital information systems (HISs) and radiology information systems

(RISs). The Veterans Health Administration launched a nationwide teleradiology network in 2009 that interfaces with its EHR (VistA). All images will be sent to a server in California that all VA radiologists can access. The Department of Defense is planning for a similar multi-facility PACS solution in the near future.

PACS has become mainstream among most large healthcare systems because of ease of use, popularity among physicians, speed of image retrieval and flexibility of the imaging platform. It is estimated that about 90% of large teaching hospitals have PACS but usage by small community hospitals is lower. [4-5] According to a June 2010 report on PACS adoption, replacement is now the most common reason for purchase of a system, with only 15% of purchases occurring for the first time. In this report 84% of hospitals of 100 + beds surveyed had implemented PACS in multiple locations outside the hospital, with the remainder of hospitals having single implementations. Additionally, 59% of hospitals also had Cardiac PACS that could store and display cardiology studies such as cardiac catheterization. [6]

One of the future challenges of teleradiology will be to share PACS images among disparate healthcare organizations.[7] SuperPACS is a new concept that would allow a radiology group that serves multiple sites with different PACS, radiology information systems (RISs) and hospital information systems (HISs) to view the sites as a single entity.[8] PACS was initially associated with expensive work stations ($50K) using thick-client technology. Now the trend is for thin or smart clients that permit clinicians to access PACS via a web browser from the office or home. [9] Health Information Organizations (see chapter on health information exchange) are beginning to link to web-based PACSs so images from different organizations can be viewed and shared.

PACS Basics

PACS is made possible by faster processors, higher capacity disk drives, higher resolution monitors, more robust hospital information systems, better servers and faster network speeds. PACS is also frequently integrated with voice recognition systems to expedite report turnaround. PACS usually has a central server that serves as the image repository and multiple client computers linked with a local or wide area network. Images are stored using the DICOM data standard. Input into PACS can also occur from a DICOM compliant CD or DVD brought from another facility or teleradiology site via satellite. Most diagnostic monitors are still grayscale as they have better resolution (3-5 megapixels), compared to color. Newer "medical monitors" have 2,048 x 2,560 pixel resolution and can display 1000+ shades of grey instead of the 250 shades of grey seen on a standard desktop monitor.

PACS Key Components (Figure 19.1)

- **Digital acquisition devices**: the devices that are the sources of the images. Digital angiography, fluoroscopy and mammography are the newcomers to PACS
- **The Network**: ties the PACS components together
- **Database server:** high speed and robust central computer to process information
- **Archival server:** responsible for storing images. A server enables short term (fast retrieval) and long term (slower retrieval) storage. HIPAA requires separate back up
- **Radiology Information system (RIS)**: system that maintains patient demographics, scheduling, billing information and interpretations
- **Workstation or soft copy display:** contains the software and hardware to access the PACS. Replaces the standard light box or view box
- **Teleradiology:** the ability to remotely view images [10]

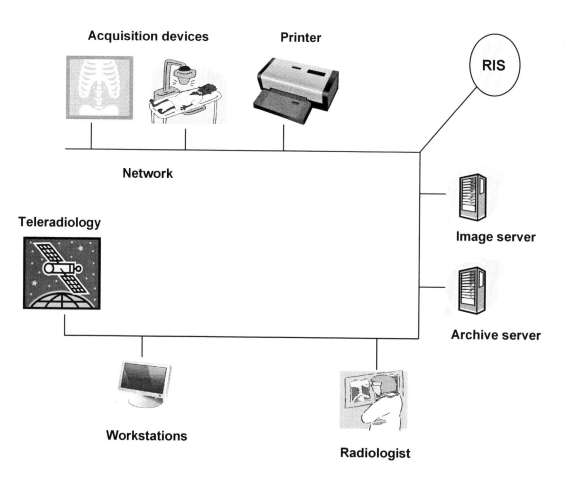

Figure 19.1 PACS Components

Types of digital detectors

1. **Computed radiography (CR)**: after x-ray exposure to a special cassette, a laser reader scans the image and converts it to a digital image. The image is erased on the cassette so it can be used repeatedly.[10] (Fig.19.2)

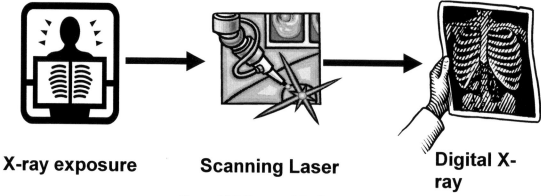

Figure 19.2 Computed Radiography

2. **Digital radiography (DR):** does not require an intermediate step of laser scanning.[10]

As noted, a PACS should interface with both the HIS and RIS. Typically, the patient is indentified in the HIS and an order created that is sent to the RIS via an HL7 protocol. Orders will go to the imaging device via the DICOM protocol and the image is created and sent to the PACS server. A diagnostic report is generated by the radiologist and stored on the PACS server. Diagnostic reports are sent back to the HIS via HL7 messages so they can be viewed on a DICOM viewer.

It is important to point out that many facilities with digital systems or PACS still print hard copies or have some non-digital services. This could be due to physician resistance, lack of resources or the fact that it has taken longer for certain imaging services such as mammography to go digital. *Full PACS* means that images are processed from ultrasonography (US), magnetic resonance imaging (MRI), positron emission tomography (PET), computed tomography (CT), routine radiography and endoscopy. *Mini-PACS*, on the other hand, is more limited and processes images from only one modality.[11]

Example of PACS Available on a Hospital Desktop Computer

The AGFA IMPAX 6.3 PACS is an example of a client-server based system used by the US Navy.[12] The PACS receives HL7 messages from the hospital information system (HIS) and provides diagnostic reports and other clinical notes along with the patient's images. Although resolution is slightly better with special monitors, the quality of the images on the standard desktop monitor is very acceptable for non-diagnostic viewing. Any physician on the network can rapidly retrieve and view standard radiographs, CT scans and ultrasounds. The desktop program is intuitive with the following features:

- Zoom-in feature for close-up detail
- Ability to rotate images in any direction
- Text button to see the report
- Mark-up tool that does the following to the image:
  o Adds text
  o Has a caliper to measure the size of an object
  o Has a caliper to measure the ratio of objects: such as the heart width compared to the thorax width
  o Measures the angle: angle of a fracture
  o Measures the square area of a mass or region
  o Adds an arrow
  o Right click on the image and short cut tools appear
- Export an image to any of the following destinations:
  o Teaching file
  o CD-ROM
  o Hard drive, USB drive or save on clipboard
  o Create an AVI movie

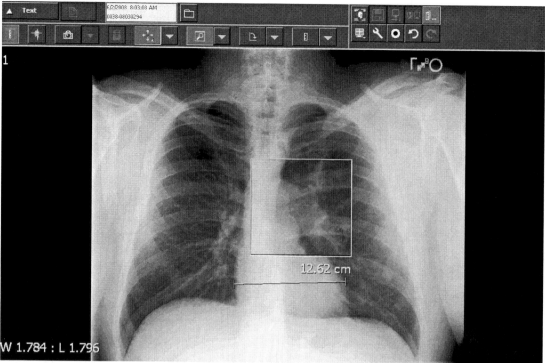

Figure 19.3 Chest X-ray viewed in PACS

The following are two scenarios that point out how practical PACS can be for the average primary care physician:

Scenario #1: An elderly man is seen in the emergency room at the medical center over the weekend for congestive heart failure and is now in your office on a Monday morning requesting follow up. Your practice is part of the Wonderful Medicine Health Organization, so you pull up his chest x-ray on your office PC and determine how severe his heart failure was.

Scenario #2: You are seeing a patient visiting your area with a cough and on his chest x-ray you note the patient has a mass in his left lung. You download this image on a CD (or USB drive) for the patient to take to his distant PCM where he will receive a further work up.

Open Source PACS

The commercial Web-PACS vendor Heart Imaging Technologies uses the open source platform known as ClearCanvas. Modules that can be downloaded include
- Workstation DICOM PACS viewer software
- Radiology Information System for scheduled workflow, etc
- Image Server software
- Platform (based on Microsoft .NET) and software developers kit (SDK) so plug-ins can be developed [13-14]
- OsiriX is a DICOM PACS open source viewer for the MAC OS. In addition to viewing images on an Apple computer, they can be viewed on an iPhone and iPad (figure 19.4) [15]

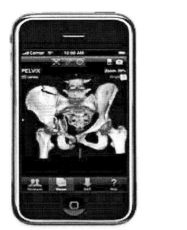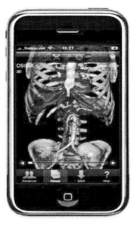

Figure 19.4 OsiriX PACS viewer on iPhone (courtesy OsiriX)

PACS Advantages and Disadvantages

PACS Advantages

- Replaces a standard x-ray film archive which means a much smaller x-ray storage space. Space can be converted into revenue generating services and it reduces the need for file clerks
- Allows for remote viewing and reporting; to also include teleradiology
- Expedites the incorporation of medical images into an electronic health record
- Images can be archived and transported on portable media, e.g. USB drive and Apple's *i*Phone [15]
- Other specialties that generate images may join PACS such as cardiologists, ophthalmologists, gastroenterologists and dermatologists
- PACS can be web based and use "service oriented architecture" such that each image has its own URL. This would allow access to images from multiple hospitals in a network
- Unlike conventional x-rays, digital films have a zoom feature and can be manipulated in innumerable ways
- Improves productivity by allowing multiple clinicians to view the same image from different locations
- Rapid retrieval of digital images for interpretation and comparison with previous studies
- Fewer "lost films"
- Radiologists can view an image back and forth like a movie, known as "stack mode"
- Quicker reporting back to the requesting clinician
- Digital imaging allows for computer aided detection (CAD)
    - Using artificial intelligence, CAD identifies mammogram abnormalities
    - CAD appears to be about as accurate as the interpretation by a radiologist
    - One study confirmed that experienced radiologists used CAD <u>after</u> they reviewed the images and 50% of lesions missed without CAD were detected with CAD [16]
    - More detail about CAD is available at E-medicine [17]

- Increased productivity. Several studies have shown increased efficiency after converting to an enterprise PACS. In a study by Reiner, inpatient radiology utilization increased by 82% and outpatient utilization by 21% after transition to a film-less operation, due to greater efficiency. [18] In another study conducted at the University of California Davis Health System, transition to digital radiology resulted in: a decrease in the average image search time from 16 to 2 minutes (equivalent to more than $1 million savings annually in physician's time); a decrease in film printing by 73% and file clerk full time equivalents (FTEs) dropped by 50% (equivalent to more than $2 million savings annually).[19] The Health Alliance Plan implemented PACS at Henry Ford Health Systems in 2003. Results indicate: turnaround time for film retrieval dropped from 96 hours to 36 minutes; net savings of $15 per film and key players noted significant time savings [20]

PACS Disadvantages and Barriers to Adoption

- Cost, although innovations such as open source and "rental PACS" are alternatives [14,21]
- New legislation cutting reimbursement rates for certain radiology procedures, thus decreasing capital that could be used to purchase a PACS [22]
- Expense and complexity to integrate with hospital and radiology information systems and EHRs
- Bandwidth limits may require network upgrades
- Different vendors may use different DICOMS tags to label films
- Viewing digital images a little slower than routine x-ray films
- Workstations may require upgrades if high resolution monitors are necessary

Key Points

- PACS is the logical result of digitizing x-rays, developing better monitors and improving medical information networks

- PACS is well accepted by radiologists and non-radiology physicians because of the ease of retrieval, quality of the images and flexibility of the platform

- PACS is a type of teleradiology, in that, images can be viewed remotely by multiple clinicians on the same network

- Cost and integration are the most significant barriers to the widespread adoption of PACS

Conclusion

PACS and digital imaging result from a predictable technological evolution beyond traditional film. For that reason, PACS has become a mainstream technology for moderate to large healthcare organizations. Like electronic health records (EHRs) PACS is an expensive technology to implement, but unlike EHRs, there is greater acceptance by clinicians. EHRs and Health Information Organizations will benefit by being interoperable with PACS. Healthcare organizations will be looking for ways to interpret and distribute a wide range of images to the entire organization. The technology is moving closer to thin client or zero client web based PACS for maximum flexibility and interoperability for the enterprise.

References

1. Vidar corp. http://www.filmdigitizer.com/about/news/glossary.htm (Accessed April 14 2006)

2. Hood MN, Scott H. Introduction to Picture Archive and Communication Systems. J Radiol Nurs. 2006;25:69-74
3. Wiley G. The Prophet Motive: How PACS was Developed and Sold http://www.imagingeconomics.com/library/tools/printengine.asp?printArticleID=200505-01 (Accessed April 14 2006)
4. Oosterwijk HT. PACS Fundamentals 2004 Aubrey Tx, Otech, Inc http://www.psqh.com/janfeb05/pacs.html (Accessed February 20 2006)
5. Matthews M. The PACS Picture. March 2009. Imaging Economics www.imaging economics.com (Accessed October 1 2009)
6. Miliard M. PACS adoption has reached the mature stage. June 25 2010. www.healthcareitnews.com (Accessed June 28 2010)
7. Moore J. Imaging meets the network. Government Health IT November 2008. pp 26-28
8. Benjamin M, Aradi Y, Shreiber R. From shared data to sharing workflow: Merging PACS and teleradiology. Eur J Rad 2010;73:3-9
9. Valenza T. Thin Wins. Imaging Economics March 2009 www.imagingeconomics.com (Accessed October 1 2009)
10. Samei, E et al. Tutorial on Equipment selection: PACS Equipment overview. Radiographics 2004; 24:313-34
11. Bucsko JK. Navigating Mini-PACS Options. Set sail with Confidence. Radiology Today. http://www.radiologytoday.net/archive/rt_071904p8.shtml (Accessed January 11 2007)
12. AGFA Healthcare. IMPAX 6. http://www.agfa.com/en/he/products_services/all_products/impax_60.jsp (Accessed June 4 2006)
13. Heart Imaging Technologies www.heartit.com (Accessed June 1 2010)
14. ClearCanvas www.clearcanvas.ca/dnn (Accessed June 1 2010)
15. Osirix http://www.osirix-viewer.com (Accessed June 1 2010)
16. Krupinski J. Digital Issues E-medicine www.emedicine.com (Accessed April 22 2006)
17. Ulissey MJ. Mammography-Computer Aided Detection. E-medicine. January 26 2005 www.emedicine.com (Accessed January 7 2007)
18. Reiner BI et al. Effect of Film less Imaging on the Utilization of Radiologic Services. Radiology 2000;215:163-167
19. Srinivasan M et al. Saving Time, Improving Satisfaction: The Impact of a Digital Radiology System on Physician Workflow and System Efficiency. J Health Info Man 2006;21:123-131
20. Innovations in Health Information Technology. AHIP. November 2005. www.ahipresearch.org (Accessed January 10 2007)
21. Vasco, C. Rental PACS Brings Digital Imaging to Rural Hospital http://www.imagingeconomics.com/issues/articles/2008-06_07.asp (Accessed June 18 2008)
22. Phillips J et al. Will the DRA Diminish Radiology's Assets? http://new.reillycomm.com/imaging/article_detail.php?id=380 (Accessed June 18, 2008)

Bioinformatics

ROBERT E. HOYT

Learning Objectives

After reading this chapter the reader should be able to:

- Define bioinformatics and how it interfaces with medical informatics
- State the importance of bioinformatics in future medical treatment
- Describe the Human Genome Project and its many important implications
- List private and governmental bioinformatics databases
- Describe the application of bioinformatics in genetic profiling of individuals and large populations

A commonly quoted definition of bioinformatics is:

"the field of science in which biology, computer science and information technology merge to form a single discipline"[1]

In the past two decades the field of bioinformatics has been involved with the creation of biological databases that helped evaluate DNA sequences and other genetic proteins. This allowed scientists to study the genetic information in the databases or add new information. The process of interpreting genetic data is referred to as **computational biology** that uses algorithms and artificial intelligence to: find the genes of various organisms, predict the structure and/or function of newly developed proteins, develop protein models and examine evolutionary relationships. [2-3] There are other bioinformatics terms worth defining:

- **Genomics**: the field that analyzes genetic material from a species
- **Proteomics**: the study of gene expression at the level of proteins
- **Pharmacogenomics**: the study of genetic material to look for drug targets

Why is bioinformatics important?

Besides diagnosing the 3000-4000 hereditary diseases that exist today, bioinformatics may be helpful to discover more targets for future drugs, develop personalized drugs based on genetic profiles and develop gene therapy to treat diseases such as cancer. The most common way to achieve this is to use genetically altered viruses that carry human DNA. This approach, however, has not been proven to be helpful and not approved by the FDA. Microbial genome alterations potentially could be used for energy production (bio-fuels), environmental cleanup, industrial processing and waste reduction. Genetically engineered plants could be made to be drought and disease resistant.

In spite of these interesting areas, it is estimated that less than 0.01% of microbes have been cultured and characterized. As an exception, the complete genome for the common human parasite Trichomonas vaginalis was reported in the January 2007 issue of the journal Science.[4] NIH will now embark on the Human Microbiome Project that will identify the genomes for 600 microbes that are related to health and disease. [5]

Genomic Primer

The human body has about 100 trillion cells and each one contains genetic information (chromosomes) in the nucleus; exceptions are eggs, sperm and red blood cells. Humans have a pair of 23 chromosomes in each cell that includes an X and Y chromosome for males and two Xs for females. Offspring inherit one pair from each parent. Chromosomes are listed approximately by size with chromosome 1 being the largest and chromosome 22 the smallest. Different organisms have different numbers of chromosomes, e.g. chimpanzees have 24 pairs. Chromosomes consist of double twisted helices of deoxyribonucleic acid (DNA). The DNA is composed of only four building blocks (nucleotides) also known as base pairs (adenine, thymine, cytosine and guanine) in varying sequences. Genes are regions on the chromosomes that provide instructions, largely by producing molecules such as ribonucleic acid (RNA), proteins and enzymes. (Figure 20.1) Humans have about 20,000 genes and 99.9% are the same among different people. Variations in the base pairs are known as single-nucleotide polymorphisms (SNPs) (pronounced snips) and are very common. There are usually three types of alterations: single base-pair changes, insertions or deletions of nucleotides and reshuffled DNA sequences. Although SNPs are common, their significance is complex and unpredictable. [6-8]

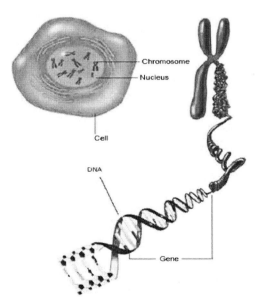

Figure 20.1 Image of a cell, gene, DNA and chromosome
(Courtesy of National Insititute of General Medical Sciences)

Bioinformatics Programs and Centers

The Human Genome Project (HGP)

One of the greatest accomplishments in medicine and the field of Biology is the Human Genome Project. This international collaborative project, sponsored by the US Department of Energy and the

National Institutes of Health, was started in 1990 and finished in 2003. In the process of evaluating the human genome (complete set of DNA sequences for the 23 chromosomes), investigators compared the human DNA results with those of the fruit fly, the bacterium E.coli and the house mouse. By mid-2007 they had identified about 3 million differences in the sequence (SNPs). In addition, the HGP addressed the ethical, legal and social issues associated with the project. Now that the HGP has been completed, it will take many years to analyze and learn from the databases.[9-11] Figure 20.2 displays the DNA sequencing of just chromosome number 12. Huge relational databases are necessary to store and retrieve this information. New technologies such as DNA arrays (gene chips) help speed the analysis and comparison of DNA fragments.[12] By 2010, a single gene chip can detect over a million variations in the base-pairs in a genome, in a few hours, costing only several hundred dollars. [6]

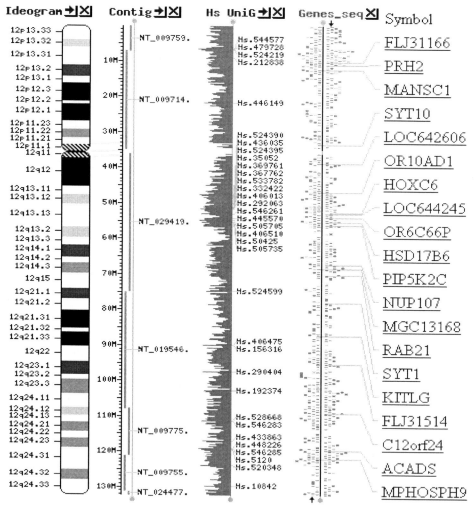

Figure 20.2 DNA sequences from chromosome 12 (Courtesy National Library of Medicine)

Human Variome Project

This Australian initiative began in 2006 with the goal to create systems and standards for storage, transmission and use of genetic variations to improve health. Rather than catalogue "normal" genomes they focus on the abnormalities that cause disease. Another aspect of their vision is to provide free public access to their databases.[13]

National Center for Biotechnology Information

The NCBI was created in 1988 and is part of the National Library of Medicine and the National Institutes of Health. They host 10 different genetic databases and thereby are considered the world's largest biomedical research center. The NCBI provides access to the complete genomes of over 1,000 organisms. Genomes represent both completely sequenced organisms and those for which sequencing is still in progress. NCBI databases are listed in figure 20.3.

If you access the Genome project you can do a search for specific genes or proteins from different species. Figure 20.4 demonstrates the result of an Entrez Gene search for a tumor protein (TP53).

The NCBI site also includes the search engine BLAST (basic local alignment search tool) that compares nucleotide or protein sequences to sequence databases and calculates the statistical significance of matches.[14]

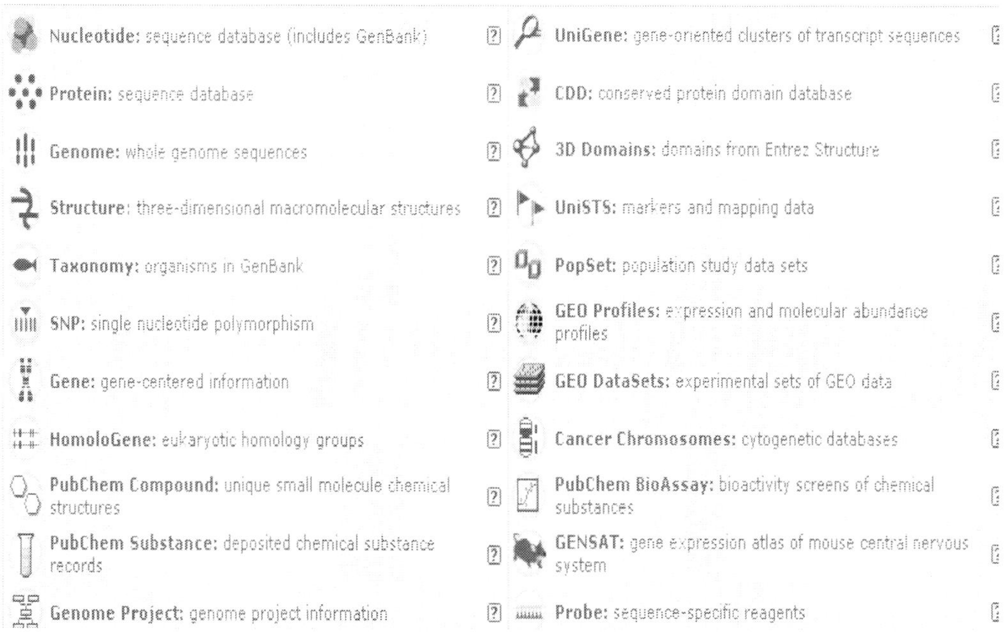

Nucleotide: sequence database (includes GenBank)

Protein: sequence database

Genome: whole genome sequences

Structure: three-dimensional macromolecular structures

Taxonomy: organisms in GenBank

SNP: single nucleotide polymorphism

Gene: gene-centered information

HomoloGene: eukaryotic homology groups

PubChem Compound: unique small molecule chemical structures

PubChem Substance: deposited chemical substance records

Genome Project: genome project information

UniGene: gene-oriented clusters of transcript sequences

CDD: conserved protein domain database

3D Domains: domains from Entrez Structure

UniSTS: markers and mapping data

PopSet: population study data sets

GEO Profiles: expression and molecular abundance profiles

GEO DataSets: experimental sets of GEO data

Cancer Chromosomes: cytogenetic databases

PubChem BioAssay: bioactivity screens of chemical substances

GENSAT: gene expression atlas of mouse central nervous system

Probe: sequence-specific reagents

Figure 20.3 NCBI Databases (Courtesy National Library of Medicine)

National Genome Research Institute

This institute is part of the National Institute of Health (NIH) and helps organize current research, available grants, educational resources, careers and policies & ethics. They lead the Human Genome Project for the NIH.[15]

GenBank

This database was established in 1982 and is the NIH genetic sequence database that is a collection of all publicly available DNA sequences (100 gigabases), the largest in the world. Interestingly, many medical journals now require submission of sequences to a database prior to publication and this can be done with NCBI tools such as BankIt.[16]

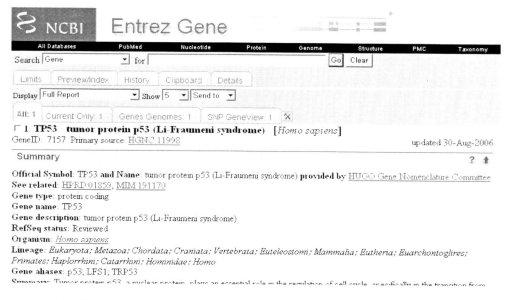

Figure 20.4 Search for specific protein (Courtesy National Library of Medicine)

Merck Gene Index

Private industry has recognized the tremendous potential of bioinformatics in research. In 1995 Merck Research Laboratory in collaboration with Washington University released to the public 15,000 gene sequences. Their ongoing releases will go to GenBank for international genetic researchers to develop future therapeutic agents.[17]

The Human Gene Mutation Database

The HGMD is a British site that collates human genetic mutations that cause inherited diseases. This has practical significance for clinicians, researchers and genetic counselors.[18]

The Online Mendelian Inheritance in Man

This is another NCBI database of genetic data and human genetic disorders. It is sponsored by Johns Hopkins University and Dr. Victor McKusick, a pioneer in genetic metabolic abnormalities. It includes an extensive reference section linked to PubMed that is continuously updated.[19]

World Community Grid

This project was launched by IBM in 2004 and simply asked people to donate idle computer time. By 2007 over 500,000 computers were involved in creating a super-computer used in bioinformatics. Projects include Help defeat Cancer, Fight AIDS@Home, Genome Comparison and Human Proteome Folding projects. This grid will greatly expedite biomedical research by analyzing complex databases more rapidly as a result of this grid.[20-21]

Pharmacogenomics Knowledge Base

This organization created by Stanford University looks at the relationships between genetics, disease and drugs. There are sections on drugs, medical literature, variant genes, pathways, diseases and phenotypes that are searchable.[22]

Framingham Heart Study SHARe Genome-Wide Association Study

In 2007, the Framingham Heart Study began a new phase by genotyping 17,000+ subjects as part of the FHS SHARe (SNP Health Association Resource) project. The SHARe database is located at NCBI's dbGaP and will contain 550,000 SNPs and a vast array of phenotypical (combined characteristics of the genome and environment) information available in all three generations of FHS subjects. These will include measures of the major risk factors such as systolic blood pressure, total, LDL and HDL cholesterol, fasting glucose, and cigarette use, as well as anthropomorphic measures such as body mass index, biomarkers such as fibrinogen and C-reactive protein (CRP) and electrocardiography (EKG) measures such as the QT interval.[23]

Cancer Biomedical Informatics Grid (CaBIG)

The Cancer Biomedical Informatics Grid is an IT project sponsored by the National Cancer Institute. The architecture is known as CaGrid and is an open source service oriented architecture (SOA). The infrastructure will support the collection and analysis of data from disparate systems to promote biomedical research. The core software and associated tools can be downloaded from their web site.[24-25]

For more information on bioinformatics and genetic databases, we refer you to the classic textbook by Shortliffe and Cimino.[26]

What is the future direction of bioinformatics?

At least three trends are appearing in the field of bioinformatics: 1) integration of genomic information into electronic health records 2) personal genetic services 3) population-based genetic data.

Integration with electronic health records: Eventually, the patient's genetic profile will be one more data field in the electronic health record. Recently, gene variants have been identified for diabetes, Crohn's disease, rheumatoid arthritis, bipolar disorder, coronary artery disease and multiple other diseases.[27] In late 2006 the Veterans Affairs healthcare system began collecting blood to generate genetic data that it will link to its EHR. The goal is to bank 100,000 specimens as a pilot project and link this information to new drug trials.[28] Similarly, Kaiser Permanente has created the Research Program on Genes, Environment and Health. In the first phase, 2 million members will be surveyed to determine their medical history, exercise and eating habits. The second phase will require the voluntary submission of genetic material.[29] SNOMED is making changes to its codes to include genetic information and the National eHealth Initiative is developing "use cases" for family history and genetics so standards can be created by organizations like the Health Information Technology Standards Panel. Organizations such as Partners Health, IBM, Cerner and multiple data mining vendors are all gearing up to add genetic information to what we currently know about patients and integrate that with electronic health records.[30]

The Agency for Healthcare Research and Quality (AHRQ) is developing computer-based clinical decision support tools to help clinicians use genetic information to treat breast cancer. The tools that could be integrated into EHRs are: whether women with a family history of breast cancer need BRCA1/BRCA2 testing and which women who already have breast cancer need genetic testing.[31]

It is surprising that family history is often overlooked by clinicians and that it usually does not exist as computable data for analysis. To our knowledge, no electronic health record collects this information in a computable format and uses it for clinical decision support. Data standards have been developed so family history can be part of EHRs and PHRs, in order to be shared.[32] There is a 2009 government sponsored free web tool available for the public to record their family history using the newest data standards. In this way, the results can be saved as a XML file and shared by EHRs and PHRs. The site, *My Family Health Portrait,* is available for English or Spanish speaking patients, is easy to use and does not store any patient information on the site. Instead, patients can store the XML file on their personal computers.[33] The program is open source and downloadable from this site.[34]

Personal Genetics: Patients will want to know their own genetic profile, even if the consequences are uncertain. Companies such as *Celera Genomics* will take advantage of the genomics project to offer genetic mapping services and pharmacogenomics.[35] *DNA Direct* is another company that offers online genetic testing and counseling. They do offer both patient and physician education and have staff genetic counselors.[36] *Decode Genetics Corporation* will collect disease, genetic and genealogical data for the entire population of Iceland. Their goal is to develop better drugs based on genetic profiles. They currently have three profiles: Complete analysis for 39 diseases, traits and ancestry for $1000, Cardio Scan for 6 cardiovascular diseases for $195 and Cancer Scan for 7 common cancers for $225. A simple mouth wash provides the DNA needed for analysis.[37] *23andMe* is a direct-to-consumer online genetic testing company. For $399-499 they will send a testing kit to homes based on analyzing saliva with a turnaround time of 4-6 weeks. Currently, they look for 81 diseases, 24 carrier states and 15 drug response conditions. They also offer an analysis of ancestry based on the genetic profile.[38] Google's co-founder Sergey Brin has funded a project through this company to study the genetic inheritance of Parkinson's disease. They hope to recruit 10,000 subjects from various organizations and offer a discount price for complete analysis.[39]

Multiple labs such as Affymetrix and Pacific Biosciences have the technology to genotype with microarrays with the goal of producing a faster and less expensive genetic analysis.[30]

However, as pointed out by Dr. Harold Varmus, personal genetics *"is not regulated, lacks external standards for accuracy, has not demonstrated economic viability or clinical benefit and has the potential to mislead customers".*[40]

Population Studies: *Oracle Corporation* will partner with the government of Thailand to develop a database to store medical and genetic records. This initiative was undertaken to offer individualized "tailor made" medications and to offer bio-surveillance for future outbreaks of infectious diseases such as avian influenza.[41] Harvard University has developed a new program "Informatics for Integrating Biology at the Bedside" to analyze 2.5 million patient records to look for links between DNA and illnesses such as asthma. It is known that certain patients respond poorly to standard asthma medications and the root may be genetic. Artificial intelligence will be used to search medical records for terms such as asthma and smoking.[42]

Genetic Prediction Obstacles

In order for genetics to enter the mainstream, new technologies and specialties will need to be developed and numerous ethical questions will arise. Just finding the abnormal gene is the starting point. Genetic tests will have to be highly sensitive and specific to be accepted. In general, patients will not be willing to undergo e.g. a prophylactic mastectomy or prostatectomy to prevent cancer unless the genetic testing is nearly perfect.

Additionally, the Genetic Information Nondiscrimination Act of 2008 was passed to protect patients against discrimination by employers and healthcare insurers based on genetic information. Specifically, the Act prohibits health insurers from denying coverage to a healthy individual or charging that person higher premiums based solely on genetic information and bars employers from using individuals' genetic information when making decisions related to hiring, firing, job placement, or promotion. [43]

Many obstacles face the routine ordering of genetic tests by the average patient. Ioannidis et al. points out that in order for genetic testing to be reasonable several facts must be true. The disease you are interested in must be common. Even with breast cancer, when you evaluate 7 established genetic variants, they only explain about 5% of the risk for the cancer. If the disease (example Crohn's disease) is rare, then the test must be highly predictive. In order for genetic testing to be relevant you should have an effective treatment to offer, otherwise there is little benefit. The test must be cost effective, as many currently are too expensive. As an example, screening for sensitivity to the blood thinner warfarin (Coumadin) makes little sense at this time due to cost. [44]

A May 2010 Lancet commentary also warned of additional concerns. Whole-genome sequencing will generate a tremendous amount of information that the average physician and patient will not understand without extensive training. At this point we lack adequate numbers of geneticists and genetic counselors. Patients will need to sign an informed consent to confirm that many of the findings will have unclear meaning. They will have to deal with the fact that they may be found to be carriers of certain diseases which would have impact on childbearing, etc. Genetic testing may cause many further tests to be ordered, thus leading to increased healthcare expenditures. As we gain more information about whole-genome sequencing, more patients will desire it but who will pay for it? [45]

Two other recent articles drive home additional practical points. When the risk of cardiovascular disease based on the chromosome 9p21.3 abnormality was evaluated in white women, it only slightly improved the ability to predict cardiovascular disease above standard, well-accepted risk factors. [46] Meigs et al. looked at whether multiple genetic abnormalities associated with Type 2 diabetes would be predictive of the disease. They found that the score based on 18 genetic abnormalities only slightly improved the ability to predict diabetes, compared to commonly accepted risk factors. [47]

Key Points

- Bioinformatics will likely introduce a treasure trove of genetic information into the field of medicine over the next few decades

- At this point bioinformatics seems like a field remote from medicine, but that will change with time

- Many organizations worldwide are beginning to collect and collate genetic information

- Electronic health records will incorporate genetic profiles in the future

- At this time, direct to consumer genetic testing can not be recommended

Conclusion

The Human Genome Project and bioinformatics will seem foreign to most clinicians. When they can access data that tells them who should be screened for certain cancers and which drugs are effective in which patients, these developments will be part of their day to day practices. In the meantime, scientists and biomedical companies will continue to add to the many genetic databases, develop genetic screening tools and get ready for one of the newest revolutions in medicine. The American Health Information Community (AHIC) recommended in 2008 that the federal government should prepare for the storage and integration of genetic information into many facets of healthcare.[48] Their recommendations will initiate the necessary dialogue that must take place to prepare for bioinformatics to align with the practice of medicine. But, as pointed out by Dr. Varmus *"the full potential of a DNA-based transformation of medicine will be realized only gradually, over the course of decades".*[40]

References

1. NCBI. A Science Primer. www.ncbi.nlm.nih.gov/About/primer/bioinformatics.html (Accessed July 1 2006)
2. Biotech: Bioinformatics: Introduction www.biotech.icmb.utexas.edu/pages/bioinform/BIintro.html (Accessed July 10 2006)
3. Bioinformatics Overview. Bioinformatics Web www.geocities.com/bioinformaticsweb/?200630/ (Accessed July 6 2006)
4. Carlton JM et al. Draft genome sequence of the sexually transmitted pathogen Trichomonas vaginalis. Science 2007;315:207-212
5. NIH Roadmap http://nihroadmap.nih.gov/hmp/initiatives.asp (Accessed January 4 2008)
6. Feero WG, Guttmacher AE, Collins FS. Genomic Medicine—An Updated Primer. NEJM 2010;362:2001-2011
7. National Institute of General Medical Sciences. The New Genetics. http://publications.nigms.nih.gov/thenewgenetics (Accessed June 24 2010)
8. Genome: The autobiography of a species in 23 chapters. Matt Ridley. Harper Perennial. 2006
9. Human Genome Project www.ornl.gov/sci/techsources/Human_Genome/project/info.shtml (Accessed July 5 2006)
10. Human Genome Project www.genome.gov (Accessed January 3 2008)
11. NCBI Human Genome Resources www.ncbi.nlm.nih.gov/genome/guide/human/ (Accessed July 19 2006)
12. DNA Arrays http://en.wikipedia.org/wiki/Dna_array (Accessed December 5 2006)
13. Human Variome Project. www.humanvariome.org (Accessed June 27 2009)
14. NCBI BLAST http://www.ncbi.nlm.nih.gov/BLAST/ (Accessed June 27 2009)
15. National Genome Research Institute www.genome.gov (Accessed June 27 2009)
16. GenBank www.ncbi.nlm.nih.gov/Genbank/ (Accessed June 27 2009)
17. Merck Gene Index www.bio.net/bionet/mm/bionews/1995-February/001794.html (Accessed December 11 2006)
18. Human Gene Mutations Database http://www.hgmd.cf.ac.uk/docs/new_back.html (Accessed June 20 2009)
19. The Online Mendelian Inheritance in Man. http://www.ncbi.nlm.nih.gov/entrez/query.fcgi?db=OMIM (Accessed June 18 2009)
20. World Community Grid www.worldcommunitygrid.org (Accessed June 25 2009)
21. Massive computer grid expedites medical research March 14 2007 www.ihealthbeat.org (Accessed March 14 2007)
22. Pharmacogenomics Knowledge Base http://www.pharmgkb.org/ (Accessed June 20 2009)
23. Framingham SNP Health Association Resource http://www.ncbi.nlm.nih.gov/projects/gap/cgi-bin/study.cgi?id=phs000007 (Accessed June 26 2009)

24. CaBIG http://cabig.cancer.gov/ (Accessed June 20 2009)
25. Oster S et al. caGrid 1.0: An Enterprise Grid Infrastructure for Biomedical Research. JAMIA 2008;15:138-149
26. Shortliffe E and Cimino J (eds) Biomedical Informatics, Computer Applications in Health Care and Bioinformatics. 3rd edition. 2006. Springer Science and Media, LLC. New York, New York
27. Pennisi, E. Breakthrough of the Year: Human Genetic Variation. Science 2007;318 (5858):1842-1843
28. Ferris N. VA to launch large-scale genetic data collection. Dec. 27 2006 www.govhealthit.com (Accessed January 3 2007)
29. Kaiser Seeks Member's Genetic Info for Database. www.ihealthbeat.org February 15 2007 (Accessed February 16 2007)
30. Kmiecik T, Sanders D. Integration of Genetic and Familial Data into Electronic Medical Records and Healthcare Processes. http://www.surgery.northwestern.edu/dos-contact/infosystems/Kmiecik%20Sanders%20Article.pdf (Accessed June 28 2009)
31. AHRQ Launches Project on Computer-Based Genetic Tools. September 23 2008 www.ihealthbeat.org (Accessed September 24 2008)
32. Ferro WG. New tool makes it easy to add crucial family history to EHRs. Perspectives. ACP Internist May 2009 p. 6
33. Family History http://familyhistory.hhs.gov (Accessed June 27 2009)
34. National Cancer Institute Gforge http://gforge.nci.nih.gov/projects/fhh (Accessed June 28 2009)
35. Celera Genomics http://www.celera.com (Accessed June 20 2009)
36. DNA Direct www.dnadirect.com (Accessed June 15 2009)
37. DeCODE genetics http://www.decodeme.com (Accessed May 10 2009)
38. 23andMe www.23andme.com (Accessed June 27 2009)
39. Google Co-Founder To Back DNA Database Study on Parkinsons. March 12 2009. www.ihealthbeat.org (Accessed March 12 2009)
40. Varmus H.Ten Years On—The Human Genome and Medicine. NEJM 2010;362:2028-2029
41. Oracle and Thai Government to build medical and genetic database www.ihealthbeat.org July 13 2005 (Accessed August 1 2005)
42. Cook G. Harvard Project to scan millions of medical files. The Boston Globe www.boston.com July 3 2005 (Accessed September 1 2005)
43. Hudson, KL, Holohan JD, Collins FS. Keeping Pace with the Times—the Genetic Information Nondiscrimination Act of 2008. NEJM 2008;358:26612663
44. Ioannidis JPA. Personalized Genetic Prediction: Too Limited, Too Expensive or Too Soon? Editorial. Annals of Internal Medicine 2009;150(2):139-141
45. Samani NJ, Tomaszewski M, Schunkert H. The personal genome—the future of personalized medicine? Lancet 2010;375:1497-1498
46. Paynter NP, Chasman DI, Buring JE et al. Cardiovascular Disease Risk Prediction With and Without Knowledge of Genetic Variation at Chromosome 9p21.3. Annals of Internal Medicine 2009;150(2):65-72
47. Meigs JB, Shrader P, Sullivan LM et al. Genotype Score in Addition to Common Risk Factors for Prediction of Type 2 Diabetes. NEJM 2008;359(21):2208-2219
48. HHS considers adding genetic information to EHRs. HealthImagingNews June 12 2008. www.healthimaging.com (Accessed June 12 2008)

Public Health Informatics

JUSTICE MZIBO
ROBERT E. HOYT

Learning Objectives

After reading this chapter the reader should be able to:

- Define the scope and goals of public health informatics
- State the significance of the Public Health Information Network
- Identify the various disparate public health informatics programs
- Describe the current biosurveillance programs
- State the significance of syndromic surveillance for early detection of bioterrorism and natural epidemics

Public health informatics is a growing and increasing important field that promises to revolutionize public health practice. Developing practical methods to detect the outbreak of disease, both natural and bioterroristic diseases has important cost and human benefits to the public health system and indeed the economy. [1] Public health tracks trends in the health of populations with the goal of preventing disease or detecting it early enough to initiate treatment. Integral to public health system's response to disease outbreaks is surveillance. At its core, public health surveillance serves to:

- Estimate magnitude of the problem
- Determine geographic distribution of illness
- Portray the natural history of a disease
- Detect epidemics/define a problem
- Generate hypotheses, stimulate research
- Evaluate programs and control measures
- Monitor changes in infectious agents
- Detect changes in health practices and behaviors
- Facilitate planning

Traditionally, public health surveillance programs monitor diseases using pre-specified case definitions and employ data collection, human decision making and manual data entry with timely dissemination of these data to those who need them. [2-3] In order to study a large population we need information technology such as networks, databases and reporting software, thus current electronic surveillance systems employ complex information technology and embedded statistical methods to gather and process large amounts of data and display the information for decision makers, first responders in the event of terrorist attack. There are several definitions of public health informatics such as the frequently cited 1995 definition of public Health Informatics:

"the systematic application of information and computer science and technology to public health practice, research and learning to the key public health functions of disease surveillance, detection and response in a timely manner" [4]

There are several types of surveillance systems:

- ***Sentinel surveillance systems*** are informatics systems that collect and analyze data by designated institutions selected for their geographic location, medical specialty, and ability to accurately diagnose and report high quality data. They include health facilities or laboratories in selected locations that report all cases of a certain condition to indicate trends in the entire population. These types of surveillance systems are useful when there are limited resources to monitor suspected health problems. Examples include networks of health providers reporting cases of influenza or a laboratory-based system reporting cases of certain bacterial infections among children. Sentinel surveillance systems are less reliable in assessing the magnitude of health related events on a national level since data collection is limited to specific geographic locations

- ***Household surveys*** are population-based, selecting a sufficiently large random sample of households representative of the whole population to make generalizabilitymore feasible. Their utility to monitor diseases is especially enhanced if the surveys are consistent and repeated periodically, e.g. every three to five years. Examples include Demographic and Health Surveys in developing countries and the Behavioral Risk Factor Surveillance System in the United States

- ***Laboratory-based surveillance*** are tools used to detect and monitor infectious diseases based on standard methods for identifying and reporting the genetic makeup of specific disease-causing agents. [5] An example is PulseNet, developed and maintained by the Centers for Disease Control and Prevention. The same system is also used in Canada, Europe, Latin America and recently with the emergence of SARS and Avian Influenza its use has been expanding in Asia and the Pacific resulting in PulseNet Asia Pacific, a network of 14 countries

- ***Integrated disease surveillance and response*** (IDSR) incorporate data from health facilities and laboratories in systems designed to monitor communicable diseases. The IDSR strategy aims to improve the availability and use of surveillance and laboratory data for control of priority infectious diseases that are the leading cause of death, disability, and illness in the African region. [6] The emphasis is on integrating surveillance with response thus: 1) strengthening public health surveillance and response to priority infectious diseases at the district level, 2) integrating surveillance with laboratory support, and 3) translating surveillance and laboratory data into specific and timely public health actions. Consequently, IDSR has several core activities: detecting, registering and confirming individual cases of disease; reporting, analysis, use, and feedback of data; and preparing for and responding to epidemics

Prior to 2001, Public Health reporting consisted of hospitals and clinics sending paper reports to local health departments, who in turn forwarded information to state health departments who sent the final data to the Centers for Disease Control (CDC) via mail or fax. This system would not suffice for epidemics, natural disasters or bioterrorism. The terrorist events of September 11, 2001 only heightened the concern for the means to detect illnesses and perform bio-surveillance more rapidly and accurately. Paper-based reporting is simply inadequate to detect subtleties in symptoms and inadequate to report large volumes of data to a central data repository. With an electronic system, artificial intelligence and rules engines could detect trends and alert officials. If the United States had a mature national health information network linked to electronic health records (EHRs) in every medical facility, the reporting of data could be uniform, rapid and less difficult to analyze. For example, in 2006 the United Kingdom began using *QFlu*, a national influenza surveillance system. The system collects data on the diagnosis and treatment of flu-like illnesses on a daily basis from over 3,000 physicians. An almost universal electronic health record has greatly facilitated reporting in the UK.[7] The United States, on the other hand, has spent billions of dollars on biosurveillance and has little to show for it, largely because of disparate systems, lack of universal EHRs and lack of a functioning Nationwide Health Information Network.[8]

Public Health Information Network (PHIN)

The creation of a Public health informatics Network (PHIN) faces many of the same issues shared by the Nationwide Health Informatics Network (NHIN). Both require data standards, databases, networks, tight security and decision support. The sources of public health data are also very disparate, deriving from hospitals, clinics, public health offices, labs, environmental agencies and poison control centers.[9] Similarly, they both face budgetary hurdles due to their complexity, difficulty in implementation and need for a long term business plan.

The PHIN is a relatively new CDC concept with its roots starting in 2004. The goal is to link together the players involved with US Public Health, using well established data standards. At the core of the PHIN are a set of programs to ensure interoperability between local, state and national public health entities. The vision is to improve disease surveillance, health status indicators, data analysis, monitoring, intervention, prevention, decision support, knowledge management, alerting and the official public health response. Integral to this vision is interoperability with EHRs, as part of Meaningful Use.

Like the NHIN, the PHIN would not be a separate network, but instead, would consist of standards, services and policies.[10-11] Historically, its first program was the PHIN Preparedness Initiative. This initiative was funded for $849 million dollars in 2004 to improve preparedness in all states and US territories. Approximately 25% of funds went towards information technology. The goals of the initiative were as follows: define the functional requirements including early event detection, outbreak management and connection of lab systems, identify and use industry wide data standards like HL7 and LOINC and make software solutions available for public health partners.[12]

National Center for Public Health Informatics (NCPHI) is one of 11 Centers within the Centers for Disease Control and Prevention (CDC) in Atlanta, Georgia. Established in 2005, it has the goal of providing national leadership to transform public health through informatics. The Center consists of 5 divisions to lead in areas of emergency preparedness and response, integrated surveillance systems, informatics, alliance management and consultation and knowledge management services. Several of the programs we will discuss in this section fall under the Center's divisions. They also support 5 Centers of Public Health Excellence at the University of Washington, University of Utah, Johns Hopkins University, New York City Health Department and Harvard Medical School.[13] NCPHI supports the National Notifiable Diseases Surveillance Program (NNDS) which consists of the following programs:

- **The National Electronic Disease Surveillance System (NEDSS)** is a major component of the PHIN that will create an interoperable surveillance system between federal, state and local networks. Specifically it will connect public health, laboratories and clinicians to support disease surveillance. The system will replace several older systems and gather as well as analyze data [14]
- **The National Notifiable Diseases Surveilllance System (NNDSS)** is a state based surveillance system for infectious conditions deemed to be nationally notifiable. All 50 states and 5 US territories report data

The CDC's Outbreak Management System (OMS) program is web based and capable of being used in the field on a laptop computer during an actual outbreak.[15-16]

The Health Alert Network (HAN) function as the PHIN's health alert component disseminating alerts and advisories via the Internet at the state and local levels. By late 2006 all 50 states were connected to the HAN and are funded for continuation of the initiative.[17] The CDC has also proposed a new system that would create actionable public health alerts that could be consumed by EHRs and that would allow for bidirectional exchange of information between public health and the private sector. Furthermore, they are investigating the potential role of health information exchanges (HIEs) in Biosurveillance.

Epi-X is a highly secure communications network that ties together select Public Health officials (about 4,200 users) around the United States. Users must be designated by health agencies in order to use the program. The system allows for rapid reporting, alerts and discussions about possible disease outbreaks. Since its inception in 2000 over 6,700 disease reports have been posted to include sentinel events such as the 2002 West Nile virus outbreak.[18]

Epi Info is a public domain software suite developed by the CDC (NCPHI) for public health officials and researchers that creates a database that can be analyzed along with graphs and maps. Users can develop a questionnaire for epidemiology studies and visualize an outbreak by using "geographic information systems" (GIS). Features also include: compatibility with Microsoft Access, SQL and ODBC databases, a report tool, EpiMap, a nutrition/anthropometry program and statistical tools. Version 3.5 was released in 2008 as a free downloadable program from the web site.[19]

TransStat is a software program developed by the National Institute of Health as part of the Models of Infectious Disease Agent Study (MIDAS). It is available as a free download for public health officials. Data such as age, sex, time of onset of symptoms, contacts, etc will be entered if there is a people-to-people or animal-to-animal epidemic. The software helps to determine if an individual contacted the illness from another individual, the rate an infection could spread and other epidemiologic information.[20]

Biosurveillance Programs

Biosurveillance. The CDC is not the only federal agency engaged with biosurveillance activities. The Department of Homeland Security (DHS) established the National Biosurveillance Integration System that will combine data from the CDC, US Department of Agriculture and environmental monitoring from the program BioWatch to improve pandemic and bioterrorism detection and response.[21] BioWatch is a Homeland Security Department program that monitors bioterrorism sensors in major US cities. The sensors are equipped and co-located with EPA air quality sensors.[22]

Syndromic surveillance. An important new public health function is syndromic surveillance defined as *"surveillance using health-related data that precede diagnosis and signal a sufficient probability of a case or an outbreak to warrant further public health response"*[23] It means that symptoms are monitored (like diarrhea or cough) before an actual diagnosis is made. If, for example, multiple individuals complain of stomach symptoms over a short period of time, you can assume there is an outbreak of gastroenteritis. In addition to the obvious sources of health data, public health officials can also monitor and analyze: unexplained deaths, insurance claims, school absenteeism, work absenteeism, over the counter medication sales, Internet based health inquiries by the public and animal illnesses or deaths.[24]

Initially, public health officials were very interested in detecting trends or epidemics in infectious diseases such as severe acute respiratory syndrome (SARS) and avian influenza. After the terrorist attacks and anthrax outbreak in 2001, they have had to improve biosurveillance to detect bioterrorism. The objective is to "*identify illness clusters early, before diagnoses are confirmed and reported to public health agencies and to mobilize a rapid response, thereby reducing morbidity and mortality*"[25] The challenge is to develop elaborate systems that can sort through the information and reduce the signal to noise ratio. The syndrome categories that are most commonly monitored are:

- Botulism-like illnesses
- Febrile (fever) illnesses (influenza-like illnesses)
- Gastrointestinal (stomach) symptoms
- Hemorrhagic (bleeding) illnesses
- Neurological syndromes
- Rash associated illnesses
- Respiratory syndromes
- Shock or coma

Ambulatory electronic health records (EHRs) are a potentially rich source of data that can be used to track disease trends and biosurveillance. EHRs contain both structured (ICD-9 coded) data as well as narrative free text. Hripcsak et al. assessed the value of outpatient EHR data for syndromic surveillance. Specifically, they developed systems to identify influenza-like illnesses and gastrointestinal infectious illnesses from Epic® EHR data from 13 community health centers. The first system analyzed structured EHR data and the second used natural language processing (MedLEE processor) of narrative data. The two systems were compared to influenza lab isolates and to a verified emergency room (ER) department surveillance system based on "chief complaint". The results showed that for influenza-like illnesses the structured and narrative data correlated well with proven cases of influenza and ER data. For gastrointestinal infectious diseases, the structured data correlated very well but the narrative data correlated less well. They concluded that EHR structured data was a reasonable source of biosurveillance data. [26]

The Electronic Surveillance System for the Early Notification of Community Based Epidemics (ESSENCE) is part of the Department of Defense Global Emerging Infections System (DOD-GEIS). It began in the national capital area in 1999 and by 2001 it was in place at military treatment facilities (MTFs). The national capital area was selected due to its increased risk of bioterrorism. Over the past three years data has been collected from 121 Army, 110 Navy and 80 Air Force installations worldwide. ESSENCE receives and analyzes data for about 90,000 outpatient and emergency room visits daily for the Department of Defense facilities. The syndromic surveillance data comes from outpatient encounters (standardized ambulatory data record) that include patient demographics and ICD-9 diagnostic codes. The data is sent to a centralized server in Denver, Colorado. Every 8 hours data related to the syndromes described above is downloaded and graphed to compare daily trends with historical data. A current limitation is that it takes several days for the data to arrive at the central server. In spite of this delay, there have been several instances where the surveillance network has identified early outbreaks before local authorities were aware. Newer versions have evolved due to collaborative efforts with Johns Hopkins University Applied Physics Laboratory and the Division of Preventive Medicine at the Walter Reed Army Institute of Research. ESSENCE II incorporated civilian data. ESSENCE IV is the most current version with the following features:

- Chemical-Biological detectors in limited distribution
- Data from civilian emergency rooms
- Prescription data

- Data from the Veterans system
- National insurance claims data
- Over the counter drug sales
- Standard reportable diseases such as TB, meningitis, etc [27]

Real-Time Outbreaks Detection System (RODS). The RODS system was initially developed by researchers at the University of Pittsburg and was the first real-time detection system for outbreaks. RODS collected patient chief complaint data from eight hospitals in a single health-care system via Health Level 7 (HL7) messages in real time, categorized these data into syndrome categories by using a classifier based on *International Classification of Diseases, Ninth Revision* (ICD-9) codes, aggregated the data into daily syndrome counts and analyzed the data for anomalies possibly indicative of disease outbreaks. Much like the ESSENCE system, RODS system started with a set of mutually exclusive and exhaustive categories of eight syndromic categories. However, as the program has gone through revisions and refinement, the categories have been reduced to seven as follows: respiratory, gastrointestinal, botulinic, constitutional, neurologic, rash and hemorrhagic.

Figure 21.1 shows the daily counts of respiratory cases for Washington County, PA in the period June-July 2003.

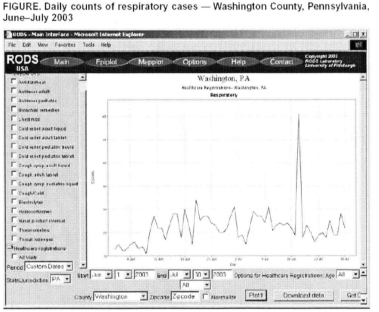

FIGURE. Daily counts of respiratory cases — Washington County, Pennsylvania, June–July 2003

Source: Real-Time Outbreak and Disease Surveillance project.
* The June 2003 increase corresponds to new hospitals being added to the system.
† The sudden increase on July 18, 2003, was caused by 60 persons reporting to one emergency department within 4 hours for carbon monoxide exposure.

Figure 21.1 Daily count of respiratory diseases using RODS

In order to increase the adoption of the RODS system, the University of Pittsburg started offering the software free of charge to public health departments. In 2003 the software was offered under an open source license and since then many more agencies have adopted the software for their use. [28]

BioSense. This is a CDC national web based program to improve disease detection, monitoring and situational awareness for healthcare organizations in the United States by reporting emergency room data. Participants include DOD, VA and civilian hospitals. The program addresses identification, tracking and management of naturally occurring events as well as bioterrorism.

Through the BioIntelligence Center, the CDC assists in the analysis of almost real-time data using advanced algorithms, statisticians and epidemiologists. This program will be part of the PHIN and use data standards such as HL7, SNOMED and LOINC. They currently have about 800 users from 570 non-governmental hospitals and 1300 Department of Defense and VA hospitals. In 2008 the program looked at the utility of connecting HIOs and testing the program during the seasonal influenza epidemic. [29]

GOARN. The Global Outbreak Alert and Response Network was established in 1997 and is supported by the World Health Organization (WHO). The severe acute respiratory syndrome (SARS) outbreak in 2003 was the first opportunity for the GOARN to be utilized. Since 2000 they have responded to more than 50 events worldwide. Features of the system include: alerts the international community about outbreaks, technical collaboration to pool human resources for rapid identification and response to outbreaks [30]

FluNet is another WHO initiative that is part of the WHO Communicable Disease Global Atlas. The goal is to collect and analyze infectious disease data from a country and global perspective. The Atlas will collect demographic and epidemiologic data so it can be used for queries, disease mapping and access to resources.[31]

Google Flu Trends is a web-based program launched by the philanthropic section of Google. This program analyzes web searches on influenza-related topics to estimate influenza frequency and maps of different regions of the US. Their results correlated strongly with data from the CDC and it is felt that Google trends might precede CDC results by about one week. Currently, they are studying trends from the US, Australia, New Zealand and Mexico. For the United States, users or consumers can also look at trends by state.[32-34] Figure 21.2 shows trends in the 2008-2009 flu season, while figure 21.3 shows yearly peaks using Google versus CDC data.

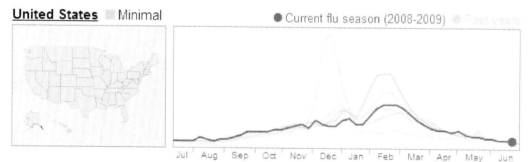

Figure 21.2 2008-2009 Flu trends (Courtesy Google Flu trends)

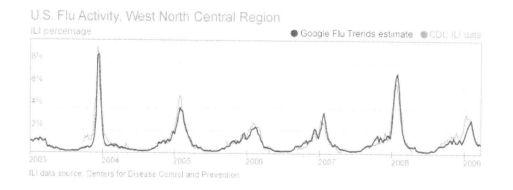

Figure 21.3 Annual Flu spikes; CDC data versus Google (Courtesy Google Flu Trends)

eLEXNET (electronic laboratory exchange network) is a web based information network that provides real time access to food safety analysis. The site has tools to look for trends, geographical locations, etc. Network allows food safety experts to collaborate. [35]

Global Public Health Intelligence Network (GPHIN) is an earlier Canadian initiative that monitors news media on the Internet in seven languages from around the world and reports on emerging disease threats. Of interest is the fact that the GPHIN detected the 2002-2003 SARS outbreaks with this technology. The information is automatically collated and analyzed by public health officials. There is a cost to subscribe to the service.[36]

Open Source Utah Disease Tracking System is the first open-source web-based infectious disease tracking and management system in the US. The software (TriSano) is part of the Collaborative Software Initiative. It will track and manage infectious diseases, biohazards and bioterrorism. The program can be downloaded from the TriSano web site for public use and customization. [37-38]

Geographic Information Systems (GISs)

As early as 1855 Dr. John Snow created a simple map to show where patients with cholera lived in London in relation to the drinking water source in the Soho District of London. Using his hand drawn map and basic epidemiological investigation techniques, much of which is still used today, he determined the epidemic was caused by a common water pump. We have come a long way since then thanks to the Internet and other technologies. Modern GIS software systems use digitized maps from satellites or aerial photography. Virtually all of the biodetection systems we have discussed have a GIS component that allows for the mapping of disease outbreak events giving public health practitioners the ability to timely deploy resources to control the outbreak and prevent further spread. Variables can be inputted by zip code, latitude, longitude, etc. GIS is a system of hardware, software and data used for the mapping and analysis of geographic data. A GIS provides: access to large volumes of data, the ability to select and query data, merge and spatially analyze data and visually display data through maps. GIS can provide geographic locations, trends, conditions and spatial patterns. Spatial data has a specific location such as longitude-latitude, whereas attribute data is the database that describes a feature on the map. Each layer on a GIS map has an attribute table that describes the layer. GIS maps are created by adding layers (features) and the data used can be of two types. *Vector* data appears as points (spots), lines (linear feature or boundary) or polycons (enclosed areas that have a perimeter like parcels of land). *Raster* data utilizes aerial photography and satellite imagery as a layer. Using GPS and mobile technology, field workers can enter epidemiologic data that can populate a GIS. This geospatial visualization is helpful to track infectious diseases, natural disasters and bioterrorism. Many more GIS details are available at these resources. [39-41]

HealthMap is a global project to integrate infectious disease news and visualization using an Internet geographic map. This program classifies alerts by location and disease. For example you can select "Salmonella" and the "United States" and see if there were any reported cases in the past 30 days (figure 21.4).[42] "Mouseover" an icon and you will see what is being reported in that area. The program was developed by the Harvard-MIT Division of Health Sciences and Technology and a more detailed explanation of the system and architecture is provided at this reference.[43]

Zyxware Health Monitoring System uses Google Maps as the GIS platform to track communicable diseases. This is an open source software program written in PHP/MySQL and available for download from www.sourceforge.net . [44]

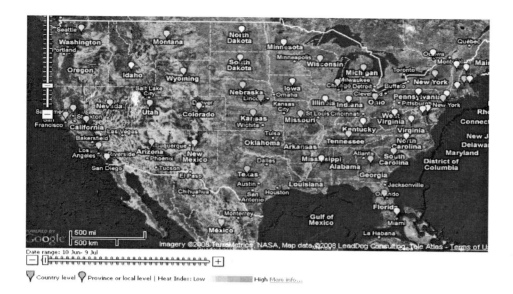

Figure 21.4 HealthMap search results for Salmonella (Courtesy HealthMap)

Microsoft Bing Health Maps and Google Fusion Tables are participating in the Community Health Data Initiative (covered in the first chapter). The Department of Health and Human Services (HHS) is making a huge health data set available to the public and developers to help with health planning, disaster management, disease management, public health, etc. [45] One of the data sets, County Health Rankings allows a user to explore multiple health indicators by county in the GIS mode. [46] In June 2010 Microsoft announced partnership with HHS and the Institute of Medicine (IOM) to make Bing maps available with data overlays from HHS. At this time they offer correlation with birth measures, death measures and risk factors. [47] Figure 21.5 shows a map of Florida with the overlay of the indicator, coronary heart disease.

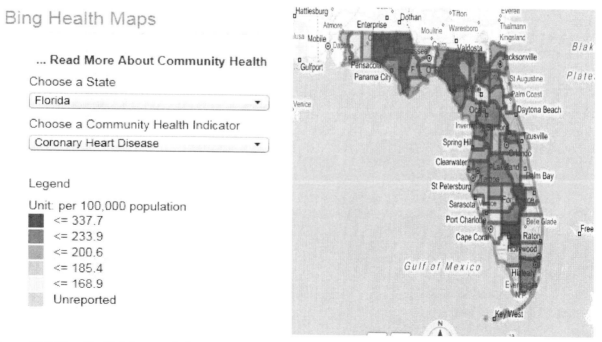

Figure 21.5 Bing Health Map correlating coronary heart disease and location (Courtesy Microsoft)

Key Points

- Public health is concerned with the health of populations, instead of individuals

- In order to study large populations and track trends in health and bioterrorism, new high speed networks must exist. Paper-based reporting is no longer tenable

- Creating a Public Health Information Network (PHIN) will not be easy due to multiple disparate systems nationwide

- The creation of the PHIN will face the same issues as the NHIN: high cost and issues of privacy, security and interoperability

Conclusion

The potential of a Public Health Information Network is great but may not be achievable in the immediate future due to high cost and disparate systems. Although the PHIN is critical for biosurveillance data reporting, it would also be important for reporting routine public health information such as influenza outbreaks and immunization status. In early 2008 the CDC awarded contracts worth $38 million to three health information exchange recipients to develop data sharing plans for public health information. [48] In September 2009 three states were able to report H1N1 data from emergency rooms to the CDC by using CONNECT and the NHIN. [49] Many details about standards and architecture are undetermined but it appears that "service oriented architecture" (discussed in chapter on architecture of information systems) will play a major role. [50]

Geographic information systems, free or commercial, are now commonplace to correlate public health epidemiology data with geography to produce dramatic visual representation of a variety of conditions.

Acknowledgement: we would like to thank Anita Ramesch at the University of West Florida for her contributions to the GIS section of this chapter

References

1. Levin, J.E. & Raman, S. (2004) Early Detection of Rotavirus Gastrointestinal Illness Outbreaks by Multiple Data Sources and Detection Algorithms at a Pediatric Health System. AIMA 2005 Symposium Proceedings: 445- 449
2. Stoto MA. Public Health Surveillance in the Twenty-First Century: Achieving Population Health Goals While Protecting Individual Privacy and Confidentiality. The Georgetown Law Journal, 96:703 703-917 (Accessed July 1, 2010)
3. Wallace R. (1998) 14th edition. Public Health and Preventive Medicine. Appleton & Lange
4. Friede A, Blum HL, McDonald M. Public Health Informatics: How Information Age Technology Can Strengthen Public Health. Ann Rev Pub Health 1995;16:239-252
5. Public Health Surveillance: The Best Weapon to Avert Epidemics. Fogarty International Center of the U.S. National Institutes of Health The World Bank World Health Organization Population Reference Bureau , Bill & Melinda Gates Foundation http://www.dcp2.org/file/153/dcpp-surveillance.pdf (Accessed June 25, 2010)

6. Integrated Disease Surveillance and Response. Centers for Disease Control and Prevention http://www.cdc.gov/idsr/ (Accessed June 30 2010)
7. QFLU: new influenza monitoring in UK primary care to support pandemic influenza planning. http://riskobservatory.osha.eu.int (Accessed November 17 2006)
8. The biosurveillance money pit. Government Health IT. November 16 2006 www.governmenthealthit.com (Accessed November 20 2006)
9. Yasnoff WA et al. Public Health Informatics: Improving and Transforming Public Health in the Information Age. J Pub Health Man 2000;6:67-75
10. PHIN: Overview CDC. http://www.cdc.gov/phin/index.html (Accessed January 4 2008)
11. Morris T. PHIN and NHIN. CDC. http://www.ncvhs.hhs.gov/060727p11.pdf (Accessed January 10 2007)
12. Loonsk JW et al. The Public Health Information Network (PHIN) Preparedness Initiative. JAIMA 2006;13:1-4
13. National Center for Public Health Informatics www.cdc.gov/ncphi (Accessed June 26 2009)
14. National Electronic Disease Surveillance System www.cdc.gov/nedss (Accessed October 4 2009)
15. PHIN: Outbreak Management System. http://www.cdc.gov/phin/software-solutions/oms/index.html (Accessed January 4 2008)
16. Public Health Informatics Institute www.phii.org (Accessed January 4, 2009)
17. Health Alert Network. CDC. http://www2a.cdc.gov/han/Index.asp (Accessed January 4 2008)
18. Epi-X http://www.cdc.gov/epix/ (Accessed January 4 2008)
19. Epi Info http://www.cdc.gov/epiinfo (Accessed July 1 2008)
20. TransStat. https://www.epimodels.org/midas/transtat.do (Accessed December 8 2007)
21. Beckner C. National Biosurveillance Integration System Moves Forward. May 12 2006. www.hlswatch.com (Accessed November 17 2006)
22. Shea DA, Lister SA. The BioWatch Program: Detection of Bioterrorism. www.fas.org/sgp/crs/terror/RL32152.html (Accessed November 17 2006)
23. Centers for Disease Control and Prevention http://www.cdc.gov/EPO/dphsi/syndromic.htm (Accessed September 20 2009)
24. Henning K. What is syndromic surveillance? MMWR September 24 2004 www.cdc.gov (Accessed September 18 2006)
25. Bioterrorism Preparedness and Response: Use of Information Technologies and Decision Support Systems www.ahrq.gov/clinic/epcsums/bioitsum.htm (Accessed September 21 2006)
26. Hripcsak G, Soulakis, ND, Li L et al. Syndromic Surveillance Using Ambulatory Electronic Health Records. JAMIA 2009;16:354-361
27. ESSENCE II www.geis.fhp.osd.mil/GEIS/SurveillanceActivities/ESSENCE/ESSENCE.asp (Accessed January 4 2008)
28. Espino J. et al. (2004) Removing a Barrier to Computer-Based Outbreak and Disease Surveillance –The RODS Open Source Project. MMWR, Supplement 53: 32-39
29. BioSense http://0-www.cdc.gov.pugwash.lib.warwick.ac.uk/biosense/ (Accessed January 4 2009)
30. Global Outbreak Alert and Response Network www.who.int/csr/outbreaknetwork/en/ (Accessed January 4 2008)
31. FluNet http://gamapserver.who.int/GlobalAtlas/home.asp (Accessed March 2 2007)
32. Google Flu Trends www.google.org/flutrends (Accessed June 24 2009)

33. Monegain B. Google Flu. November 12 2008. www.ihealthbeat.org (Accessed November 14 2008)
34. Ginsberg J, Mohebbi MH, Patel RS et al. Detecting influenza epidemics using search engine query data. Nature 2009;457:1012-1014
35. eLEXNET www.elexnet.com (Accessed January 4, 2008)
36. Global Public Health Intelligence Network. http://www.phac-aspc.gc.ca/media/nr-rp/2004/2004_gphin-rmispbk_e.html (Accessed March 20 2008)
37. Hayes HB. Open-sourcing public health. Government Health IT. November 2008. p32-33
38. TriSano. www.trisano.org (Accessed June 26 2009)
39. Geographic Information Systems. Wikipedia. www.wikipedia.com (Accessed June 27 2009)
40. ESRI. www.esri.com (Accessed June 27 2009)
41. Geographic Information Systems www.gis.com (Accessed June 27 2009)
42. HealthMap. http://www.healthmap.org/en (Accessed March 2 2008)
43. Freifeld CC. Health Map: Global Infectious Disease Monitoring through Automated Classification and Visualization of Internet Media Reports. JAMIA 2008;15:150-157
44. Zyxware www.zyxware.com (Accessed July 6 2009)
45. Community Health Data Initiative. www.hhs.gov/open/datasets/communityhealthdata.html (Accessed June 5 2010)
46. County Health Rankings www.countyhealthrankings.org (Accessed June 5 2010)
47. Bing Health Maps www.bing.com (Accessed June 5 2010)
48. CDC Grants Aim at Using RHIOs to Share Public Health Data. March 13 2008. www.ihealthbeat.org (Accessed March 13 2008)
49. Mosquera M. CDC begins to collect flu data through NHIN. Government Health IT September 25 2009. www.govhealthit.com (Accessed October 1 2009)
50. Arzt NH. Service-Oriented Architecture in Public Health. Interoperability Case Studies. JHIM 2010;24 (2):45-52

E-Research

ROBERT E. HOYT

Learning Objectives

After reading this chapter the reader should be able to:

- Identify the multiple ways information technology can improve research
- State the general benefits of research automation
- Describe the benefits of electronic collaborative web sites
- Describe the specific benefits of electronic case research forms
- Compare and contrast the pros and cons of mobile technology-based research forms

One of the definitions of medical informatics cited in chapter one includes medical research:

*"Medical informatics is the application of computers, communications and information technology and systems to all fields of medicine - medical care, medical education and **medical research**"*[1]

Like most areas in medicine, automation and digitization are making inroads, and research is no exception. At this point there is very little written about e-research in the medical literature, yet numerous commercial and home grown solutions now exist.[2] Ironically, there have been multiple advances in technology that could potentially make research automated, seamless and paperless.

Potential of information technology to improve research:

- Enhanced information retrieval through search engines such as Google and PubMed
- Automation of patient information
- Online registration
- Online surveys
- Online recruitment of subjects
- EHR recruitment of subjects [3]
- Electronic grant submission
- Data analysis with software programs such as Statistical Package for the Social Sciences (SPSS)[4] and Matlab[5]
- Software programs like LabVIEW can control medical devices, collate all data into a Microsoft Access database and display data real-time on a monitor during a study [6]
- E-collaboration web sites
- E-forms
- Research interfaced with electronic health records and health information exchanges
- Service oriented architecture (SOA) can connect universities, pharmaceutical companies and other partners with major disparate databases and services worldwide

Paperless Research

Multiple problems exist with paper forms in regards to data collection, validation, entry and storage:

- Data collection and transcription errors may be unrecognized until after the study has been completed
- Paper forms can be time-consuming to create and then to enter, store, retrieve and analyze the data. More time required means more money spent
- Data storage takes up valuable space that could be used for more profitable ventures
- Data stored in filing cabinets limits collaboration within and outside an organization

The British market analyst Datamonitor estimated that large life science companies could save approximately $10-$12 million dollars annually by transitioning to electronic data capture and clinical trials management systems.[7] Several reports in the literature have suggested greater efficiency and cost savings by converting from paper to electronic case report forms (e-CRF).[8-10] Electronic forms are also effective for data validation and creating an audit trail. With the widespread approval of electronic signatures, there is little reason to rely on paper.

Forms of every description can be stored on a web site with data fields mapped to a back-end database. With data validation tools, this should result in fewer data entry and transcription errors. An alert would notify the person filling out the form that correction or completion of information is needed before the form can be submitted. Web based forms would require Internet access by the researcher using a PC or laptop computer in a wired or wireless mode. Additionally, patients can be recruited via e-mail and referred to forms on the web site. Some complicated web forms may require programming time to add necessary features. Excellent back up is mandatory.

Commercial-Off-the-Shelf (COTS) Forms for Research

An example of a comprehensive e-form generating program would be OneForm Designer Plus. Customizable html forms can be created for web pages with JavaScript coding or a PDF "fillable" form. Forms can then be hosted on a web site and data automatically sent to a database.[11]

Microsoft InfoPath is an e-forms generating software program that is part of the Microsoft Office Suite 2003 and 2007. The electronic form is usually created on a personal or laptop computer and then uploaded to a web site. InfoPath forms can be hosted on a web site, sent via e-mail or used on mobile devices. All data is written in XML (extensible markup language). Office SharePoint Server 2007 is a portal where InfoPath forms can be uploaded and managed. In that way users do not need to have InfoPath on their personal computers to complete a form. Forms creation is easy (drag-and-drop) and includes many shortcuts such as drop down menus. Forms can be created by converting Microsoft Word or Excel files. Incorrect and missing data can be detected with data validation tools. Figure 22.1 provides an example of a study form used for a drug trial.[12]

IBM purchased PureEdge forms in 2005 and they are now known as Lotus 3.0 forms. This comprehensive form generating program is available to use offline or hosted on a web page. The program uses XForms, the latest XML format and offers a mobile solution as well. The US Army now has more than 2000 forms available using this technology.[13]

The Effects of Modafinil on Motion Sickness

Subject Progress Log

All the information that you provide in this questionnaire is strictly confidential and will become part of your medical record.

Date: [_____] 📅

Subject Information - Intake Day

Subject Number: Last 4 of Social Security Number	Sex: M ○ F ○	Date of Birth: (mm/dd/yyyy) / /

Race: ○ White ○ Black ○ Asian/Oriental ○ Multiracial ○ Other

List of Foods and Behaviors given: ☐ YES ☐ NO	Appointment Date given: ☐ YES ☐ NO
Consent Process: The subject was given the IRB approved informed consent form (ICF) with ample time to read and ask questions regarding the document. The subject signed and dated the ICF and was given two copies of the document. ☐ Yes	Date Consent Obtained 📅

Figure 22.1 InfoPath e-form

Mobile e-forms

Handheld computers have been used for more than a decade in clinical trials.[14] Studies have suggested that this technology is accurate and fast.[15-16] The main advantages they offer in research are their mobile nature, low cost and small form factor. Data can be collected anywhere, including in the field and later synchronized to a computer and uploaded to a database. The disadvantages are a small screen and relatively short battery life. Additionally, a study reported in 2009 compared electronic data collection using a PDA with a laptop computer and concluded that handheld technology was twice as slow for data entry and were associated with more errors.[17] PDA forms can be created by a programmer at considerable expense, but recently, commercial products exist that allow the average user to create forms. One of the products is Pendragon Forms that can build a form using 23 common field types that includes images and signatures. Form creation does not require any programming experience and a two week free download trial is available. The collected data uploads to an Access database on the computer, although an enterprise edition can synchronize data to a remote server.[18]

Electronic Case Report Forms (eCRFs)

Multiple vendors can produce electronic forms used in research to record patient data and events. Frequently, electronic forms reside on the local computer and are stored on a server; the client-server model. Because other aspects of research can now be electronic, automated and integrated, it has become more common to see electronic case report forms (eCRSs) integrated with the other functions as a whole package. Researchers have created programs just for single trials using the client server model.The reality is that web-based programs offer more functionality and interoperability for multiple researchers located at multiple different sites. Electronic data capture is becoming more common. According to a recent Canadian study 41% of respondents to a survey used eCRFs to conduct research.[19] In the next section we will discuss additional comprehensive solutions that include more functions than just eCRFs.

The Need to Collaborate and Integrate

Whether it is within or between organizations, communication is very important. A recent paper from the Mayo Clinic stated " *as we are in the midst of a fundamental investment by government and the stimulus package on health data standardization, enhanced interoperability and patient-centered care models, it seems self-evident that cooperation around an intellectual commons for developing shared, standards-based infrastructure for clinical and research data organization must emerge".* [20] Traditionally, people meet face to face to discuss how they might partner to write a grant or paper or analyze data sets. This is relatively simple if the collaborators work in the same building or organization but difficult if they work in different states. What is needed therefore is a means to communicate asynchronously and securely. The Internet provides the network and space to allow collaboration. Harris et al. listed the key desirable features to promote effective biomedical e-research: 1) data access among different institutions 2) user authentication and role-based security 3) eCRFs 4) data validation and data quality checks 5) audit trails 6) protocol management 7) central data storage and backup 8) data export to statistical packages 9) data import capability from different sources. [21]

An appropriate web site for research can be home grown as described by Marshall and Haley in a 2000 article in the Journal of the American Medical Association. They enumerated the 10 major steps to create a secure collaborative web site and estimated its cost to be $20-$35,000 with annual maintenance costs of $2500. The article also pointed out that data in a digital format allows for more rapid uploading and analysis. [22] Avidan reported the use of a web based platform serving 37 medical centers in 17 European countries for a study of decision making in intensive care units. The article discussed the importance of data validation or the means to alert researchers when data is missing or incorrect. They point out four ways data can be validated to include client-sided or server-sided validation or both. The importance of local and remote validation is stressed to prevent missing data. Their solution used both commercial off the shelf (COTS) products and Java Script programming. [23] Other authors have also published their home grown web-based solutions. [24-26]

Commercial Collaborative Solutions

Simplified Clinical Data Systems hosts a web-based solution that includes more than eCRFs. Some of the features of this research platform are as follows:
- Web site serves as a repository for all data collection tools, data storage and all documents related to the study
- Remote data entry via the Internet
- Online subject randomization
- Electronic case report forms (eCRFs) that are fee-based
- Data validation and audit trail
- Electronic signatures
- Automated real-time (e-mail) notifications for enrollment, adverse events, protocol deviations, subject visits, etc
- Integration with a wide variety of databases: Oracle, SQL Server, MS Access, etc
- Customized reports
- 128 bit SSL encryption for all system transactions
- Collaborators must log on with 3 types of information to include a 6 digit number contained on a key fob that changes every minute [27]

ClickCommerce is another commercial comprehensive administrative research software program. It includes additional tools for Institutional Review Board (IRB) requirements, grants and contracts management, e-forms library and several other research functions.[28]

Velos eResearch is a clinical research information system that supports account management, protocol management, patient management, IRB review and monitoring, project planning, study design, data safety monitoring and adverse event reporting. This web-based program integrates all of the research clinical data management functions with the administrative management. They are HL7 compliant and have interfaces for labs, devices and EHRs.[29]

OpenClinica is an open-source software application for research offered as a free download. Because it is web-based and not client-server based it provides a collaborative platform for researchers from different institutions. Its features include: protocol configuration, design of case report forms (CRFs), electronic data capture (EDC), retrieval and clinical data management. The application allows for importing and exporting of data sets to SPSS, CDISC, ODM and XML. It is capable of supporting regulatory guidelines and is built upon a modern architecture based on common standards. In 2009 they had 5000 community members from 76 countries.[30] Figure 22.3 demonstrates OpenClinica with color coding of records.

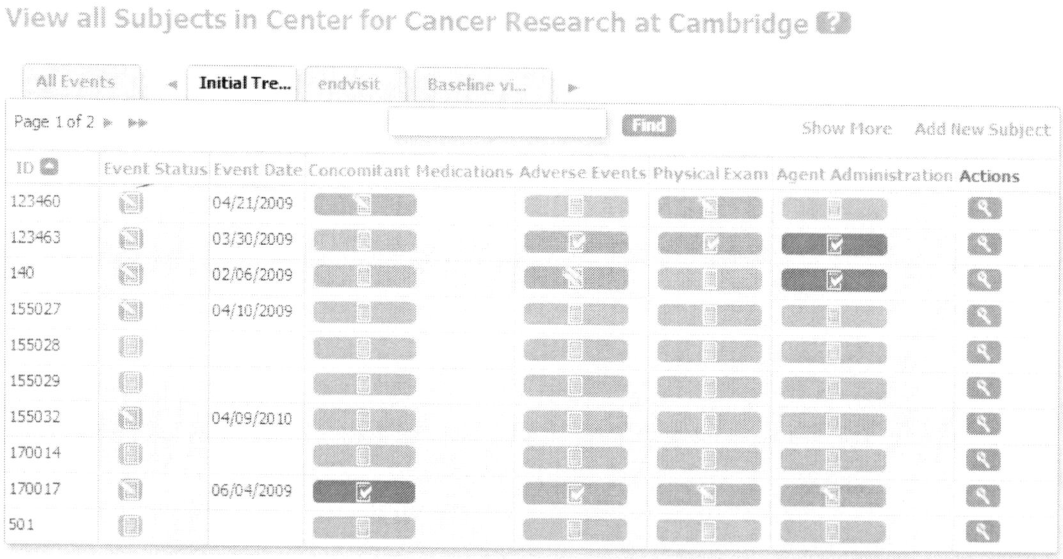

Figure 22.3 Sample research with OpenClinica (Courtesy OpenClinica)

REDCap (research electronic data capture) is another free web-based research platform actually consists of two web-based programs REDCap and REDCap Survey. It facilitates secure data capture, data validation, an audit trail and the ability to download seamlessly to statistical programs such as SPSS, SAS and Stata. It is currently available in English, Spanish and Japanese. The project is based on PHP and JavaScript programming with a MySQL database for data storage and management. The REDCap Survey tool is an online survey generating application used in research. Fifty-three organizations use this platform and to date it has been used for 650 studies. While this program is free to institutions, it is not open source so the code cannot be customized and an "end-user license agreement" must be signed.[19, 31]

A 2004 article in The Journal of Urology by Lallas outlined how a commercial web-based product improved efficiency and collaboration among twenty-one participating institutions. He concluded that the program became more cost effective as the number of enrolled subjects increased. For instance, once eCRFs are created it doesn't cost any more to use them for more subjects. 83% of participants rated the new way to collaborate as satisfactory to excellent.[32]

Enterprise-Level Research

For large civilian and federal research organizations there is a growing trend to use web services and service oriented architecture (SOA) (see chapter on the architectures of information systems) to connect disparate data and organizations. We will present several examples of large E-research initiatives. Many other institutions such as Intermountain Health, Vanderbilt, Duke and Massachusetts General Hospital have robust datawarehouses, but won't be discussed.

National Institute of Health (NIH). Perhaps the best example of enterprise research is the National Institutes of Health, a huge research organization that consists of 27 Institutes and Centers. In a statement by its Director, he noted that the new vision was to *"explore a standard clinical research informatics strategy, which will permit the formation of Nation-wide communities of clinical researchers made up of academic researchers, qualified community physicians and patient groups".*[33]

In order to execute the vision, the National Cancer Institute (NCI) created the cancer Biomedical Informatics Grid (CaBIG), also discussed in the chapter on Bioinformatics. This initiative was part of the National Cancer Institute Center for Biomedical Informatics and Technology (CBIT). The concept was to create a platform for all aspects of cancer research and for all participants, researchers, clinicians and patients. As of 2008, 46 NCI-designated Cancer Centers and 16 Community Cancer Centers have connected to the infrastructure and its set of tools. The architecture is known as CaGrid 1.3 and is an open access, open-source federated environment. More than 1000 software developers from 200 organizations contributed. Although intended for cancer research, this platform could be used for non-cancer research and they are poised to connect to EHRs and genomic databases. A list of CaBIG tools is available on their web site.[34]

Although smaller than CaBIG, other organizations have also created a network using SOA to facilitate collaboration and data exchange in genomics [35] and orthopedics.[36]

Mayo Clinic Enterprise Data Trust. The Mayo Clinic created an enterprise information system to collect and analyze data from patient care, education, research and administrative systems. Their datawarehouse was created with the assistance of IBM in 2005. The Data Trust connects to billing and their electronic health record. They have also incorporated an open source natural language processing system from IBM for data extraction and business analytics (intelligence). More details of their program are available in a 2010 article.[20]

Electronic Health Record (EHR) Based Research. While it is largely theoretical that an EHR could assist with research Greenway Medical EHR has created a new module PrimeResearch™ that is completely dedicated to clinical research. Patients can be recruited based on the study protocol and patient characteristics. The eCRF can be autopopulated with existing data to make the process much easier for participating clinicians with minimal disruption of workflow. Forms can be submitted to the appropriate private and public research organizations. This process also establishes an office as a participating site for future research.[37] Figure 22.4 shows how EHR-based research interacts with the other members of the research team.

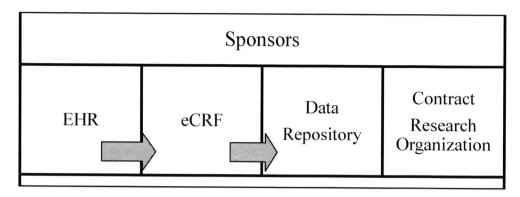

Figure 22.4 EHR-based research

Distributed Practice-Based Network for Observational Comparative Effectiveness Research (DARTNet) is a federated network of electronic data designed for comparative effectiveness research and medication and device outcomes research. The network links separate databases so that single queries can return results from other databases in the system. This system currently receives data from 8 organizations representing more than 500 physicians and 400,000 patients. DARTNet was created from grants from the Agency for Healthcare Research and Quality. The system is capable of receiving 150 data elements patient-level from multiple different EHR systems. [38]

Clinical Research Informatics: An Emerging Sub-Domain

Another convergence has arisen between biomedical informatics and research, primarily due to the multiple new technologies listed above. This emerging new sub-domain of biomedical informatics would be known as clinical research informatics. A proposed definition is a subdomain *"concerned with the development, application and evaluation of theories, methods and systems to optimize the design and conduct of clinical research and the analysis, interpretation and dissemination of the information generated"*. This is in keeping with the NIH Road map to re-engineer how bench research gets translated into clinical practice. Also supporting this new direction is the Clinical Research Informatics Working Group that is part of the American Medical Informatics Association. Further dialogue about this emerging domain can be found in an article in a 2009 article by Emby and Payne in the Journal of the American Medical Informatics Association. [39]

Key Points

- Traditionally, research has been paper based but that is quickly changing

- New technologies are now available to automate the research process from start to finish

- New web based collaborative sites exist to help organize and expedite research within one or between many research organizations

- Research networks will combine medical data from multiple diverse institutions to make scientific conclusions more valid

Conclusion

The evidence points towards improved productivity and accuracy using electronic collaboration tools and forms. It also seems likely that information technology can reduce full time equivalents (FTEs) due to fewer steps in data entry and data storage. The literature also suggests that data integrity is enhanced by automated data entry and validation. Multiple commercial products now exist to move research into the information age.

As we increase our adoption of electronic health records and create more health information organizations we will generate more medical data, ripe for mining. Large organizations have developed a variety of federated and non-federated data warehouses to aggregate and mine data. Health and Human Services plans to create a new database for all payors that is based on claims data that will support robust analysis.[40] Others have suggested that we need a National Distributed Health Data Network for the same reasons: to conduct multi-instituitional research to examine population health issues, compartative effective research, etc.[41]

References

1. MF Collen. Preliminary announcement for the Third World Conference on Medical Informatics. MEDINFO 80, Tokyo
2. E-Research: Methods, Strategies and Issues. 2002. Anderson T and Kanuka H. Ally & Bacon Publishers
3. Embi PJ et al. Effect of a Clinical Trial Alert System on Physician Participation in Trial Recruitment. Arch Int Med 2005;165:2272-2277
4. Statistic Package for the Social Sciences www.spss.com (Accessed January 25 2007)
5. Matlab www.mathworks.com (Accessed February 20 2007)
6. LabVIEW www.ni.com (Accessed January 4 2009)
7. Pizzi R. Millions could be saved with paperless clinical trials. August 8 2008. www.healthcareitnews.com (Accessed August 8 2008)
8. Eisenstein EL, Collins R, Cracknell BS et al. Sensible approaches for reducing clinical trial costs. Clin Trials 2008;5(1):75-84
9. Lopez-Carrero C, Arriaza E, Bolanos E et al. Internet in clinical research based on a pilot experience. Contemp Clin Trials Apr 2005;26(2):234-43
10. Pepine CJ, Handberg-Thurmond e, Marks RG et al. Rationale and design of the international verapamilSR/Trandolapril Study (INVEST): An Internet based randomized trial in coronary artery patients with hypertension. J Am Coll Cardiol 1998;32(5):1228-37
11. Amgraf Software Technology www.amgraf.com (Accessed July 16 2009)
12. Microsoft InfoPath. www.microsoft.com (Accessed January 4, 2008)
13. IBM Lotus forms http://www-306.ibm.com/software/lotus/forms/ (Accessed June 20 2009)
14. Koop A, Mosges R. The use of handheld computers in clinical trials. Control Clin Trials 2002;23:469-80
15. Lal SO et a. Palm computer demonstrates a fast and accurate means of burn data collection. J Burn Care Rehab 2000;21:559-61
16. Rivera, ML et al. Prospective, randomized evaluation of a personal digital assistant-based research tool in the emergency department. BMC Med Info Dec Making 2008;8:1-7
17. Haller G, Haller DM, Courvoisier DS et al. Handheld vs Laptop Computers for Electronic Data Collecton in Clinical Research: A Crossover Randomized Trial. J Am Med Assoc 2009:16:651-659
18. Pendragon Forms www.pendragon-software.com (Accessed June 27 2009)

19. El Emam K, Jonker E, Sampson M et al. The Use of Electronic Data Capture Tools In Clinical Trials: Web-Survey of 259 Canadian Trials. JMIR. 2009;11 (1) www.jmir.org (Accessed December 29 2009)

20. Chute CG, Beck SA, Fisk TB et al. The Enterprise Data Trust at Mayo Clinic: a semantically integrated warehouse of biomedical data. J Am Med Inform Assoc 2010;17:131-135

21. Harris PA, Taylor R, Thielke R et al. Research electronic data capture (REDCap)—a metadata-driven methodology and workflow process for providing translational research informatics support. J of Biomed Inf 2009;42:377-381

22. Marshall WW, Haley RW. Use of a Secure Internet Web Site for Collaborative Medical Research. JAMA 2000;284:1843-1849

23. Avidan A, Weissman C and Sprung CL. An Internet Web Site as a Data Collection Platform for Multicenter Research. Anesth Analg 2005;100:506-11

24. Sippel H, Eich HP, Ohmann C. Data collection in multicenter clinical trials via Internet: a generic system in Java. Medinfo 1998:9:93-97

25. Mezzanaa P, Madonna Terracina FS, Valeriani M. Use of a web site in a multicenter plastic surgery trial: a new option for data acquisition. Plast Recontr Surg 2002;109:1658-61

26. Marks R, Bristol H, Conlon M, Pepine CJ. Enhancing clinical trials on the Internet: lessons from INVEST.Clin Cardiol 2001;24:V17-23

27. Simplified Clinical Data Systems. www.simplifiedclinical.com (Accessed January 4, 2008)

28. ClickCommerce. www.clickcommerce.com (Accessed June 24 2009)

29. Velos www.velos.com (Accessed June 20 2009)

30. OpenClinica www.openclinica.org (Accessed June 20 2009)

31. REDCap www.project-redcap.org (Accessed June 29 2009)

32. Lallas CD et al. Internet Based Multi-Institutional Clinical Research: A Convenient and Secure Option. J Urol 2004;171:1880-1885

33. Zerhouni EA. A new Vision for the National Institutes of Health. Editorial. J of Biomed Biotech 2003;3:159-160

34. CaBIG http://cabig.nci.nih.gov (Accessed June 27 2009)

35. Stark K, Schulte J, Hampel T et al. GATiB-CSCW, Medical Research Supported by a Service-Oriented Collaborative System. Lecture Notes in Computer Science. Advanced Information Systems Engineering. 2008. Springer Berlin/Heidelberg.

36. Makola D, Sim YW, Wang C et al. A Service-Oriented Architecture for a Collaborative Orthopaedic Research Environment. http://eprints.ecs.soton.ac.uk/12898/ (Accessed June 28 2009)

37. Greenway Medical PrimeResearch www.greenwaymedical.com (Accessed June 25 2010)

38. Pace WD, Cifuentes M, Valuck RJ et al. An Electronic Practice-Based Network for Observational Comparative Effectiveness Research. Ann Int Med 2009;151:338-340

39. Embi PJ, Payne PRO. Clinical Research Informatics: Challenges, Opportunities and Definition for an Emerging Domain. JAMIA 2009;16:316-327

40. Mosquera M. HHS to build universal claims database for health research. Government HealthIT. December 17 2009. www.govhealthit.com (Accessed December 29 2009)

41. Maro JC, Platt R, Holmes JH et al. Design of a National Distributed Health Data Network. Ann of Int Med 2009;151:341-344

Emerging Trends in Health Information Technology

ROBERT E. HOYT
ROBERT W. CRUZ
FRED TROTTER

Learning Objectives

After reading this chapter the reader should be able to:

- Identify the features of successful technology innovations
- Describe some of the future prediction by national experts
- State the significance of increased artificial intelligence in medicine
- List the innovations found at the 100 most wired hospitals that will likely permeate the healthcare system in the future
- Enumerate several new Web innovations

"Computers of the future may weigh no more than 1.5 tons"
Popular Mechanics 1949

"Any sufficiently advanced technology is indistinguishable from magic"
Arthur C. Clarke

Technology continues to evolve at a rate faster than our ability to understand and assimilate it into healthcare. In this chapter we will discuss some of the more significant emerging trends in technology that have applicability in the field of medicine. What determines the long term success or failure of a technology trend is often unclear but seems to be partly related to the features listed in Table 23.1.

Table 23.1 Features and examples of successful technology trends

Features	Examples
Unique new concept	VoIP, voice recognition, digital images, RFID
Saves time or money	Voice recognition, VoIP
National mandate	E-prescribing and EHRs for Medicare physicians
Affordable	Wireless capability, cell phones, USB memory devices
Convenient form factor	PDAs, smartphones, USB memory devices
Ease of operation	USB memory devices, smartphones, PACS

Additionally, in the 2005 RAND monograph *The Diffusion and Value of Healthcare Information Technology*, Bower describes the following features that predict innovative success: [1]

- Relative advantage over earlier innovations
- Compatibility with existing values or past experiences
- Complexity; the lower the better
- External influence by vendor marketing

- Peer pressure to try an innovation
- Network realities; the more who join, the more practical the innovation
- Degree of specialization; the broader the better
- Government policy; financial support always helps

A pivotal paper was written by Bower and Christensen in 1995 in The Harvard Business Review, entitled *Disruptive Technologies: Catching the Wave*. In their work they defined *disruptive technologies* as overturning the dominant standard-of-care technologies. These technologies catch the market by surprise. They maintained that the very technologies customers enjoy may hold back advancement and market share. Initially disruptive technologies are likely to have at least one attribute that is considered undesirable, like price, but in a short time these issues are overcome and they soon dominate the market. *Sustaining technologies*, on the other hand, are those that evolve slowly with the same attributes. Examples might be digital cameras replacing film photography or flash drives replacing floppy discs. Also, consider how the iPhone has succeeded compared to devices based on the Palm operating system, even though the latter was the originator of handheld technology. At this point we have several disruptive technologies in the field of medicine such as digital radiology replacing film-based radiology. Solid state drives will likely replace standard hard drives in the not too distant future. As electronic health records evolve they may also be considered a disruptive technology.[2-3]

Medical informatics is heavily influenced not only by the need to solve problems in the field of medicine, but by what new technologies are available. For instance, picture archiving and communication systems (PACS) are possible today solely because of innovations in monitors, servers, digital images and processors. Another example would be the increasing capabilities of the Internet made possible by greater bandwidth. Table 23.2 demonstrates the advances in personal computer technology.[4] Integrating the latest developments in technology into the field of medicine will require that more healthcare workers become formally. Healthcare administrators and chief information officers (CIOs) will also need additional education to understand new systems such as EHRs, PACS and HIOs, prior to implementation.

Table 23.2 Computer components in 1999 compared to 2005 and 2010

Component	1999	2005	2010
Random Access Memory (RAM)	64 megabytes	1 gigabyte	4 gigabytes
Processor speed (instructions per second)	400 million	7 billion	12 billion
Circuit density (total transistors)	7.5 million	125 million	500 million
Hard Drive	8 gigabytes	135 gigabytes	360 gigabytes
Internet speed (bits per second)	56,000	1 million	30 million

What Do the Experts Predict?

Health and Healthcare 2010 Study by the Institute for the Future

In this book future predictions are listed as "stormy", "long and winding" or "sunny". Some of the predicted information technology advances are as follows:

- Faster microprocessor speed
- Better data storage, warehousing and mining
- More wireless applications
- Better bandwidth

- More use of artificial intelligence
- Better encryption
- Internet 2
- Smaller, more accurate and less expensive sensors
- Drugs designed by computers
- Home telemonitoring
- Pharmacogenomics
- Improved imaging
- Mini-MRIs with much smaller magnets
- Electron beam CT instead of x-ray CT
- More 3-D images
- Better resolution, contrast and display
- PET scans that use tumor specific markers instead of glucose making them more accurate
- Computer aided diagnostic interpretations [5]

Health Care in the 21st Century

In this interesting 2005 article by Senator Bill Frist, published in the New England Journal of Medicine, changes in medical care expected to be seen by the year 2015 were:

- Highly sophisticated EHRs that are integrated with patient portals and national information networks
- Combination medications that you only have to take daily, weekly or monthly, known as "polypills"
- Implantable computer chips that monitor vital signs and blood chemistries
- Injectable nanorobots that correct problems such as blood vessel blockages
- Automatic transmission of hospital data to insurance company so bills are paid before the patient leaves the hospital [6]

Emerging Trends

In this section we will take a look at some of the emerging technological trends that have already impacted the field of medicine or will likely do so in the future. This is only a partial enumeration of emerging trends. Some of those presented may fail and others may appear that were not predicted.

Artificial Intelligence in Medicine (AIM)

AIM has gone from a vague concept thirty years ago to new technologies such as neural networks that can aid medical diagnosis and treatment. Swartz in 1970 was very optimistic in thinking that computers would radically change the delivery of healthcare by the year 2000. He stated *"It seems probable that in the not too distant future the physician and the computer will engage in frequent dialogue, the computer continuously taking note of history, physical findings, laboratory and the like...."*.[7]

Over the past thirty years universities have used artificial intelligence to develop software programs to assist medical care or what is now referred to as clinical decision support. Neural networks, a form of artificial intelligence, are data modeling tools that capture complex relationships and learn over time. An example of a neural network would be optical character recognition (OCR). A document is scanned and the image is converted to a format such as a Microsoft Word document. Each group of pixels scanned produces a value that the OCR software recognizes and converts to text. Other examples of neural networks include: target recognition, computer aided diagnosis, voice recognition and financial forecasting.[8-11]

The majority of patient information is stored as free text (to include voice recognition). This makes it difficult to extract information for coding and data mining. Companies such as Language and Computing are using natural language processing (NLP) and natural language understanding (NLU) to extract meaningful data from free text. They have entered into an agreement with Kaiser Permanente's Southern California region to develop a solution for evaluation and management (E&M) coding.[12] In Australia, students at Sydney University have developed the Clinical Data Analytics Language that converts physician notes to SNOMED CT using NLP. This should facilitate excellent data mining based on diagnoses.[13] As more regional and national health information networks and databases are developed it is likely that artificial intelligence will automatically analyze these data and generate reports and alerts to physicians, insurers and the government. This could eventually be more cost effective than standard chart reviews.

Better Imaging

All imaging devices have improved greatly in the past decade. In spite of the fact that imaging has not become less expensive over time, devices are smaller, faster and with better resolution. For example, Siemen's Somatom Sensation™ 64 CT scanner is one of the fastest scanners of its generation, circling the patient in 1/3 second and producing 64 images per rotation with a resolution of 0.4 millimeters.[14] One can assume that all imaging will continue to improve in the future: ultrasounds, mammograms, CT scans, MRIs, etc. As an example, ultrasound machines are now also available as portable devices with impressive features. In figures 23.1 the new SonoSite M-Turbo has full features to include color flow patterns and faster speed, yet weighs only 7 pounds.[15] As a result of portability, this model is now used frequently in emergency rooms and in deployed military hospitals. Figure 23.2 shows a new GE VScan portable ultrasound, the size of a cellphone. Some speculate that this will replace the stethoscope for cardiologists and others.[16]

Figure 23.1 SonoSite M-Turbo   **Figure 23.2** Color flow (Courtesy SonoSite)

Monitors will have higher resolution and be available in color as well as black and white. Internet-2 and Lambda Rail will merge to provide a much faster Internet. This will facilitate teleradiology and permit radiologists to simultaneously compare side-by-side old and new CTs and

MRIs. With the larger pipe, images will be available on information networks for others to view remotely.

Hospitals of the future

As a rule, the more affluent hospitals are able to purchase state-of-the-art technology as a marketing edge over the competition. Every year a survey is conducted in order to nominate the United State's 100 Most Wired hospitals. Current achievements by these cutting edge hospitals will likely predict future trends in health information technology for the average hospital:

- 90% provide access to the EHR online
- 69% offer online access to nurses notes
- 88% offer lab results online
- 90% offer radiology reports online
- Most have physicians (60%) and nurses (95%) involved in IT planning and training
- Roughly 60% offer IT training as CME
- Most offer multiple patient services online such as pre-registration and disease specific self-triage
- More than 75% offer wireless access to clinical information
- Much higher adoption of bar coding and RFID [17]

Perhaps the best known of the "most wired" hospitals is the Indiana Heart Hospital built in 2003. They spent $25 million on technology, (total budget of $65 million), yet they expect a return on investment in only 6.6 years. A portion of their estimate, however, presumes decreased medical errors and legal costs as a result of going digital. One area of cost savings is from the installation of IP telephony (VoIP). This was less expensive than installing a formal private branch exchange (PBX) system. The hospital selected GE Medical Solutions throughout, to include Centricity as their EHR solution. Due to the completely digital (paperless and filmless) and wireless environment, there are no nursing stations, as nurses are located primarily in patient rooms where they have a computer for all nursing functions. [18]

Voice over IP (VoIP)

Voice over IP (VoIP) refers to a family of transmission technologies (also known as IP telephony) that deliver voice communications over IP networks such as the Internet, rather than a public switched telephone network. The analog voice signal is converted to a digital format that is transmitted and compressed over the Internet. A user can place a phone call by connecting to the Internet through a computer or directly to the Internet with a VoIP phone or device using WiFi. This technology has the advantage of decreasing communication and infrastructure costs, compared to standard telephony.

There are many applications for VoIP including communication within a hospital over wireless area networks. The goal is to decrease the dependence on phone lines and pagers. The Social Security Administration is in the process of replacing its phone system with VoIP for its 1500+ field offices and 63,000 employees. The 3 year roll out will be centrally managed and run on its data network. [19] Vocera offers a hands-free device and Avaya and Cisco offer typical handsets. Vocera is a wireless device worn around the neck that follows commands thru voice recognition. The system operates on a wireless 802.11b/g network and the badge device can also send and receive telephone calls. Other options include call waiting and call forwarding, call recognition by name, function or group membership. The system can also include a nurse call button integration feature so a patient can communicate with their nurse. In 2008 a Vocera T1000 phone was added that also uses 802.11b/g with the same function as the Vocera WiFi badge. [20] In 2010 Richardson and Ash reported an exploratory study on the use of "hands free communication" at an academic center and a community hospital. They concluded that this technology can improve communication access but may disrupt workflow unless there is excellent training.[21]

Figure 23.3 Vocera badge (courtesy Vocera)

Voalte´ is a new program that uses an existing hospital infrastructure to communicate with hospital staff carrying smartphones (iPhones and BlackBerries) using VoIP. Staff can receive voice calls, text messages, physiologic monitoring output, nurse call messages, smart infusion pump data and alarms. [22]

Voice (speech) recognition

As previously noted, voice recognition (VR) is a form of artificial intelligence that is catching on in healthcare. The accuracy is said to be currently in the range of 98% under ideal circumstances (i.e. by an experienced VR dictator). No longer does a clinician need to practice dictating for an extended period of time for the VR software to recognize the speech pattern. A training period of about 15 minutes using the existing software is adequate in many cases. The speed at which you can dictate has also improved dramatically and is in the 100+ words per minute range. However, in 2010 Hoyt and Yoshihashi reported a failure rate of 31% for a large scale deployment of VR at a military facility. They pointed out that training must be well planned and follow on training considered as well. [23] A separate medical vocabulary program must be purchased in addition to the basic VR software. The technology has evolved to include natural language processing, templates and macros to make the process much more robust and user-friendly. A template can structure a note in logical sections, whereas a macro can add blocks of commonly used text. Programming script can be developed to give the computer simple commands such as *"open Outlook"*. Most clinicians dictate using VR into their PC with the obvious disadvantage of not being able to use the program on other PCs. This can be overcome by hosting the software on a central server or using a portable tape recorder and later syncing with the PC or using a wireless (Bluetooth) headset. There are currently only three VR vendors that are major players: IBM, Dragon and Phillips. The most recent Dragon Naturally Speaking medical version in 2010 is 10.1 that has 80 medical vocabularies and can network with Citrix thin clients.[24-25] Many smartphones also now incorporate voice recognition.

Smartcards

France determined that it needed a universal identification card for healthcare. In 1997 they developed a smartcard that has been deployed to over 57 million consumers. A smartcard is also given to all healthcare professionals to allow access to a patient data warehouse. This card is also used for visits to dentists and pharmacies. As a result of the smartcard, reimbursement has been shortened from 6 weeks to 2-3 days. Spain, Germany, the Czech Republic and Russia are also experimenting with this type of health transaction card. Each card has a microprocessor that allows them to store, process and exchange basic medical data that can be used for emergencies and authentication. They carry enough information (32K) for an emergency and unlike most credit cards, they can be read without contact. Cards were updated in 2009 to include better encryption and a patient photo. Figure 23.4 shows the newer Vitale 2 card.[26-27]

Figure 23.4 Vitale 2 smart card (courtesy Vitale)

Medical smart cards have been slow to enter the American healthcare scene. Traditionally they have been used solely for identification and authentication. The US military depends on similar smart cards for authentication, but to date, the cards do not store medical information. With improved memory and other features we can expect them to find new niches. Smart cards are also being used at the Queens Health Network and University of Pittsburgh Medical Center and the St. Lukes Episcopal Health System.[28] Florida eLife Card offers 4 MB of memory with the ability to store a complete medical history to include living wills.[29] If smart cards can be shown to expedite and improve the accuracy of insurance claims payments then adoption can't be too far behind.

Improved Memory

Memory in all formats continues to improve. It is estimated that over 130 million USB flash drives were sold in 2007.[30] With the rapidly increasing memory capabilities of flash, solid state drives (SSDs) are now available for use in many computers. They are currently more expensive than standard hard drives but are smaller, faster, lighter, use less current, have non-volatile memory and have no moving parts. The technology behind SSDs has advanced to a point that options are now available ranging from smaller 16-32GB drives for consumer products such as slates and netbooks, to enterprise models such as the 1TB SSD from OCZ technology.[31] Corsair has released a ruggedized 32 GB flash drive that is water resistant and compatible with all Windows operating systems, so it could function as a hard drive. Operating systems such as Windows or Linux could be installed on USB memory drives along with software programs.[32] Nantero plans to introduce solid state memory composed of carbon nanotubes (one carbon atom thick) that will be incredibly fast and light. This new technology would permit instant "booting up" of a laptop. The memory will be called Nano RAM or NRAM.[33]

USB 3.0

The newest USB standard was released in 2008 and USB hardware is now available. Transfer speeds of 10 times the current USB 2.0 standard are promised. This high performance is only possible when two USB 3.0 devices are connected with a USB 3.0 cable. New USB 3.0 cables contain 8 wires, instead of the usual 4 wires and they are backwards compatible with USB 2.0.[34]

Radio Frequency Identification (RFID)

RFID continues to grow in popularity, particularly as prices drop and tags are miniaturized. The RFID industry is predicted to grow to $ 8.25 billion by 2014.[35] RFID *passive* (no battery) tags are now much cheaper ($.10-$.50).[36] In the non-medical arena RFID is being used on passports, used for toll booth identification and added to credit cards for contactless identification. In the field of medicine they are primarily used for tracking of patients, medications and assets. A new twist is that *active* RFID devices can be connected to Zigbee networks. These low cost wireless personal area networks (WPANs) transmit each other's messages, thus bypassing the wired network.[37] Although *active* RFID tags are not cheap, they are durable with long battery life. Many authorities believe RFID tags will slowly replace bar coding for most common applications. Passive RFID tags will undoubtedly have more memory in the future and therefore find new indications for use. RFID receivers can be linked to existing WiFi networks which reduces cost. Hospitals have begun to use active RFID tags to track staff, patients and assets with a system known as real-time locating systems (RTLS).[38] Newer RFID tags will integrate sensors capable of transmitting temperature, decibels, etc. The University of Southern California will use RFID RTLS tags that contain temperature sensors (-28° C to + 90° C) to alert the hospital if sterilization of instruments falls outside ideal temperature ranges.[39] The future of injectable subcutaneous RFID devices for patient tracking is less certain. Although this strategy has become a standard way to identify pets, it is in its infancy with humans. Horizon Blue Cross/Blue Shield of New Jersey plans a pilot study to implant VeriChips into patients with chronic diseases to access medical information. The chip stores an identification number that correlates with an online patient medical information database. They are in the process of seeking 280 volunteers for the program.[40-41] RFID is also covered in the patient safety chapter.

Cell phones

It seems predictable that future cell phones will be so robust that they will become a major means of pushing and pulling information, in addition to telephonic communication. In the hospital setting they have the potential to replace pagers and access patient information from the EHR or central data repository and view patient monitoring real time. They will likely connect with the computer at home and at work and be capable of paying bills. Is it possible physicians will dictate into them using voice recognition and the files are automatically sent to an EHR or PC. Security issues, however, will remain a rate limiting factor. Another area that cell phones will assist is remote disease reporting. Mobile phones have been used to track diseases in remote areas like Rwanda. Data can be sent to a central database via cell phone, PDA or the web.[42] As a result of this successful trial and the fact that 60% of Africa has cell phone coverage, the "Phones-for-Health" project began in 2007 to fight the spread of AIDS.[43] Cell phones are also being used to manage chronic diseases such as hypertension in the United States. In one small study, blood pressures were significantly reduced in those patients who reported blood pressure readings using their cell phones via Bluetooth. Alerts were sent to doctors and patients if the blood pressure was too high.[44]

The smartphone has continued its proliferation as a "jack of all trades" gadget for the individual, accounting for 17.3 percent of mobile phone sales in the first quarter of 2010. These devices typically run on conceptually pared down versions of PC operating systems. The top 6 in quarter one 2010 (in order of units sold) were the Symbian operating system, Research In Motion's (RIM) blackberry OS, Apple's iPhone OS, Google's Android, Microsoft Windows Mobile, and variations on Linux.[45] Smartphones are commonly interfaced with by users through either software or hardware keyboards. Touchscreen capabilities are a key navigation scheme implemented in various operating system dependent manners.

The maturation of a handheld touchscreen device, coupled with an increasingly savvy populace, and an "always online" capability through the growth of our wireless infrastructure is changing the way people are capable of interacting with computable data. As a result of these devices, software vendors have been developing small, intuitive, focused applications that provide users with detail oriented mechanisms for interacting with data stored either on the device or in the cloud. Healthcare applications on a mobile platform could provide a range of functional benefits from being able to provide a physician with real time updates on lab results to giving a patient full access to their PHR without having to be attached to a desk or carry around a laptop. More information about smartphones is available in the chapter on mobile technology.

Laptop, Tablet and Slate Computers

Laptop computers have gotten smaller, lighter and faster with more than adequate RAM and processor speed. Battery life has been a limiting factor and healthcare workers may seek more functionality. There are multiple new laptops, notebooks and tablet PCs available in the healthcare field with expanding feature sets. A recent laptop, Motion Computing C5, is worth mentioning as a new platform to be used in healthcare. It was designed by Intel and Motion Computing specifically for the healthcare field. Several of its notable features are as follows:

- Handwriting recognition via a digital pen
- Built-in 2.0 megapixel camera to document wounds, etc
- Bar coding and RFID capability
- Wireless and Bluetooth capability
- Biometric fingerprint reader
- Programmable buttons
- Weight of 3 lbs
- Hardened chassis and hard drive
- Ability to be wiped clean for disinfection
- 3 hour battery life with ability to swap batteries rapidly [46]

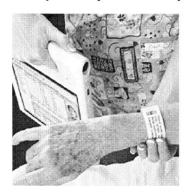

Figure 23.5 Motion Computing C5 laptop computer (Courtesy Motion Computing)

A medical laptop from Panasonic, known as the ToughBook CF-H1 Mobile Clinical Assistant, has similar features to the Motion Computing C5 laptop. Panasonic also manufactures the MDWD Wireless monitor which is a small monitor that wirelessly connects to a remote PC. It has an 8.4 inch screen, a touch screen and 300 foot range but a battery life of only 2 hours. [47] Netbook computers may be attractive to certain healthcare workers due to a smaller form factor, less expensive price tag and longer battery life.

While smartphones can provide us with a quicker onramp to a "compute anywhere" world, their ultra portability also lends to limitations as far as ease of use over the long term is concerned. To that end, manufacturers have begun to scale up devices (known as slates) into the tablet and netbook tier. Slate computers are capable of running portable versions of desktop operating systems, much like netbooks, but also have been seen with the same operating systems that run on the smartphone platform. Slate computers are interfaced solely through software keyboards manifested on touchscreen displays, much like tablets, only in a thinner footprint. Slate computers are the result of lessons learned from the problems that were seen in the first iterations of the tablet computer, as a fork of the laptop. A lot has been learned about providing an easy intuitive touch interface during the lifecycle of the smartphone, and it is the marriage of this knowledge to the application of the tablet that has birthed the slate.

The iPad was released in 2010 from Apple as the first mass market next generation slate computer. Early adopters are already discussing the merits of this device in the medical field but one should be aware of the advantages and disadvantages. [48-49] Expect more innovative tablet PCs (slate computers) in 2010. Some will be based on the NVIDIA's new Tegra mobile processor. New thin tablets or slate computers will be offered by HP (Windows OS and Palm Web OS) and Dell (Android OS). The Dell slate computer, known as Streak, will have Wi-Fi, Bluetooth, and will handle e-mail, texting, instant messages and voice calls on a 3G network. It will have a 5 inch screen along with a 1GHz Snapdragon processor, a 5-megapixel camera with dual LED flash, and 2 Gigabytes of internal storage (expandable up to 32GB with the Micro SD card). [50]

Thin Client Technology

In the early stages of computer deployment to organizations information technology staff would often provide the user with a "dumb terminal" that would link multiple users to a single computer, in order to save money. As the price of computers dropped and the operating features improved everyone in the organization wanted a desktop personal computer, loaded with their favorite software (fat client). Slowly this trend is reversing for some large organizations, including healthcare, that are looking to save money and simplify their operations. Thin client technology means that you have a software program on your PC that directs what you want to do on a distant server where all of your files and software programs are stored. You no longer have a hard drive, standard Microsoft Office software, etc on your PC. Your desktop station will include a keyboard, monitor, mouse, USB drive, printer and the means to connect to the network. This technology has the following advantages:

- Desktop device can be much smaller, will require less power and be less expensive to purchase and maintain
- Security is much easier as there is nothing to attack on the desktop
- Deployment and maintenance is easier and less expensive
- Data is centralized so it can be accessed from any thin client device
- Less likely to be stolen and more rugged in harsh environments
- Less network bandwidth
- Easier to upgrade hardware and software [51]

Hybrid thin-client technology exists such as "ultra thin client" and "zero client" technologies. In the zero client model there is no software or operating system on the desktop device. It does, however, require a small desktop hardware device and a separate management server. [52]

Thin client technology is being used in healthcare organizations such as the Veterans Integrated Systems Network (VISN) 23 in the northern United States that uses thin clients about 50% of the time. It is a reasonable alternative when you require a large number of computers and the same suite of software. VISN 23 reported that 80% of trouble tickets were for fat client and 20% were for thin client technology. [53] Wireless tablet PCs can also utilize the thin client model which means the device is much lighter and cooler to operate.

Web 2.0

Although there is no strict definition of Web 2.0, most people would say it is the new use of the Internet for collaborative purposes. Others argue it means using newer, richer web development applications. Rather than one person accessing one web site, it is multiple individuals taking advantage of multiple free web tools and web pages. Examples would be Wikipedia, Flickr and Clinical Informatics Wiki. Adding to this new movement would be the appearance of social networks, blogs, podcasts, vodcasts and RSS feeds.[54-56]

Clearly, Web 2.0 is beginning to affect the field of medicine. The following are examples of medical programs that are taking advantage of new Internet applications and philosophies:

- As noted in the chapter on consumer health informatics multiple new medical services, blogs and podcasts are appearing [57-60]
- Massachusetts Institute of Technology has developed an "opencourseware" concept that results in posting all of its course materials free to educators and learners worldwide [64]
- Universities can now post online courses using the "course management system software" known as Moodle [61]
- SugarStats is a diabetic program that permits uploading of blood sugar results with Twitter. Multiple health related web pages such as iHealthbeat and organizations such as the CDC and the FDA can be followed on Twitter. The reality is that more uses of this program are appearing for both clinicians and patients. New ventures are appearing constantly [62]

Web 3.0 or the Semantic Web

There is a tremendous amount of government and civilian data on the Internet but it often is stored in formats such as pdf that are largely non-computable. The Semantic Web will find and interpret the data or create a common framework for data sharing. Data will need to be tagged with metadata tags (data that describes data) and known as "linked data". The World Wide Web Consortium (W3C) has promoted the notion of Resource Description Framework (RDF) as the means to describe documents and images. Another specification will be Web Ontology Language. Better definitions will produce better search results. It will also allow for applications run on the Internet to receive and understand data from another application. Sir Timothy Berners-Lee, considered to be the father of the Internet, now promotes the concept of linked data as part of the RDF. He points out that currently you must have application programming interfaces (APIs) and programs like Excel and PDF to interpret data. If the data was linked and encoded by RDF standards, the extra steps would not be necessary. Slowly, organizations such as BestBuy, eBay, BBC and Data.gov have begun participation in Web 3.0. [63-64]

Cloud Computing

Just when people are comfortable with having their software located on their computers, along comes cloud computing. In essence, it means that the software is hosted on a remote server or "software as a service" (SaaS). A good example of SaaS is the EHR ASP model discussed in chapter 2. As we move closer to large scale interoperability with regional health exchanges and the NHIN we can expect to see more cloud computing. Online scheduling, consult referrals, e-prescribing and other medical processes become easier.

The advantages of cloud computing are as follows:

- No need to install or update the software. This is done for you so less IT support is needed
- Process was designed for better sharing; you can invite others to join your files on the web which cannot be done with files on your PC
- Better for mobile computing so you can access your files on the go. As we progress beyond 3G speeds, access to the Internet will become much more rapid and make cloud computing more attractive
- Deployment is faster and costs less because of decreased IT support
- Software is often free

- It is operating system independent
- It fits with the Web 2.0 strategy and service oriented architecture
- This movement creates new competition and innovations

As with every new technological movement there are disadvantages:

- Privacy may be more of an issue
- Someone else is hosting your data
- Aspects of your data might be sold without your knowledge

Multiple large scale examples of cloud computing now exist:

- Amazon's Elastic Compute Cloud
- Google Docs
- Rackspace Cloud
- IBM Blue Cloud
- Microsoft Office Live and Windows Azure (an operating system for the cloud)
- Sun Microsystems Open Cloud Platform [65-69]

Virtualization

Many large organizations have reduced the number of servers needed in their data centers by leveraging virtualization technology, allowing a single set of hardware to run multiple separate operating system instances (each with its own virtualized environment). This trend also extends to the desktop environment, where a user may not need or want to be tied to a single device. Since virtualization allows a single machine (server or desktop) to act like many, the case of the virtualized desktop encompasses the user's programs and applications being stored on a remote server so they can be accessed by any secure client device. Virtualization results in higher utilization rates, lower power consumption, lower operating costs and reduced space. Operating systems being managed under a system known as a hypervisor are aware of their virtualized environment, and therefore can request extra resources from the virtual machine manager as needed. This allows for lower average operating costs while still providing for a system that can scale based on load.[70]

Open Source Software in Medicine

Open source software such as Apache for web servers, Sun OpenOffice.org, the GNU/Linux operating system and Firefox browser are considered mainstream programs in terms of use and acceptance. Open source has been slower to enter the healthcare field, with the exception of the VistA EHR, but clearly is making a significant presence. Open source has often been used for a low cost alternative for community clinics and "third world" countries but many feel that open source will soon become very competitive with proprietary products. In the chapters on EHRs, interoperability, telemedicine, bioinformatics, public health informatics and e-research we have mentioned open source applications. Although criticized by some proprietary vendors, open source is created by a collaborative team so frequently the end product is well thought out and more secure. The American Medical Informatics Association has an open source working group and they have published a November 2008 monograph on open source for healthcare available on their web site. [71]

Another use of open source is in the development of standards based frameworks to be used as a component of a larger application or system of applications. When developing a system with a need to serve web pages to a browser, a vendor could implement the entire server infrastructure from scratch or they could plug the apache web server in and benefit from the project's community. The same analogy holds true to the implementation of healthcare information technology standards. Open eHealth is an open source initiative whose Integration Platform project has been used in IHE Connectathon testing to evaluate interoperability among different proprietary EHR vendors. [76] Mirth Corporation's Mirth Connect product utilizes a myriad of open source projects to provide a system for integrating disparate healthcare systems. [73] Openhealthtools (OHT) is an organization, forked from the Eclipse foundation, whose goal is creating a network of contributors and open software geared towards solving software problems in healthcare. [74] The OHT iheprofiles project is used by many EMR vendors to provide connectivity of their systems to IHE compliant HIOs (as discussed in the chapter on health information exchange) without having to integrate the necessary standards into their systems from scratch. Government contracted open source has seen an upsurge recently in healthcare as well with projects such as MITRE's pophealth (for aggregating and submitting quality reporting to CMS) being funded for early stage development, prototyping, and then release to the general open source community for the remainder of the project lifecycle. [75]

Wikipedia has an extensive list of over 100 healthcare software applications that covers many of the same categories as chapters in this textbook and we highly recommend this site for browsing. [76]

Readers will frequently see the acronym FOSS that stands for free open source software which requires further explanation. Free is meant to imply freedom, not price. "Free" software licenses will define the ability to copy, run and distribute the software and are determined by the Free Software Foundation. Licenses that are open source are determined by the Open Source Initiative. Public domain software has no license so it can be sold or modified and later licensed as proprietary, free or open source. For more detail on these important distinctions we refer you to these references. [70,77]

Faster Ethernet

The Institute of Electrical and Electronics Engineers (IEEE) is expected to release new standards for a 100 Gigabit Ethernet network in June 2010. It is anticipated that this network will be used by researchers and others like the Department of Energy (ESnet) who need advanced speed. Verizon recently deployed a 100 Gigabit network in Europe as the first commercial network of its type. There is already talk about bundling 100 Gigabit pipes to create a Terabit Ethernet.

For most people the next move in networks will be to adopt 10 Gigabit Ethernet standards. Higher speeds make virtualization easier. While this standard has been around since 2002, it has only recently become cost-effective and increasingly adopted. The primary cost will be for switches and interface cards. This standard can operate on both fiber-optic cable and twisted-pair cable. To add to the choices, 40 Gigabit networks will be offered in late 2010. [38] Networks are discussed in more detail in a separate chapter.

Internet2 and LambdaRail

Internet2 is a not-for-profit networking consortium of more than 200 universities, government agencies, researchers and business groups developing applications and a network for the future. The current network is known as *Abilene* and it operates at 10 gigabits per second (100-1000 times faster than Internet1). They have deployed 13,500 miles of dedicated fiber optics as the backbone of the system. They anticipate future transmission speeds of 40-100 Gbps (7,000 times faster than a T1 line).

National LambdaRail (NLR) also connects universities (150+) across the nation through fiber optic networks. This unique network connects 28 American cities. Members benefit from using the faster Internet to communicate and from the development of interesting middleware. Research is underway to develop programs to support digital video, authentication and security. [78-80]

In 2010 the National Telecommunications and Information Administration (NTIA) Broadband Technology Opportunities Program, funded the United States Unified Community Anchor Network (U.S. UCAN) that will be a nationwide advanced network infrastructure that, together with state and regional network partners (Internet2, LambadaRail, Cisco and others) will connect anchor institutions such as schools, libraries, community colleges, health centers and public safety organizations. This will make possible advanced services such as telemedicine and distance learning for all community anchor institutions. The new funding will permit upgrades to existing networks to create a 100 Gbps backbone. [81]

Biometrics

Security continues to be an ongoing issue with all technologies that store or communicate personal health information (PHI). Numerous laptop computers now offer finger print identification in lieu of signing on with a password.[82] While this is a reasonable solution for many people, certain conditions make finger prints unreliable. Retinal scanning is felt to be the most accurate biometric measure but comes with a much higher price tag. Iris imaging has also been shown to be highly accurate and has been used for authentication in the Netherlands and United Arab Emirates since 2001. Other biometric measures being used for authentication are face and hand geometry and speech recognition.[83-84] All technologies will require high tech scanners and a subject database for verification.

In February 2009 Sony announced a new biometric testing technique based on finger veins known as mofiria that scans veins below the skin.[85] Figure 23.6 demonstrates the technology. Fujitsu has similar biometric technology known as PalmSecure™ that scans the veins in the palm.[86]

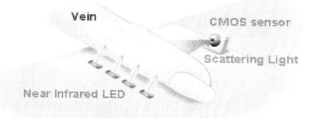

Figure 23.6 Vein scanning (Courtesy Technology Experts)

Key Points

- Technology is advancing faster than our ability to incorporate it into the field of medicine

- Expect smarter processes with natural language processing and smart cards

- Expect faster processes with Internet2 and LambdaRail

- Expect improved laptops, imaging, RFID, flash memory and battery life

- Anticipate more open source initiatives

Conclusion

We live in a very exciting time in terms of rapidly improving technology. It will require highly trained clinicians who are also information technology advocates (informaticians) to embrace and successfully implement innovations into healthcare. More research is needed to critically appraise new information technologies before they are recommended on a large scale. Better methods will be necessary to train busy clinicians and their staff. Improved productivity, patient safety and medical quality continue to be the promise. Security, privacy, high cost and human resistance, however, will continue to be the rate limiting factors for years to come.

References

1. Rand. The Diffusion and Value of Healthcare Information Technology. 2005. http://www.rand.org/pubs/monographs/2006/RAND_MG272-1.pdf (Accessed October 3 2007)
2. Bower JL, Christensen CM. Disruptive Technologies: Catching the Wave. Harvard Business Review Jan-Feb 2005 pp 43-53
3. Disruptive Technology www.wikipedia.com (Accessed June 28 2009)
4. University of Toledo http://homepages.utoledo.edu/akunnat/chap10pt2.ppt (Accessed September 10 2005)
5. Health and Healthcare 2010: Institute for the Future Wiley John & Sons Inc or http://www.iftf.org/docs/SR-794_Health_&_Health_Care_2010.pdf (Accessed November 15 2005)
6. Frist WH. Shattuck Lecture: Health Care in the 21st Century. NEJM 2005;352:267-272
7. Swartz WB. Medicine and the Computer: The Promise and Problems of Change. NEJM 1970;1257-1264
8. Artificial Intelligence Systems in Routine Clinical Use http://www.coiera.com/ailist/list-main.html (Accessed September 10 2005)
9. What is a neural network? Neuro Solutions. http://www.nd.com (Accessed October 1 2005)
10. Szolovits P. Artificial Intelligence and Medicine. Westview Press, Boulder, Colorado 1982
11. Coiera E. Guide to Health Informatics, 2nd Edition Oxford University Press 2003
12. Language and Computing www.landcglobal.com (Accessed February 22 2007)
13. Dearne, K. Processor Cleans Up doctor's Notes. Nov 13 2007. www.australianit.news.com. (Accessed December 3 2007)
14. Siemens www.siemens.com (June 20 2009)
15. Sonosite www.sonosite.com (Accessed July 6 2009)
16. GE VScan. http://www.gereports.com/vscan-pocket-sized-ultra-smart-ultrasound-unveiled (Accessed May 1 2010)
17. The 100 Most Wired Vroom www.hhnmag.com July 15 2004 (Accessed July 1 2005)
18. Nash K S Indiana Heart Hospital: Real Time ER www.baselinemag.com May 4 2005 (Accessed July 5 2005)
19. Jackson W. SSA goes big on VoIP. Government Computer News June 1 2009 p 24-25
20. Vocera www.vocera.com (Accessed January 1 2009)
21. Richardson JE, Ash JS. The effects of hands-free communication device systems: communication changes in hospital organization. J Am Med Inform Assoc 2010;17:91-98
22. Voalte www.voalte.com (Accessed May 10 2010)
23. Hoyt RE, Yoshihashi AK. Lessons Learned from Voice Recognition for Documentation. Perspectives in Health Information Management Winter 2010. http://perspectives.ahima.org (Accessed February 4 2010)

24. Bergeron B. Voice Recognition and Medical Transcription 8/24/2004 www.medscape.com (Accessed November 25 2005)

25. Nuance www.nuance.com (Accessed January 1 2009)

26. French Sesam Vitale Smartcards http://www.smartcardalliance.org/pdf/about_alliance/user_profiles/French_Health_Card_Profile.pdf (Accessed May 5 2006)

27. New Health Card Introduced Across France. www.frenchentree.com (Accessed February 20 2009)

28. Smart Card Alliance www.smartcardalliance.org (Accessed February 1 2009)

29. EMIDASI www.emidasi.com (Accessed January 4, 2008

30. USB Flash Drive Market http://www.u3.com/platform/ (October 23 2005)

31. OCZ Technology www.ocztechnology (Accessed May 16 2010)

32. Corsair. www.corsair.com (Accessed February 12 2008)

33. Nantero. www.nantero.com (Accessed February 12 2008)

34. 10 Technologies to Watch in 2010. Government Computer News. January 18 2010. P:16-25

35. RFID Market to reach 5.35 Billion www.abiresearch.com/press/1618-RFID+Market+to+Reach+$5.35+Billion+This+Year March 5 2010 (Accessed June 6 2010)

36. RFID's Second Wave www.businessweek.com August 9 2005 (Accessed August 10 2005)

37. Zigbee networks http://en.wikipedia.org/wiki/ZigBee (Accessed January 24 2007)

38. Awarix http://www.awarix.com/products.html (Accessed February 20 2007)

39. RFID News www.rfidnews.org June 12 2009 (Accessed July 6 2009)

40. Agovino T. Insurers to test implantable microchip. USA Today www.usatoday.com July 16 2006 (Accessed July 20 2006)

41. Verichip www.verichipcorp.com (Accessed July 6 2009)

42. Voxiva. www.voxiva.net (Accessed October 19 2006)

43. Project Taps Cell Phones To Fight AIDS in Africa February 13 2007 www.ihealthbeat.org (Accessed February 13 2007)

44. Logan AG et al. Mobile Phone-Based Remote Patient Monitoring System for Management of Hypertension in Diabetic Patients. Am J Hyper 2007;20 (9):942-948

45. Gartner http://www.gartner.com/it/page.jsp?id=1372013 May 19 2010 (Accessed June 6 2010)

46. Motion Computing. www.motioncomputing.com (Accessed January 1 2008)

47. Panasonic www.panasonic.com (Accessed July 6 2009)

48. Bertmann J. Tech Talk. Will the Apple iPad and other Next-generation Devices Make 2010 the Year of the Tablet PC? www.hcplive.com/endocrinology . March 2010 (Accessed April 10 2010)

49. Soto C. 10 Reasons Feds Should Be Eager. Government Computer News Lab. April 19 2010. P16-19

50. Dell Streak. C-Net Reviews http://reviews.cnet.com/tablets/dell-streak/4505-3126_7-34117657.html (Accessed June 5 2010)

51. Thin Client http://en.wikipedia.org/wiki/Thin_client (Accessed July 1 2008)

52. Pano Logic www.panologic.com (Accessed July 1 2008)

53. Gerber C. Thin Client Military Medical Technology online. http://www.military-medical-technology.com/print_article.cfm?DocID=1428 2006;10 (3) (Accessed July 1 2008)

54. Tapscott D, Williams A. Wikinomics. How Mass Collaboration Changes Everything. Penguin Books. 2006. London, England

55. Web 2.0 www.wikipedia.org (Accessed January 4 2007)

56. Health Care 2.0 Government HealthIT April 2007 vol 2. No. 2. p 22-29

57. Sermo. www.sermo.com (Accessed April 1 2007)

58. Survey: Consumers Read, Post Info on Health Blogs January 10 2007 (Accessed January 10 2007)
59. Savel, RH et al. The iCritical Care Podcast: A Novel Medium for Critical Care Communication and Education. JAMIA. 2007;14:94-99
60. Opencourseware. http://ocw.mit.edu (Accessed January 6 2007)
61. Moodle. www.moodle.org (Accessed January 8 2008)
62. SugarStats www.sugarstats.com (Accessed June 17 2009)
63. Moore J. Waiting for Web 3.0. Federal Computer News. July 20 2009. www.fcw.com (Accessed August 3 2009)
64. Jackson J. The Web's next act. Government Computer News November 9 2009. p.20-29
65. Robinson B. Gathering Storm. Federal Computer Week January 12 2009 p.28-29
66. Microsoft offers to host agency data. Government Computer News May 18 2009 p.9
67. Future Forecast. Cloud Computing. CMIO April 2009 p. 28-30
68. Wollman D. The Future of Cloud Computing. Laptop Magazine. August 2008 p. 24-26
69. Cloud Computing at a Higher Level. www.cloudsrus.net/upload/cloud_computing_primer.pdf (Accessed March 10 2010)
70. Virtualization. www.wikipedia.org (Accessed June 4 2010)
71. Valdes I. Free and Open Source Software in Healthcare 1.0. AMIA Open Source Working Group White Paper November 2008 www.amia.org (Accessed July 4 2009)
72. Open eHealth www.openehealth.org (Accessed July 5 2009)
73. Mirth Corp www.mirthcorp.com (Accessed June 6 2010)
74. Open Health Tools www.openhealthtools.org (Accessed June 6 2010)
75. Project Pop Health www.projectpophealth.org (Accessed June 6 2010)
76. List of Open Source Healthcare Software www.wikipedia.org (Accessed July 2 2009)
77. What Does Free Mean? Or what do you mean by Free Software? www.debian.org/intro/free (Accessed May 15 2009)
78. McGill MJ. An Internet Upgrade www.healthcare-informatics.com Jan 12 2005 (Accessed April 20 2006)
79. Internet2. www.internet2.edu (Accessed January 20 2008)
80. National LambdaRail. www.nlr.net (Accessed January 29 2008)
81. Broadband Funding Grant: Implications for Research & Education. July 2 2010. www.nlr.net (Accessed July 23 2010)
82. Toshiba www.toshiba.com (Accessed August 6 2006)
83. Biometrics www.wikipedia.com (Accessed August 6 2006)
84. Arun R, Prabhakar S, Jain A. An Overview of Biometrics. http://biometrics.cse.msu.edu/info.html (Accessed August 6 2006)
85. Sony's New Biometric Tech more than a little vein http://technologyexpert.blogspot.com February 21 2009 (Accessed July 6 2009)
86. Fujitsu http://www.fujitsu.com/us/services/biometrics/palm-vein/healthcare.html (Accessed July 22 2009)

Index